Doing Meta-Analysis with R

Doing Meta-Analysis with R

A Hands-On Guide

Mathias Harrer
Pim Cuijpers
Toshi A. Furukawa
David D. Ebert

CRC Press is an imprint of the
Taylor & Francis Group, an **informa** business
A CHAPMAN & HALL BOOK

First edition published 2022
by CRC Press
6000 Broken Sound Parkway NW, Suite 300, Boca Raton, FL 33487-2742

and by CRC Press
2 Park Square, Milton Park, Abingdon, Oxon, OX14 4RN

CRC Press is an imprint of Taylor & Francis Group, LLC

Library of Congress Cataloging-in-Publication Data

Names: Harrer, Mathias, author.
Title: Doing meta-analysis with R : a hands-on guide / Mathias Harrer [and three others].
Description: First edition. | Boca Raton : CRC Press, 2022. | Includes bibliographical references and index.
Identifiers: LCCN 2021017096 (print) | LCCN 2021017097 (ebook) | ISBN 9780367610074 (hardback) | ISBN 9780367619770 (paperback) | ISBN 9781003107347 (ebook)
Subjects: LCSH: Meta-analysis. | R (Computer program language)
Classification: LCC R853.M48 H37 2022 (print) | LCC R853.M48 (ebook) | DDC 610.727--dc23
LC record available at https://lccn.loc.gov/2021017096
LC ebook record available at https://lccn.loc.gov/2021017097

ISBN: 9780367610074 (hbk)
ISBN: 9780367619770 (pbk)
ISBN: 9781003107347 (ebk)

DOI: 10.1201/9781003107347

Typeset in Alegreya
by KnowledgeWorks Global Ltd.

The problems are solved, not by giving new information,
but by arranging what we have known since long.

– **Ludwig Wittgenstein**, *Philosophical Investigations*

Contents

Preface xiii

About the Authors xxiii

List of Symbols xxv

I Getting Started 1

1 Introduction 3
- 1.1 What Are Meta-Analyses? 4
- 1.2 "Exercises in Mega-Silliness": A Historical Anecdote 6
- 1.3 Apples and Oranges: A Quick Tour of Meta-Analysis Pitfalls 8
- 1.4 Problem Specification, Study Search & Coding 11
  - 1.4.1 Defining the Research Question 12
  - 1.4.2 Analysis Plan & Preregistration 16
  - 1.4.3 Study Search 18
  - 1.4.4 Study Selection 21
  - 1.4.5 Data Extraction & Coding 24
- 1.5 Questions & Answers 26
- 1.6 Summary 26

2 Discovering R 29
- 2.1 Installing R & R Studio 29
- 2.2 Packages 31
- 2.3 The *{dmetar}* Package 34
- 2.4 Data Preparation & Import 36
- 2.5 Data Manipulation 38
  - 2.5.1 Class Conversion 39
  - 2.5.2 Data Slicing 42
  - 2.5.3 Data Transformation 44
  - 2.5.4 Saving Data 46
- 2.6 Questions & Answers 48
- 2.7 Summary 49

II Meta-Analysis in R 51

3 Effect Sizes 53

3.1 What Is an Effect Size? 54
3.2 Measures & Effect Sizes in Single Group Designs 59
3.2.1 Means 59
3.2.2 Proportions 60
3.2.3 Correlations 61
3.3 Effect Sizes in Control Group Designs 64
3.3.1 (Standardized) Mean Differences 64
3.3.2 Risk & Odds Ratios 70
3.3.3 Incidence Rate Ratios 76
3.4 Effect Size Correction 80
3.4.1 Small Sample Bias 80
3.4.2 Unreliability 81
3.4.3 Range Restriction 84
3.5 Common Problems 87
3.5.1 Different Effect Size Data Formats 87
3.5.2 The Unit-of-Analysis Problem 88
3.6 Questions & Answers 90
3.7 Summary 90

4 Pooling Effect Sizes 93

4.1 The Fixed-Effect & Random-Effects Model 94
4.1.1 The Fixed-Effect Model 95
4.1.2 The Random-Effects Model 99
4.2 Effect Size Pooling in R 105
4.2.1 Pre-Calculated Effect Size Data 108
4.2.2 (Standardized) Mean Differences 112
4.2.3 Binary Outcomes 115
4.2.4 Correlations 127
4.2.5 Means 130
4.2.6 Proportions 132
4.3 Questions & Answers 136
4.4 Summary 136

5 Between-Study Heterogeneity 139

5.1 Measures of Heterogeneity 140
5.1.1 Cochran's Q 140
5.1.2 Higgins & Thompson's I^2 Statistic 145
5.1.3 The H^2 Statistic 147
5.1.4 Heterogeneity Variance τ^2 & Standard Deviation τ 147
5.2 Which Measure Should I Use? 149
5.3 Assessing Heterogeneity in R 150
5.4 Outliers & Influential Cases 153

5.4.1 Basic Outlier Removal 153
5.4.2 Influence Analysis 156
5.4.3 GOSH Plot Analysis 163
5.5 Questions & Answers 170
5.6 Summary 170

6 Forest Plots **173**
6.1 What Is a Forest Plot? 173
6.2 Forest Plots in R 174
6.2.1 Layout Types 177
6.2.2 Saving the Forest Plots 178
6.3 Drapery Plots 180
6.4 Questions & Answers 182
6.5 Summary 182

7 Subgroup Analyses **183**
7.1 The Fixed-Effects (Plural) Model 184
7.1.1 Pooling the Effect in Subgroups 184
7.1.2 Comparing the Subgroup Effects 185
7.2 Limitations & Pitfalls of Subgroup Analyses 187
7.3 Subgroup Analysis in R 190
7.4 Questions & Answers 194
7.5 Summary 194

8 Meta-Regression **197**
8.1 The Meta-Regression Model 198
8.1.1 Meta-Regression with a Categorical Predictor 198
8.1.2 Meta-Regression with a Continuous Predictor 200
8.1.3 Assessing the Model Fit 201
8.2 Meta-Regression in R 203
8.3 Multiple Meta-Regression 206
8.3.1 Interactions 207
8.3.2 Common Pitfalls in Multiple Meta-Regression 209
8.3.3 Multiple Meta-Regression in R 212
8.4 Questions & Answers 224
8.5 Summary 225

9 Publication Bias **227**
9.1 What Is Publication Bias? 228
9.2 Addressing Publication Bias in Meta-Analyses 230
9.2.1 Small-Study Effect Methods 231
9.2.2 P-Curve 254
9.2.3 Selection Models 272
9.3 Which Method Should I Use? 281
9.4 Questions & Answers 283
9.5 Summary 283

III Advanced Methods **285**

10 "Multilevel" Meta-Analysis **287**
10.1 The Multilevel Nature of Meta-Analysis 287
10.2 Fitting Three-Level Meta-Analysis Models in R 291
10.2.1 Model Fitting 293
10.2.2 Distribution of Variance across Levels 295
10.2.3 Comparing Models 296
10.3 Subgroup Analyses in Three-Level Models 298
10.4 Questions & Answers 301
10.5 Summary 301

11 Structural Equation Modeling Meta-Analysis **303**
11.1 What Is Meta-Analytic Structural Equation Modeling? 304
11.1.1 Model Specification 304
11.1.2 Meta-Analysis from a SEM Perspective 307
11.1.3 The Two-Stage Meta-Analytic SEM Approach 308
11.2 Multivariate Meta-Analysis 309
11.2.1 Specifying the Model 311
11.2.2 Evaluating the Results 313
11.2.3 Visualizing the Results 315
11.3 Confirmatory Factor Analysis 316
11.3.1 Data Preparation 317
11.3.2 Model Specification 319
11.3.3 Model Fitting 324
11.3.4 Path Diagrams 326
11.4 Questions & Answers 328
11.5 Summary 328

12 Network Meta-Analysis **329**
12.1 What Are Network Meta-Analyses? 330
12.1.1 Direct & Indirect Evidence 330
12.1.2 Transitivity & Consistency 332
12.1.3 Network Meta-Analysis Models 334
12.2 Frequentist Network Meta-Analysis 335
12.2.1 The Graph Theoretical Model 336
12.2.2 Frequentist Network Meta-Analysis in R 338
12.3 Bayesian Network Meta-Analysis 356
12.3.1 Bayesian Inference 356
12.3.2 The Bayesian Network Meta-Analysis Model 358
12.3.3 Bayesian Network Meta-Analysis in R 360
12.3.4 Network Meta-Regression 373
12.4 Questions & Answers 378
12.5 Summary 378

13 Bayesian Meta-Analysis 381

13.1 The Bayesian Hierarchical Model 381
13.2 Setting Prior Distributions 383
13.3 Bayesian Meta-Analysis in R 385
13.3.1 Fitting the Model 386
13.3.2 Assessing Convergence 387
13.3.3 Interpreting the Results 389
13.3.4 Generating a Forest Plot 391
13.4 Questions & Answers 395
13.5 Summary 395

IV Helpful Tools 397

14 Power Analysis 399

14.1 Fixed-Effect Model 401
14.2 Random-Effects Model 404
14.3 Subgroup Analyses 406

15 Risk of Bias Plots 407

15.1 Data Preparation 408
15.2 Summary Plots 410
15.3 Traffic Light Plots 411

16 Reporting & Reproducibility 413

16.1 Using R Projects 413
16.2 Writing Reproducible Reports with R Markdown 415
16.3 OSF Repositories 417
16.3.1 Access Token 417
16.3.2 The *{osfr}* Package & Authentication 418
16.3.3 Repository Setup 418
16.3.4 Upload & Download 419
16.3.5 Collaboration, Open Access & Pre-Registration 420

17 Effect Size Calculation & Conversion 423

17.1 Mean & Standard Error 423
17.2 Regression Coefficients 424
17.3 Correlations 426
17.4 One-Way ANOVAs 427
17.5 Two-Sample t-Tests 428
17.6 p-Values 428
17.7 χ^2 Tests 430
17.8 Number Needed to Treat 430
17.9 Multi-Arm Studies 434

Appendix 437

A Questions & Answers 437

B Effect Size Formulas 445

C R & Package Information 449

Bibliography 451

Index 471

Preface

It is a trivial observation that our world is complex. Scientific research is no exception; in most research fields, we are often faced with a seemingly insurmountable body of previous research. Evidence from different studies can be conflicting, and it can be difficult to make sense out of various sources of information. *Evidence synthesis* methods therefore play a crucial role in many disciplines, for example the social sciences, medicine, biology, or econometrics. *Meta-analysis*, the statistical procedure used to combine results of various studies or analyses, has become an indispensable tool in many research areas. Meta-analyses can be of enormous importance, especially if they guide practical decision-making, or future research efforts. Many applied researchers therefore already have some meta-analysis skills in their "statistical toolbox", while others want to learn how to perform meta-analyses in their own research field. Meta-analyses have become so ubiquitous that many graduate and undergraduate students already learn how to perform one as part of their curriculum – sometimes with varying levels of enthusiasm.

The way meta-analyses can be performed, like statistical computing as a whole, has seen major shifts in the last decades. This has a lot to do with the rise of open source, collaborative statistical software, primarily in the form of the R Statistical Programming Language and Environment. The R ecosystem allows researchers and statisticians everywhere to build their own *packages*, and to make them available to everyone, at no cost. This has lead to a spectacular rise in readily available statistical software for the R language. While we are writing this, the CRAN Task View[1] lists more than 130 packages dedicated to meta-analysis alone. In R, you can do *anything* – literally. It is a full programming language, so if you do not find a function for something you want to do, you can easily write it yourself. For meta-analyses, however, there is hardly any need to do this anymore. Just a small collection of R packages already provide all the functionality you can find in current "state-of-the-art" meta-analysis programs – for free. Even more so, there are many novel meta-analysis methods that can currently *only* be applied in R. In short: the R environment gives researchers much more tools for their meta-analyses. In the best case, this allows us to draw more robust conclusions from our data, and thus better informed decision-making.

This raises the question: why isn't everyone using R for meta-analyses? We think there are two main reasons: *convenience* and *anxiety* (and sometimes a mixture of both). Both reasons are very understandable. Most meta-analysts are applied researchers, not

[1] `https://cran.r-project.org/web/views/MetaAnalysis.html`

statisticians or programmers. The thought of learning an obscure and complicated-seeming programming language can act as a deterrent. The same is true for meta-analytic methods, with their special theoretical background, their myriad analytic choices, and different statistics that need to be interpreted correctly.

With this guide, we want to show that many of these concerns are unfounded, and that learning how to do a meta-analysis in R is worth the effort. We hope that the guide will help you to learn the skills needed to master your own meta-analysis project in R. We also hope that this guide will make it easier for you to not only learn *what* meta-analytic methods to apply when, but also *why* we apply them. Last but not least, we see this guide as an attempt to show you that meta-analysis methods and R programming are not mere inconveniences, but a fascinating topic to explore.

This Book Is for Mortals

This guide was not written for meta-analysis experts or statisticians. We do not assume that you have any special background knowledge on meta-analytic methods. Only basic knowledge of fundamental mathematical and statistical concepts is needed. For example, we assume that you have heard before what are things like a "mean", "standard deviation", "correlation", "regression", "p-value" or a "normal distribution". If these terms ring a bell, you should be good to go. If you are really starting from scratch, you may want to first have a look at Robert Stinerock's statistics beginner's guide (Stinerock, 2018) for a thorough introduction including hands-on examples in R–or some other introductory statistics textbook of your choice.

Although we tried to keep it as minimal as possible, we will use mathematical formulas and statistical notation at times. But do not panic. Formulas and Greek letters can seem confusing at first glance, but they are often a very good way to precisely describe the idea behind some meta-analysis methods. Having seen these formulas, and knowing what they represent, will also make it easier for you to understand more advanced texts you may want to read further down the line. And of course, we tried our best to always explain in detail what certain symbols or letters stand for, and what a specific formula *wants to tell us*. In the beginning of this book, you can find a list of the symbols we use, and what they represent. In later chapters, especially the Advanced Methods section, we need to become a little more technical to explain the ideas behind some of the applied techniques. Nevertheless, we made sure to always include some background information on the mathematical and statistical concepts used in these sections.

No prior knowledge of R (or programming in general) is required. In the guide, we try to provide a gentle introduction into basic R skills you need to code your own meta-analysis. We also provide references to adequate resources to keep on learning.

Furthermore, we will show you how you can set up a free computer program which allows you to use R conveniently on your PC or Mac.

As it says in the title, our book focuses on the "doing" part of meta-analysis. Our guide aims to be an accessible resource which meets the needs of applied researchers, students and data scientists who want to get going with their analyses using R. Meta-analysis, however, is a vast and multi-faceted topic, so it is natural that not everything can be covered in this guide. For this book, limitations particularly pertain to three areas:

- Although we provide a short primer on these topics, we do not cover *in detail* how to define research questions, systematically search and include studies for your meta-analysis, as well as how to assess their quality. Each of these topics merits books of their own, and luckily many helpful resources already exist. We therefore only give an overview of important considerations and pitfalls when collecting the data for your meta-analysis, and will refer you to adequate resources dealing with the nitty-gritty details.
- The second limitation of this guide pertains to its level of technicality. This book is decidedly written for "mortals". We aim to show you when, how and why to apply certain meta-analytic techniques, along with their pitfalls. We also try to provide an easily accessible, conceptual understanding of the techniques we cover, resorting to more technical details only if it benefits this mission. Quite naturally, this means that some parts of the guide will not contain a deep dive into technicalities that expert-level meta-analysts and statisticians may desire. Nevertheless, we include references to more advanced resources and publications in each chapter for the interested reader.
- Contents of a book will always to some extent reflect the background and experience of its authors. We are confident that the methods we cover here are applicable and relevant to a vast range of research areas and disciplines. Nevertheless, we wanted to disclose that the four authors of this book are primarily versed in current research in psychology, psychiatry, medicine and intervention research. "Real-world" use cases and examples we cover in the book therefore concentrate on topics where we know our way around. The good news is that meta-analytic methods (provided some assumptions, which we will cover) are largely agnostic to the research field from which data stem from, and can be used for various types of outcome measures. Nonetheless, and despite our best intentions to make this guide as broadly applicable to as many applied research disciplines as possible, it may still be possible that some of the methods covered in this book are more relevant for some research areas than others.

Topics Covered in the Book

Among other things, this guide will cover the following topics:

- What a meta-analysis is, and why it was *invented.*
- *Advantages* and *common problems* with meta-analysis.
- How *research questions* for meta-analyses are specified, and how the *search for studies* can be conducted.
- How you can set up R, and a *computer program* which allows you to use R in a convenient way.
- How you can *import* your meta-analysis data into R, and how to *manipulate* it through code.
- What *effect sizes* are, and how they are calculated.
- How to *pool effect sizes* in fixed-effect and random-effects meta-analyses.
- How to analyze the *heterogeneity* of your meta-analysis, and how to explore it using *subgroup analyses* and *meta-regression.*
- Problems with *selective outcome reporting,* and how to tackle them.
- How to perform *advanced types* of meta-analytic techniques, such as "multi-level" meta-analysis, meta-analytic structural equation modeling, network meta-analysis, or Bayesian meta-analysis.
- How to *report* your meta-analysis results, and make them *reproducible.*

How to Use This Book

Work Flow

This book is intended to be read in a "linear" fashion. We recommend that you start with the first chapters on meta-analysis and R basics, and then keep on working yourself through the book one chapter after another. Jumping straight to the hands-on chapters may be tempting, but it is not generally recommended. From our experience, a basic familiarity with meta-analysis, as well as the R Studio environment, is a necessary evil to avoid frustrations later on. This is particularly true if you have no previous experience with meta-analysis *and* R programming. Experienced R users may skip Chapter 2, which introduces R and R Studio. However, it will certainly do no harm to work through the chapter anyway as a quick refresher.

While all chapters are virtually self-contained, we do sometimes make references to topics covered in previous chapters. Chapters in the Advanced Methods section in particular assume that you are familiar with theoretical concepts we have covered before.

The last section of this book contains helpful tools for your meta-analysis. This does not mean that these topics are the final things you have to consider when performing a meta-analysis. We simply put these chapters at the end because they primarily serve as reference works for your own meta-analysis projects. We link to these tools throughout the book in sections where they are thematically relevant.

Online Version

This book also has an online version[2]. On the website, click on "Read the Guide" to open it. The contents of the online version are nearly identical with the ones you will find here. However, the website does contain some extra content, including a few sections on special interest topics that we did not consider essential for this book. It also contains interactive material which can only be used via the Internet. We reference supplementary online content in the book where it is thematically relevant.

The online version of the guide also contains an additional chapter called *Corrections & Remarks*. We regularly update the online version of the book. Potential errors and problems in the printed version of the book that we, or others, may have encountered in the meantime will be displayed there.

Companion R Package

This book comes with a companion R package called *{dmetar}*. This package mainly serves two functions. First, it aims to make your life easier. Although there are fantastic R packages for meta-analysis out there with a vast range of functionalities, there are still a few things which are currently not easy to implement in R, at least for beginners. The *{dmetar}* package aims to bridge this gap by providing a few extra functions facilitating exactly those things. Secondly, the package also contains all the data sets we are using for the hands-on examples included in this book. In Chapter 2.3, the *{dmetar}* package is introduced in detail, and we show you how to install the package step by step. Although we will make sure that there are no substantial changes, *{dmetar}* is still under active development, so it may be helpful to have a look at the package website[3] now and then to check if there are new or improved functionalities which you can use for your meta-analysis. While advised, it is not essential that you install the package. Wherever we make use of *{dmetar}* in the book, we will also provide you with the raw code for the function, or a download link to the data set we are using.

[2] `www.protectlab.org/meta-analysis-in-r/`

[3] `dmetar.protectlab.org`

Text Boxes

Throughout the book, a set of text boxes is used.

General Note
General notes contain relevant background information, insights, anecdotes, considerations or take-home messages pertaining to the covered topic.

Important Information
These boxes contain information on caveats, problems, drawbacks or pitfalls you have to keep in mind.

Questions
After each chapter, this box will contain a few questions through which you can test your knowledge. Answers to these questions can be found at the end of the book in Appendix A.

***{dmetar}* Note**
The *{dmetar}* note boxes appear whenever functions or data sets contained in the companion R package are used. These boxes also contain URLs to the function code, or data set download links, for readers who did not install the package.

How Can I Report This?
These boxes contain recommendations on how you can report R output in your thesis or research article.

Conventions

A few conventions are followed throughout the book.

{packages}

All R packages are written in italic and are put into curly brackets. This is a common way to write package names in the R community.

```
R Code
```

All R code or objects we define in R are written in this monospace font.

```
## R Output
```

The same monospace font is used for the output we receive after running R code. However, we use two number signs (hashes) to differentiate it from R input.

Formula

This serif font is reserved for formulas, statistics and other forms of mathematical notation.

What to Do When You Are Stuck

Undeniably, the road to doing meta-analyses in R can be a rocky path at times. Although we think this is sometimes exaggerated, R's learning curve *is* steep. Statistics *is* hard. We did our best to make your experience of learning how to perform meta-analyses using R as painless as possible. Nevertheless, this will not shield you from being frustrated sometimes. This is all but natural. We all had to start from scratch somewhere down the line. From our own experience, we can you assure that we have never met anyone who was *not* able to learn R, or how to do a meta-analysis. It only takes practice, and the understanding that there will be no point in time when you are "done" learning. We believe in you.

If you are looking for something a little more practical than this motivational message: here are a few things you can do once you stumble upon things that this guide cannot answer.

Do Not Panic

Making their first steps in R, many people are terrified when the first red error messages start popping up. That is not necessary. *Everyone* gets error messages *all the time.* Instead of becoming panicky or throwing your computer out the window, take a deep breath and take a closer look at the error message. Very often, it only takes a few tweaks to make the error messages disappear. Have you misspelled something in your code? Have you forgotten to close a bracket, or to put something into quotation marks? Also, make sure that your output actually *is* an error message. R distinguishes between `Errors`, `Warnings` and plain messages. Only the first means that your code could not be executed. `Warnings` mean that your code did run, but that something *may* have gone awry. Messages mean that your code did run completely, and are usually shown when a function simply wants to bring your attention to something it has done for you under the hood. For this reason, they are also called *diagnostic messages*.

Google

A software developer friend once told the first author this joke about his profession: "A programmer is someone who can Google better than Average Joe". This observation certainly also applies to R programming. If you find yourself in a situation in which you cannot make sense out of an error or warning message you receive, do not hesitate to simply copy and paste it, and do a Google search. Adding "R" to your search is often helpful to improve the results. Most content on the Internet is in English; so if your error message in R is in another language, run `Sys.setenv(LANGUAGE = "en")` and then rerun your code again. There is a large R community out there, and it is very likely that someone had the same problem as you before. Google is also helpful when there is something specific you want to do with your data, but do not know what R commands you should use. Even for experts, it is absolutely normal to use Google *dozens* of times when writing R code. Do not hesitate to do the same whenever you get stuck.

StackOverflow & CrossValidated

When searching for R-related questions on Google, you will soon find out that many of the first hits will link you to a website called StackOverflow[4]. StackOverflow is a large community-based forum for questions related to programming in general. On StackOverflow, everyone (including you) can ask and answer questions. In contrast to many other forums on the Internet, answers you get on StackOverflow are usually goal-oriented and helpful. If searching Google did not help you to solve your problem, addressing it there might be a good solution. However, there are a few things to

[4] `https://stackoverflow.com/`

keep in mind. First, when asking a question, always tag your question with `[R]` so that people know which programming language you are talking about. Also, run `sessionInfo()` in R and attach the output you get to your question. This lets people know which R and package versions you are using, and might be helpful to locate the problem. Lastly, do not expect overwhelming kindness. Many StackOverflow users are experienced programmers who may be willing to point to certain solutions; but do not expect anyone to solve your problem for you. It is also possible that someone will simply inform you that this topic has already been covered elsewhere, send you the link, and then move on. Nevertheless, using StackOverflow *is* usually the best way to get high-quality support for specific problems you are dealing with. StackOverflow, by the way, is primarily for questions on programming. If your question also has a statistics background, you can use CrossValidated[5] instead. CrossValidated works like StackOverflow, but is primarily used by statisticians and machine learning experts.

Contact Us

If you have the feeling that your problem has something to do with this guide itself, you can also contact us. This particularly pertains to issues with the companion R package for this guide, *{dmetar}*. If you have trouble installing the package, or using some if its functions, you can go to our website[6], where you can find ways to report your issue. When certain problems come up frequently, we usually try to have a look at them and search for fixes. Known issues will also be displayed in the Corrections & Remarks section in the online version of the guide (see *Work Flow* section). Please do not be disappointed if we do not answer your question personally, or if takes some time to get back to you. We receive many questions related to meta-analysis and our package every day, so it is sometimes not possible to directly answer each and every one.

Acknowledgments

We would like to thank David Grubbs and Chapman & Hall/CRC Press for approaching us with the wonderful idea of turning our online guide into the printed book you are reading right now, and for their invaluable editorial support.

Many researchers and students have shared their feedback and experiences working with this guide with us since we began writing a preliminary online version of it in

[5]https://stats.stackexchange.com/
[6]www.protectlab.org/meta-analysis-in-r

late 2018. This feedback has been incredibly valuable, and has helped us considerably to tailor this book further to the needs of the ones reading it. Thanks to all of you.

We owe a great debt of gratitude to all researchers involved in the development of the R meta-analysis infrastructure presented in this guide; but first and foremost to Guido Schwarzer and Wolfgang Viechtbauer, maintainers of the *{meta}* and *{metafor}* package, respectively. This guide, like the whole R meta-analysis community, would not exist without your effort and dedication.

Furthermore, particular thanks go to Luke McGuinness, author of the gorgeous *{robvis}* package, for writing an additional chapter on risk of bias visualization, which you can find on this book's companion website. Luke, we are incredibly grateful for your continued support of this project.

Last but not least, we want to thank Lea Schuurmans for supporting us in the development and compilation of this book.

February 2021

Erlangen, Amsterdam, Kyoto and Munich

Mathias, Pim, Toshi and David

About the Authors

Mathias Harrer is a research associate at the Friedrich-Alexander-University Erlangen-Nuremberg. Mathias' research focuses on biostatistical and technological approaches in psychotherapy research, methods for clinical research synthesis and on the development of statistical software.

Pim Cuijpers is professor of clinical psychology at the VU University Amsterdam. He is specialized in conducting randomized controlled trials and meta-analyses, with a focus on the prevention and treatment of common mental disorders. Pim has published more than 800 articles in international peer-reviewed scientific journals, many of which are meta-analyses of clinical trials.

Toshi A. Furukawa is professor of health promotion and human behavior at the Kyoto University School of Public Health. His seminal research focuses both on theoretical aspects of research synthesis and meta-analysis, as well as their application in evidence-based medicine.

David D. Ebert is professor of psychology and behavioral health technology at the Technical University of Munich. David's research focuses on Internet-based intervention, clinical epidemiology, as well as applied research synthesis in this field.

List of Symbols

a, b, c, d	Events in the treatment group, non-events in the treatment group, events in the control group, non-events in the control group.
c_α	Critical value assumed for the Type I error rate α (typically 1.96).
χ^2	Chi-squared statistic.
d	Cohen's d (standardized mean difference).
δ	Non-centrality parameter (non-central t distribution).
ϵ	Sampling error.
g	Small sample bias-corrected standardized mean difference (Hedges' g).
$\int f(x)dx$	Integral of $f(x)$.
κ	True effect of an effect size cluster.
$\bar{x}$	Arithmetic mean (based on an observed sample), identical to m.
n, N	(Total) sample size of a study.
$\Phi(z)$	Cumulative distribution function (CDF), where z follows a standard normal distribution.
P(X|Y)	Conditional probability of X given Y.
β_0, β_1, β	Regression intercept, regression coefficient, Type II error rate.
$\mathcal{HC}(x_0, s)$	Half-Cauchy distribution with location parameter x_0 and scaling parameter s.
Cov(x, y)	Covariance of x and y.
D_g	Regression dummy.
d.f.	Degrees of freedom.
F	Snedecor's F statistic (used by the F-tests in ANOVAs).
I^2	Higgins' and Thompson's I^2 measure of heterogeneity (percentage of variation not attributable to sampling error).
k, K	Some study in a meta-analysis, total number of studies in a meta-analysis.
MD, SMD	(Standardized) mean difference (Cohen's d).
μ, m	(True) population mean, sample mean.
$\mathcal{N}(\mu, \sigma^2)$	Normal distribution with population mean μ and variance σ^2.
π, p	True population proportion, proportion based on an observed sample.
$\hat{\psi}$	(Estimate of) Peto's odds ratio, or some other binary effect size.

(continued)

Q	Cochran's Q measure of heterogeneity.	RR, OR, IRR	Risk ratio, odds ratio, incidence rate ratio.
$\hat{R}$	R-hat value in Bayesian modeling.	R^2_*	R^2 (explained variance) analog for meta-regression models.
ρ, r	True population correlation, observed correlation.	SE	Standard error
σ^2	(True) population variance.	t	Student's t statistic.
τ^2, τ	True heterogeneity variance and standard deviation.	θ	A true effect size, or the true value of an outcome measure.
V, v, s^2, $\widehat{\text{Var}}(x)$	Sample variance (of x), where s is the standard deviation.	w, w^*, $w(x)$	(Inverse-variance) weight, random-effects weight of an effect size, function that assigns weights to x.
z	Fisher's z or z-score.	ζ, u	"Error" due to between-study heterogeneity, random effect in (meta-)regression models.

Note. Vectors and matrices are written in bold. For example, we can denote all observed effect sizes in a meta-analysis with a vector $\hat{\boldsymbol{\theta}} = (\hat{\theta}_1, \hat{\theta}_2, \dots, \hat{\theta}_K)^\top$, where K is the total number of studies. The $\top$ symbol indicates that the vector is *transposed*. This means that elements in the vector are arranged vertically instead of horizontally. This is sometimes necessary to do further operations with the vector, for example, multiplying it with another matrix.

Part I

Getting Started

1

Introduction

Science is generally assumed to be a cumulative process. In their scientific endeavors, researchers build on the evidence compiled by generations of scientists who came before them. A famous quote by Isaac Newton stresses that if we want to see further, we can do so by standing on the "shoulders of giants". Many of us are fascinated by science *because* it is progressive, furthering our understanding of the world, and helping us to make better decisions.

At least by the numbers alone, this sentiment may be justified. Never in history did we have access to more evidence in the form of published research articles than we do today. Petabytes of research findings are produced every day all around the world. In biomedicine alone, more than one million peer-reviewed articles are published each year (Björk et al., 2008). The amount of published research findings is also increasing almost exponentially. The number of articles indexed for each year in one of the largest bibliographical databases, *PubMed*[1], symbolizes this in an exemplary fashion. Until the middle of the 20th century, only a few hundred research articles are listed for each year. These numbers rise substantially for the following decades, and since the beginning of the 21st century, they skyrocket (see Figure 1.1).

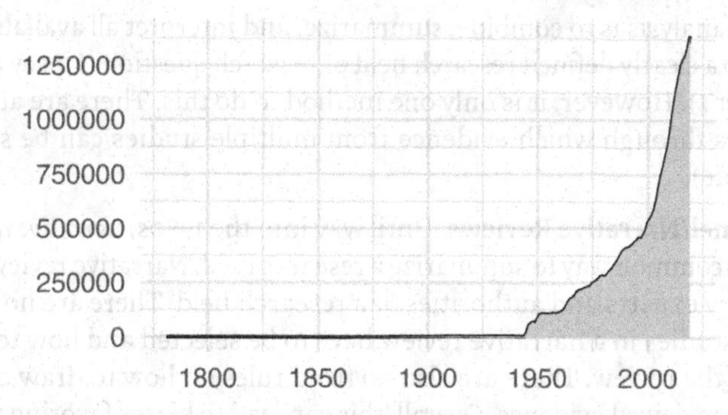

FIGURE 1.1: Articles indexed in PubMed by year, 1781-2019.

In principle, this development should make us enthusiastic about the prospects of science. If science is cumulative, more published research equals more evidence. This should allow us to build more powerful theories and to dismantle fallacies of the past. Yet, of course, it is not that easy. In a highly influential paper, John Ioannidis of

[1] pubmed.ncbi.nlm.nih.gov/

DOI: 10.1201/9781003107347-1

Stanford criticized the notion that science is automatically cumulative and constantly improving. His article has the fitting title "Why Science Is Not Necessarily Self-Correcting" (Ioannidis, 2012). He argues that research fields can often exist in a state where an immense research output is produced on a particular topic or theory, but where fundamental fallacies remain unchallenged and are only perpetuated. Back in the 1970s, the brilliant psychologist Paul Meehl already observed that in some research disciplines, there is a close resemblance between theories and fashion trends. Many theories, Meehl argued, are not continuously improved or refuted, they simply "fade away" when people start to lose interest in them (Meehl, 1978).

It is an inconvenient truth that the scientific process, when left to its own devices, will not automatically move us to the best of all possible worlds. With unprecedented amounts of research findings produced each day, it is even more important to view and critically appraise bodies of evidence *in their entirety*. Meta-analysis can be enormously helpful in achieving this, as long as we acknowledge its own limitations and biases.

1.1 What Are Meta-Analyses?

One of its founding fathers, Gene V. Glass, described meta-analysis as an "analysis of analyses" (Glass, 1976). This simple definition already tells us a lot. In conventional studies, the units of analysis are a number of people, specimens, countries, or objects. In meta-analysis, *primary studies* themselves become the elements of our analysis. The aim of meta-analysis is to combine, summarize, and interpret all available evidence pertaining to a clearly defined research field or research question (Lipsey and Wilson, 2001, chapter 1). However, it is only one method to do this. There are at least three distinct ways through which evidence from multiple studies can be synthesized (Cuijpers, 2016).

- **Traditional/Narrative Reviews**. Until way into the 1980s, *narrative reviews* were the most common way to summarize a research field. Narrative reviews are often written by experts and authorities of a research field. There are no strict rules on how studies in a narrative review have to be selected and how to define the scope of the review. There are also no fixed rules on how to draw conclusions from the reviewed evidence. Overall, this can lead to biases favoring the opinion of the author. Nevertheless, narrative reviews, when written in a balanced way, can be helpful for readers to get an overall impression of the relevant research questions and evidence base of a field.
- **Systematic Reviews**. Systematic reviews try to summarize evidence using clearly defined and transparent rules. In systematic reviews, research questions are determined beforehand, and there is an explicit, reproducible methodology through which studies are selected and reviewed. Systematic reviews aim to cover *all*

available evidence. They also assess the validity of evidence using predefined standards and present a synthesis of outcomes in a systematic way.

- **Meta-Analyses**. Most meta-analyses can be seen as an advanced type of a systematic review. The scope of meta-analyses is clearly defined beforehand, primary studies are also selected in a systematic and reproducible way, and there are also clear standards through which the validity of the evidence is assessed. This is why it is common to find studies being named a "systematic review *and* meta-analysis". However, there is one aspect which makes meta-analyses special. Meta-analyses aim to combine results from previous studies in a *quantitative* way. The goal of meta-analyses is to integrate quantitative outcomes reported in the selected studies into one numerical estimate. This estimate then summarizes all the individual results. Meta-analyses quantify, for example, the effect of a medication, the prevalence of a disease, or the correlation between two properties, *across all studies*[2]. Therefore, they can only be used for studies which report quantitative results. Compared to systematic reviews, meta-analyses often have to be more exclusive concerning the kind of evidence that is summarized. To perform a meta-analysis, it is usually necessary that studies used the same design and type of measurement, and/or delivered the same intervention (see Chapter 1.3).

Individual Participant Data Meta-Analysis

Depending on the definition, there is also a fourth type of evidence synthesis method, so called *Individual Participant Data (IPD) Meta-Analysis* (Riley et al., 2010). Traditionally, meta-analyses are based on *aggregated* results of studies that are found in the published literature (e.g. means and standard deviations, or proportions). In IPD meta-analysis, the *original* data of all studies is collected instead and combined into one big data set. IPD meta-analysis has several advantages. For example, it is possible to impute missing data and apply statistical methods in exactly the same way across all studies. Furthermore, they can make it easier to explore variables which influence the outcome of interest. In traditional meta-analyses, only so-called *study-level* variables (e.g. the year of publication, or the population used in the study) can be used to do this. However, it is often *participant-level* information (e.g. an individual person's age or gender) which may play a role as an important moderator of the results. Such variables can only be explored using IPD meta-analysis.

[2]This statement is of course only true if meta-analytic techniques were applied soundly, and if the results of the meta-analysis allow for such generalizations.

IPD meta-analysis is a relatively new method, and the overwhelming majority of meta-analyses conducted today remain "traditional" meta-analyses. This is also one reason why we will not cover IPD meta-analysis methods in this guide. This has nothing to do with traditional meta-analysis being superior–the opposite is correct. It is simply due to the fact that making all research data openly available has unfortunately been very uncommon in most disciplines until recently. While it is relatively easy to extract summarized results from published research reports, obtaining original data from all relevant studies is much more challenging. In biomedical research, for example, individual participant data can only be obtained from approximately 64% of the eligible studies (Riley et al., 2007).

1.2 "Exercises in Mega-Silliness": A Historical Anecdote

Meta-analysis was not invented by one person alone, but by many founding mothers and fathers (O'Rourke, 2007). The first attempts to statistically summarize the effects of separate, but similar studies date back around 100 years, and can be linked to two of the most important statisticians of all time, Karl Pearson and Ronald A. Fisher. Pearson, in the beginning of the 20th century, combined findings on the effects of typhoid inoculation across the British Empire to calculate a pooled estimate (Shannon, 2016). Fisher, in his seminal 1935 book on the design of experiments, covered approaches to analyze data from multiple studies in agricultural research, and already acknowledged the problem that study results may vary due to location and time (Fisher, 1935; O'Rourke, 2007).

The name "meta-analysis" and the beginning of its rise to prominence, however, can be traced back to a scholarly dispute raging in the mid-20th century. In 1952, the famous British psychologist Hans Jürgen Eysenck (Figure 1.2) published an article in which he claimed that psychotherapy (in that time, this largely meant Freudian psychoanalysis) was ineffective. If patients get better during therapy, it is because their situation would have improved anyway due to factors that have nothing to do with the therapy. Even worse, Eysenck claimed, psychotherapy would often hinder patients from getting better. The reputation of psychotherapy took a big hit, and it did not recover until the late 1970s. During that time, Gene V. Glass developed a technique he termed "meta-analysis", which allowed to pool *Standardized Mean Differences*[3] across studies. The first extensive application of his technique was in an

[3] i.e., the difference in means between two groups, for example, an intervention and control group, expressed in the units of the pooled standard deviation of both groups (see Chapter 3.3.1).

article published in the *American Psychologist*, written by Mary L. Smith and Glass himself (Smith and Glass, 1977). In this large study, results from 375 studies with more than 4000 participants were combined in a meta-analysis. The study found that psychotherapies had a pooled effect of 0.68, which can be considered quite large. Glass' work had an immense impact because it provided quantitative evidence that Eysenck's verdict was wrong. Eysenck himself, however, was not convinced, calling the meta-analysis "an abandonment of scholarship" and "an exercise in mega-silliness" (Eysenck, 1978).

FIGURE 1.2: Hans Jürgen Eysenck (*Sirswindon/CC BY-SA 3.0*).

Today we know that Smith and Glass' study may have overestimated the effects of psychotherapy because it did not control for biases in the included studies (Cuijpers et al., 2019a). However, the primary finding that some psychotherapies are effective has been corroborated by countless other meta-analyses in the following decades. Eysenck's grim response could not change that meta-analysis soon became a commonly used method in various fields of study.

The methodology behind meta-analysis has been continuously refined since that time. About the same time Glass developed his meta-analysis method, Hunter and Schmidt started crafting their own type of meta-analysis techniques putting emphasis on the correction of measurement artifacts (Schmidt and Hunter, 1977; Hunter and Schmidt, 2004). Meta-analysis first found its way into medicine through the groundbreaking work of Peter Elwood and Archie Cochrane, among others, who used meta-analysis to show that aspirin has a small, but statistically and clinically relevant preventive effect on the recurrence of heart attacks (Peto and Parish, 1980; Elwood, 2006; O'Rourke, 2007). In the mid-80s, Rebecca DerSimonian and Nan Laird introduced an approach to calculate random-effects meta-analyses (see Chapter 4.1.2) that has been in use to this day (DerSimonian and Laird, 1986). Countless other innovations have helped to increase the applicability, robustness, and versatility of meta-analytic methods in the last four decades.

The Cochrane and Campbell Collaboration

The *Cochrane Collaboration*[a] (or simply *Cochrane*), founded in 1993 and named after Archie Cochrane, has played a crucial role in the development of applied meta-analysis. Cochrane is an international network of researchers, professionals, patients, and other relevant stakeholders who "work together to produce credible, accessible health information that is free from commercial sponsorship and other conflicts of interest".

Cochrane uses rigorous standards to synthesize evidence in the biomedical field. The institution has its headquarters in London, but also has local branches in several countries around the world. The Cochrane Collaboration issues the regularly updated *Handbook for Systematic Reviews of Interventions*[b] (Higgins et al., 2019) and the *Cochrane Risk of Bias Tool*[c] (Sterne et al., 2019). Both are widely viewed as standard reference works for all technical details concerning systematic reviews and meta-analyses (see Chapter 1.4). An organization similar to Cochrane is the Oslo-based *Campbell Collaboration*[d], which primarily focuses on research in the social sciences.

[a] https://www.cochrane.org/
[b] https://training.cochrane.org/handbook
[c] https://methods.cochrane.org/bias/resources/rob-2-revised-cochrane-risk-bias-tool-randomized-trials
[d] https://campbellcollaboration.org/

1.3 Apples and Oranges: A Quick Tour of Meta-Analysis Pitfalls

In the last decades, meta-analysis has become a universally accepted research tool. This does not come without its own costs. Conducting a high-quality primary study is often very expensive, and it can take many years until the results can finally be analyzed. In comparison, meta-analyses can be produced without too many resources, and within a relatively small time. Nevertheless, meta-analyses often have a high impact and are cited frequently (Patsopoulos et al., 2005). This means that scientific journals are often very inclined to publish meta-analyses, maybe even if their quality or scientific merit is limited. Unfortunately, this creates a natural incentive for researchers to produce many meta-analyses, and scientific considerations sometimes become secondary.

Ioannidis (2016) criticized that an immense amount of redundant and misleading meta-analyses is produced each year. On some "hot" topics, there are more than 20 recent meta-analyses. Some meta-analyses may also be heavily biased by corporate interests, for example, in pharmacotherapy research (Ebrahim et al., 2016; Kirsch et al., 2002). As we have mentioned before, reproducibility is a hallmark of good science. In reality, however, the reproducibility of many meta-analyses is all too often limited because important information is not reported (Lakens et al., 2017). A common problem is also that different meta-analyses on the same or overlapping topics come to different conclusions. In psychotherapy research, for example, there has been an ongoing debate pertaining to the question if all types of psychotherapy produce equivalent outcomes. Countless reviews have been published supporting either one conclusion or the other (Wampold, 2013; Cuijpers et al., 2019c).

While some of these issues may be associated with systemic problems of the scientific process, others can be traced back to flaws of meta-analyses themselves. Therefore, we want to lead you through a quick tour of common meta-analysis pitfalls (Borenstein et al., 2011; Greco et al., 2013; Sharpe, 1997).

- **The "Apples and Oranges" problem.** One may argue that meta-analysis means combining apples with oranges. Even with the strictest inclusion criteria, studies in a meta-analysis will never be absolutely identical. There will always be smaller or larger differences between the included sample, the way an intervention was delivered, the study design, or the type of measurement used in the studies. This can be problematic. Meta-analysis means to calculate a numerical estimate which represents the results of all studies. Such an estimate can always be derived from a statistical point of view, but it becomes meaningless when studies do not share the properties that matter to answer a specific research question. Imagine the, admittedly absurd, scenario in which a meta-analyst decides to pool both studies on the effect of job satisfaction on job performance, as well as all available evidence on the effect of medication on the HbA_{1c} value of diabetic patients in one meta-analysis. The results would be pointless to organizational psychologists and diabetologists alike. Now, suppose that the same poor meta-analyst, trying to learn from previous mistakes, overcompensates and conducts a meta-analysis containing only studies published between 1990 and 1999 in which Canadian males in their sixties with moderate depressive symptoms were treated using 40mg of Fluoxetine, for exactly six weeks. The meta-analyst may proudly report the positive results of the study to a psychiatrist. However, she may only ask: "and what do I do if my patient is 45 years old and French"?
  This brings us to an important point. The goal of meta-analyses is not to heedlessly throw everything together that can be combined. Meta-analysis can be used to answer relevant research questions that go beyond the particularities of individual studies (Borenstein et al., 2011, chapter 40). The scope and specificity of a meta-analysis should therefore be based on the research question it wants to answer, and this question should be of practical relevance (see Chapter 1.4). If we want to know, for example, if a type of training program is effective across various age groups, cultural regions and settings, it makes perfect sense to put

no restriction on the population and country of origin of a study. However, it may then be advisable to be more restrictive with respect to the training program evaluated in the studies, and only include the ones in which the training had a certain length, or covered similar topics. Results of such a meta-analysis would allow us not only to estimate the pooled effect of the training but also allow us to quantify if and how much this effect may *vary* across different settings. Meta-analysis is capable to accommodate and "make sense" out of such forms of *heterogeneity*. In Chapter 5, we will have a closer look at this important concept.
To sum up, whether the "Apples and Oranges" problem is in fact an issue highly depends on the question a meta-analysis wants to answer. Variation between studies can often be unproblematic, and even insightful if it is correctly incorporated into the aims and problem specification of a meta-analysis.

- **The "Garbage In, Garbage Out" problem.** The quality of evidence produced by a meta-analysis heavily depends on the quality of the studies it summarizes. If the results reported in our included findings are biased, or downright incorrect, the results of the meta-analysis will be equally flawed. This is what the "Garbage In, Garbage Out" problem refers to. It can be mitigated to some extent by assessing the quality or *risk of bias* (see Chapter 1.4 and 15) of the included studies. However, if many or most of the results are of suboptimal quality and likely biased, even the most rigorous meta-analysis will not be able to balance this out. The only conclusion that can usually be drawn in such cases is that no trustworthy evidence exists for the reviewed topic, and that more high-quality studies have to be conducted in the future. However, even such a rather disappointing outcome can be informative, and help guide future research.
- **The "File Drawer" problem.** The file drawer problem refers to the issue that not all relevant research findings are published, and therefore missing in our meta-analysis. Not being able to integrate all evidence in a meta-analysis would be undesirable, but at least tolerable if we could safely assume that research findings are missing at random in the published literature. Unfortunately, they are not. Positive, "innovative" findings often generate more buzz than failed replications or studies with negative and inconclusive results. In line with this, research shows that in the last decades, less and less negative findings have been published in many disciplines, particularly in the social sciences and the biomedical field (Fanelli, 2012). There is good reason to believe that studies with negative or "disappointing" results are systematically underrepresented in the published literature and that there is a so called *publication bias*. The exact nature and extent of this bias can be at best a "known unknown" in meta-analyses. However, there are certain ways through which publication bias can be minimized. One pertains to the way that studies are searched for and selected (see Chapter 1.4). The other approaches are statistical methods which try to estimate if publication bias exists in a meta-analysis, and how big its impact may be. We will cover a few of these methods in Chapter 9.
- **The "Researcher Agenda" problem.** When defining the scope of a meta-analysis, searching and selecting studies, and ultimately pooling outcome measures,

researchers have to make a myriad of choices. Meta-analysis comes with many "researcher degrees of freedom" (Wicherts et al., 2016), leaving much space for decisions which may sometimes be arbitrary, and sometimes the result of undisclosed personal preferences. The freedom of meta-analysts in their *modus operandi* becomes particularly problematic when researchers are consciously or subconsciously driven by their own agenda. Meta-analyses are usually performed by applied researchers, and having extensive subject-specific expertise on the reviewed topic is a double-edged sword. On the one hand, it can help to derive and answer meaningful research questions in a particular field. On the other hand, such experts are also deeply invested in the research area they are examining. This means that many meta-analysts may hold strong opinions about certain topics, and may intentionally or unintentionally influence the results in the direction that fits their beliefs. There is evidence that, given one and the same data set, even experienced analysts with the best intentions can come to drastically varying conclusions (Silberzahn et al., 2018). The problem may be even more grave in intervention research, where some meta-analysts have a substantial *researcher allegiance* because they have helped to develop the type of intervention under study. Such researchers may of course be much more inclined to interpret outcomes of a meta-analysis more positively than indicated by the evidence. One way to reduce the researcher agenda problem is pre-registration, and publishing a detailed analysis plan *before* beginning with the data collection for a meta-analysis (see Chapters 1.4 and 16.3.5).

1.4 Problem Specification, Study Search & Coding

In the last chapter, we took some time to discuss common problems and limitations of meta-analyses. Many of these issues, such as the "Apples and Oranges" problem, the "File Drawer" problem, or the "Researcher Agenda" problem, can and should be addressed by every meta-analyst. This begins long before you start calculating your first results. No meta-analysis can be conducted without data, and this data has to come from somewhere. We first have to specify the *research question* and *eligibility criteria* of our planned meta-analysis, search for studies and select the relevant ones, extract the data we need for our calculations, and then code important information we want to report later on. There are several rules, standards, and recommendations we can or should follow during each of these steps; they can help us to create a high-quality meta-analysis. Such high-quality meta-analyses contain a comprehensive selection of all suitable evidence, are unbiased and impartial with respect to their subject, and they draw valid, justified, and practically relevant conclusions from their results.

However, even when "following all the rules", it may not always be clear which specific decision is the best to achieve this in practice. It is possible that people will disagree with the way you went about some things. This is normal and usually just fine, as long as your methodological decisions are both *transparent* and *reproducible* (Pigott and Polanin, 2020).

In this chapter, we will go chronologically through a few important building blocks needed before we can begin with our first calculations. The length of this chapter is not representative of the time this process of data acquisition takes in reality. From our experience, statistical analyses only make up a maximum of 15% of the time spent on a meta-analysis, much less compared to everything that comes before. But specifying the research question, systematically searching for studies and reliably coding extracted data is essential. It builds the basis of every good meta-analysis.

1.4.1 Defining the Research Question

When designing a study, the first thing we do is define the research question. Meta-analysis is no exception. To produce a good research question, it helps to first see it as a form of *problem specification*. To be pertinent and impactful, a meta-analysis should solve a problem. To identify such problems, some subject-specific knowledge is necessary. If you want to find a good research question for a meta-analysis, it may, therefore, be helpful to pick a research area in which you have some background knowledge and ask yourself a few basic questions first. What are the questions which are currently relevant in this particular field? Is there a gap in current knowledge on certain topics? Are there any open discussions that remain unsettled? It might also help to think about the intended target audience. What are problems that are relevant to other researchers? What issues might other people, for example, health care professionals, state agencies, schools, or human resource departments face?

Meta-analysis depends on previous research. Once you know the general direction of your research problem, it therefore helps to have a look at the current literature. Do previous primary studies exist on this topic, and how did they address the problem? What methods and outcome measures did they use? What limitations did they mention in the background and discussion section of the article? Have previous reviews and meta-analyses addressed the topic, and what issues have they left open? Cummings and colleagues (2013) have proposed a few criteria we can use to specify the problem to be covered by our meta-analysis, the FINER framework. It states that a research question should be **F**easible, **I**nteresting, **N**ovel, **E**thical, and **R**elevant.

Step by step, asking yourself these questions should make it easier to define what you want to achieve with your meta-analysis. It may also become clear that meta-analysis is *not* suitable for your problem. For example, there may simply be no relevant studies that have addressed the topic; or there may already be recent high-quality meta-analyses in the literature which address the issue sufficiently. However, if you get the feeling that your problem is relevant to one or several groups of people, that previous studies have provided data pertaining to this problem, and that previous reviews and meta-analyses have not sufficiently or adequately addressed it, you can proceed to turn it into a *research question*.

Let us give you an example of how this can be done. There is evidence suggesting that gender biases exist in medical research (Hamberg, 2008; Nielsen et al., 2017). Especially in earlier decades, many clinical trials only or largely used male participants, and results were simply assumed to generalize to women as well. This has probably lead to worse health outcomes in women for some diseases, such as heart conditions (Kim and Menon, 2009; Mosca et al., 2013)[4]. Let us imagine that you are a medical researcher. You have heard rumors that a commonly used drug, *Chauvicepine*, may have serious side effects in women that have remained largely unrecognized. You determined that this, if true, would be a highly relevant problem because it would mean that many women are prescribed with a drug that is not safe for them. A look into the literature reveals that most studies investigating Chauvicepine were randomized placebo-controlled trials. The first of these trials were conducted in populations which only or predominantly consisted of men. But you also found a few more recent trials in which the gender makeup was more balanced. Many of these trials even reported the number of negative side effects that occurred in the trial separately for men and women. You also find a recent commentary in a medical journal in which a doctor reports that in her clinic, many women have experienced negative side effects when being treated with the medication. Based on this, you decide that it may be interesting to address this problem in a meta-analysis. Therefore, you translate the issue you just discovered into a research question: *does evidence from randomized placebo-controlled trials show that Chauvicepine leads to a significant increase of negative side effects in women, compared to placebo?*

Having derived a first formulation of the research question is only the first step. We now have to translate it into concrete *eligibility criteria*. These eligibility criteria will guide the decision which studies will and will not be included in our meta-analysis. They are, therefore, extremely important and should be absolutely transparent and reproducible. A good way to start specifying the eligibility criteria is to use the PICO framework (Mattos and Ruellas, 2015). This framework is primarily aimed at intervention studies, but it is also helpful for other types of research questions. The letters in PICO stand for **P**opulation, **I**ntervention, **C**ontrol group or comparison, and **O**utcome.

- **Population**: What kind of people or study subjects do studies have to include to be eligible? Again, remember that it is important to address this questions as precisely as possible, and to think of the implications of each definition. If you only want to include studies in young adults, what does "young adults" mean? That only people between 18 and 30 were included? Can that even be determined from the published articles? Or is it just important that people were recruited from places which are usually frequented by young adults, such as universities and *Cardi B* concerts? If you only want to include studies on patients with a specific medical condition, how has that condition been diagnosed? By a trained health care professional, or is a self-report questionnaire sufficient? Many of these questions can be answered by resorting to the F and R parts of the FINER

[4]It is of note that gender bias can not only negatively affect women but also men; an example are diseases such as osteoporosis (Adler, 2014).

framework. Is it feasible to impose such a limitation on published research? And is it a relevant differentiation?

- **Intervention**: What kind of intervention (or alternatively, *exposure*) do studies have to examine? If you want to study the effects of an intervention, it is important to be very clear on the type of treatment that is eligible. How long or short do interventions have to be? Who is allowed to deliver them? What contents must the intervention include? If you do not focus on interventions, how must the *independent variable* be operationalized? Must it be measured by a specific instrument? If you study job satisfaction, for example, how must this construct be operationalized in the studies?
- **Control group** or **comparison**: To what were results of the study compared to? A control group receiving an attention placebo, or a pill placebo? Waitlists? Another treatment? Or nothing at all? It is also possible that there is no comparison or control group; for example, if you want to study the prevalence estimates of a disease across different studies, or how many specimens of a species there are in different habitats.
- **Outcome**: What kind of outcome or dependent variable do studies have to measure? And *how* must the variable be measured? Is it the mean and standard deviation of questionnaire scores? Or the number of patients who died or got sick? When must the outcome be measured? Simply after the treatment, no matter how long the treatment was? Or after one to two years?

Guidelines for Systematic Reviews and Meta-Analyses

In light of the often suboptimal quality of meta-analyses, some guidelines and standards have been established on how meta-analyses should be conducted.

If you meta-analyze evidence in biomedical research or on the effect of an intervention, we strongly advise you to follow the *Preferred Reporting Items for Systematic Reviews and Meta-Analyses*, or PRISMA (Moher et al., 2009). The PRISMA statement contains several recommendations on how nearly all aspects of the meta-analysis process should be reported. The statement can also be found online[a]. For meta-analyses of psychological and behavior research, the *American Psychological Association*'s *Meta-Analysis Reporting Standards* (Appelbaum et al., 2018), or MARS, should be followed.

[a] http://www.prisma-statement.org/

> Although these standards largely pertain to how meta-analyses should be *reported*, they also have implications on best practices when *performing* a meta-analysis. PRISMA and MARS share many core elements, and many things that we cover in this chapter are also mentioned in both of these guidelines.
>
> An even more detailed resource is the *Cochrane Handbook for Systematic Reviews of Interventions* (see Chapter 1.2), which contains precise recommendations on virtually every aspect of systematic reviews and meta-analyses. An overview of methodological standards for meta-analyses (with a focus on social science) can be found in Pigott and Polanin (2020).

While the PICO framework is an excellent way to specify the eligibility criteria of a meta-analysis, it does not cover all information that may be relevant. There are a few other aspects to consider (Lipsey and Wilson, 2001).

One relevant detail are the eligible *research designs*. In evidence-based medicine, it is common to only include evidence from randomized controlled trials (meaning studies in which participants were allocated to the treatment or control group by chance); but this is not always required (Borenstein et al., 2011, chapter 40).

It may also be helpful to specify the *cultural* and *linguistic range* of eligible studies. Most research is based on WEIRD populations, meaning western, educated, industrialized, rich, and democratic societies (Henrich et al., 2010). Especially in social science, it is very likely that certain effects or phenomena do not generalize well to countries with other societal norms. Many researchers, however, only consider publications in English for their meta-analyses, to avoid having to translate articles in other languages. This means that some evidence from different language areas will not be taken into account. Although English is the most common language for scientific publishing in most disciplines, it should be at least made transparent in the eligibility criteria that this limitation exists. If one of the goals of a meta-analysis is to examine cross-cultural differences, however, it is generally advisable to extend the eligibility criteria to other languages, provided all the other criteria are fulfilled.

Another important aspect is the *publication type* that is allowed for a meta-analysis. Sometimes, meta-analysts only include research articles which were published in peer-reviewed scientific journals. The argument is that studies taken from this source fulfill higher standards since they have passed the critical eyes of experts in the field. This justification is not without flaws. In Chapter 1.3, we already covered that the "File Drawer" problem can seriously limit the validity of meta-analysis results because positive findings are more likely to get published. A way to mitigate the risk of publication bias is therefore to also include *grey literature*. Grey literature can be defined as all types of research materials that have not been made available through conventional publication formats. This includes research reports, preprints, working papers, or conference contributions. Dissertations also often count as grey literature, although many of them are indexed in electronic bibliographic databases today (Schöpfel and Rasuli, 2018). It may be advisable to at least also include dissertations in a meta-analysis. Compared to other types of unpublished material, it is rather unlikely that the information provided in dissertations is heavily biased or downright fraudulent. Furthermore, you can still define other eligibility criteria to ensure that only studies fulfilling certain methodological requirements are included, no matter if they were published in scientific journals or not.

The last step of defining your eligibility criteria is to write them down as a list of *inclusion* and *exclusion criteria* that you will apply. Here is an example from a meta-analysis of insomnia interventions in college students showing how this can be done (Saruhanjan et al., 2020):

> *"We included: (a) RCTs [randomized controlled trials; authors' note] in which (b) individuals enrolled at a tertiary education facility (university, college or comparable postsecondary higher education facility) at the time of randomization, (c) received a sleep-focused psychological intervention, (d) that was compared with a passive control condition, defined as a control condition in which no active manipulation was induced as part of the study (wait-list, treatment as usual).*
> *For the purposes of this analysis, "sleep-focused" means that (e) effects on symptoms of sleep disturbances (global measures of sleep disturbances, sleep-onset latency [...], fatigue and daytime functionality, pre-sleep behaviour and experiences) were assessed as a (f) target outcome (by declaring a sleep outcome as the primary outcome or by stating the intervention was primarily aimed at this outcome) using (g) standardized symptom measures (objective sleep measures, standardized sleep or fatigue questionnaires, sleep diaries, items recording sleep quantity, quality or hygiene).*
> *Only studies (h) published in English or German were considered for inclusion."*

1.4.2 Analysis Plan & Preregistration

After your research question and eligibility criteria are set, it is sensible to also write an *analysis plan* (Pigott and Polanin, 2020; Tipton et al., 2019). In statistics, there is an important distinction between *a priori* and *post hoc* analyses. *A priori* analyses are specified *before seeing the data*. Post hoc, or *exploratory*, analyses are conducted *after seeing the data,* or based on the results implicated by the data. Results of *a priori* analyses can be regarded as much more valid and trustworthy than post hoc analyses. Post hoc analyses make it easier to tweak certain details about the analysis or the data itself until results support the goals of the researcher. They are therefore much more prone to the "Researcher Agenda" problem we discussed in Chapter 1.3.

In the analysis plan, we specify all important calculations we want to perform in our meta-analysis *a priori*. This serves two purposes. First, it allows others to verify that the analyses we made were indeed planned, and are not the mere result of us playing around with the data until something desirable came out. Second, a detailed analysis plan also makes our meta-analysis reproducible, meaning that others can understand what we did at each step of our meta-analysis, and try to replicate them. When using R, we can take the reproducibility of our analyses to a whole new level by writing documents which allow others to re-run every step of our analysis (see Chapter 16 in the "Helpful Tools" section). But this is relevant *after* we complete our analyses. In the analysis plan, we specify what we plan to do *before* any data has been collected.

There are a few things we should always specify in our analysis plan. We should make clear which information we will extract, and which effect size metric will be calculated for each included study (see Chapter 3). It is also recommended to decide beforehand if we will use a *fixed-* or *random-effects model* to pool results from each study, based on the amount of variation between studies we expect (see Chapter 4). An *a priori power analysis* may also be helpful to determine how many studies are required for our meta-analysis to find a statistically significant effect (see Chapter 14 in the "Helpful Tools" section). Furthermore, it is crucial to determine if we want to assess if some variables explain differences in the outcomes of included studies using a subgroup analysis (Chapter 7) or meta-regression (Chapter 8). For example, if our hypothesis states that the publication year might be associated with a study's outcome, and if we want to have a look at this association later in our meta-analysis, we must mention this in our analysis plan. If we plan to sort studies into subgroups and then have a look at these subgroups separately, we should also report the exact criteria through which we will determine that a study belongs to a specific subgroup (see Chapter 1.4.4). In Part II of this book, we will cover various statistical techniques to apply as part of a meta-analysis. Every technique we learn there and plan to apply in our meta-analysis should be mentioned in the analysis plan.

Once you are finished writing your analysis plan, do not simply bury it somewhere–make it public. There are a few excellent options for researchers to make their research documents openly available. For example, we can create a new project on the website of the *Open Science Framework* (OSF; see Chapter 16.3 in the "Helpful Tools" section) and upload our analysis plan there. We can also upload our analysis plan to a preprint server such as *medrxiv.org, biorxiv.org,* or *psyarxiv.com,* depending on the nature of our research question. Once our eligibility criteria, analysis plan and search strategy (see next chapter) are set, we should also *register* our meta-analysis. If the meta-analysis has a broadly health-related outcome, this may preferably be done using PROSPERO[5], one of the largest registries for prospective systematic reviews and meta-analyses. The preregistration service of the OSF[6] is also a good option.

In case we want to go even one step further, we can also write an entire *protocol* for our meta-analysis (Quintana, 2015). A meta-analysis protocol contains the analysis plan, plus a description of the scientific background of our study, more methodological detail, and a discussion of the potential impact of the study. There are also guidelines on how to write such protocols, such as the PRISMA-P Statement (Moher et al., 2015). Meta-analysis protocols are accepted by many peer-review journals. A good example can be found in Büscher, Torok and Sander (2019), or Valstad and colleagues (2016).

A priori analysis plans and preregistration are essential features of a well-made, trustworthy meta-analysis. And they should not make you anxious. Making the perfect choice for each and every methodological decision straight away is difficult, if not impossible. It is perfectly natural to make changes to one's initial plans somewhere down the road. We can assure you that, if you are honest and articulate about changes to your planned approach, most researchers will not perceive this as a sign of failure, but of professionalism and credibility.

1.4.3 Study Search

The next step after determining your eligibility criteria and analysis plan is to search for studies. In Chapter 1.1, we discussed that most meta-analyses are an advanced type of systematic review. We aim to find *all* available evidence on a research question in order to get an unbiased, comprehensive view of the facts. This means that the search for studies should also be as comprehensive as possible. Not only one, but several sources should be used to search for studies. Here is an overview of important and commonly used sources.

- *Review articles*. It can be very helpful to screen previous reviews on the same or similar topics for relevant references. Narrative and systematic reviews usually provide a citation for all the studies that they included in their review. Many of these studies may also be relevant for your purposes.
- *References in studies*. If you find a study that is relevant for your meta-analysis, it is sensible to also screen the articles that this study references. It is very likely that the study cites previous literature on the same topic in the introduction or discussion section, and some of these studies may also be relevant for your meta-analysis.
- *Forward search*. A forward search can be seen as the opposite of screening the references of previous primary studies and reviews. It means to take a study that is relevant for the meta-analysis as basis, and then search for other articles that have cited this study since it has been published. This can be done quite easily

[5] https://www.crd.york.ac.uk/prospero/
[6] https://osf.io/prereg/

on the Internet. You simply have to find the online entry of the study; usually, it is on the website of the journal in which it has been published. Most journal websites today have a functionality to display articles that have cited a study. Alternatively, you can also search for the study on *Google Scholar* (see Table 1.1). Google Scholar can display citing research for every entry.

- *Relevant journals*. Often, there are a number of scientific journals which are specialized in the type of research question you are focused on. Therefore, it can be helpful to search for studies specifically in those journals. Virtually all journals have a website with a search functionality today, which you can use to screen for potentially eligible studies. Alternatively, you can also use electronic bibliographical databases, and use a filter so that only results from one or several journals are displayed.
- *Electronic bibliographical databases*. The methods we described above can be seen as rather fine-grained strategies. They are ways to search in places where it is very likely that a relevant article will be listed. The disadvantage is that these approaches will unlikely uncover all evidence that is really out there. Thus, it is advisable to also use electronic bibliographic databases for one's search. An overview of important databases can be found in Table 1.1.
  One should always conduct a search in several databases, not just one. Many bibliographical databases contain an immense number of entries. Nevertheless, it is common to find that the overlap in the results of database searches is smaller than anticipated. You can select the databases you want to search based on their subject-specific focus. If your meta-analysis focuses on health-related outcomes, for example, you should at least search PubMed and CENTRAL.
  When searching bibliographic databases, it is important to develop a *search string*. A search string contains different words or terms, which are connected using operators such as AND or OR. Developing search strings takes some time and experimenting. A good way to start is to use the PICO or eligibility criteria (Chapter 1.4.1) as basis and to connect them using AND (a simplified example would be *"college student" AND "psychotherapy" AND "randomized controlled trial" AND "depression"*).
  Most bibliographical databases also allow for *truncation* and *wildcards*. Truncation means to replace a word ending with a symbol, allowing it to vary as part of your search. This is usually done using asterisks. Using *"sociolog*"* as a search term, for example, means that the database will search for "sociology", "sociological", and "sociologist" at the same time. A wildcard signifies that a letter in a word can vary. This can come in handy when there are differences in the spelling of words (for example, differences between American English and British English). Take the search term *"randomized"*. This will only find studies using American English spelling. If you use a wildcard (often symbolized by a question mark), you can write *"randomi?ed"* instead, and this will also give results in which the British English spelling was used ("randomised").
  When developing your search string, you should also have a look at the number of hits. A search string should not be too specific, so that some relevant articles are missed. For example, getting around 3000 hits for your search string is

manageable in later steps, and it makes it more likely that all important references will be listed in your results. To see if your search string is generally valid, it sometimes helps to inspect the first few hundred hits you get, and to check if at least some of the references have something to do with your research question. Once you have developed the final versions of the search strings you want to use in your selected databases, save them somewhere. It is best practice to already include your search string(s) in your preregistration. Reporting of the search string (for example, in the supplement) is required if you want to publish a meta-analysis protocol (see Chapter 1.4.1), or the final results of your meta-analysis. In conclusion, we want to stress that searching bibliographic databases is an art in and of itself, and that this paragraph only barely scratches the surface. A much more detailed discussion of this topic can be found in Cuijpers (2016) and Bramer and colleagues (2018).

TABLE 1.1: A selection of relevant bibliographical databases.

Database	Description	Website
	Core Database	
PubMed	Openly accessible database of the US National Library of Medicine. Primarily contains biomedical research.	ncbi.nlm.nih.gov/ pubmed
PsycInfo	Database of the American Psychological Association. Primarily covers research in the social and behavioral sciences. Allows for a 30-day free trial.	apa.org/pubs/ databases/psycinfo
Cochrane Central Register of Controlled Trials (CENTRAL)	Openly accessible database of the Cochrane Collaboration. Primarily covers health-related topics.	cochranelibrary.com/ central
Embase	Database of biomedical research maintained by the large scientific publisher Elsevier. Requires a license.	elsevier.com/solutions/ embase-biomedical-research
ProQuest International Bibliography of the Social Sciences	Database of social science research. Requires a license.	about.proquest.com/ products-services/ibss-set-c.html
Education Resources Information Center (ERIC)	Openly accessible database on education research.	eric.ed.gov

TABLE 1.1: A selection of relevant bibliographical databases. *(continued)*

Database	Description	Website
	Citation Database	
Web of Science	Interdisciplinary citation database maintained by Clarivate Analytics. Requires a license.	webofknowledge.com
Scopus	Interdisciplinary citation database maintained by Elsevier. Requires a license.	scopus.com
Google Scholar	Openly accessible citation database maintained by Google. Has only limited search and reference retrieval functionality.	scholar.google.com
	Dissertations	
ProQuest Dissertations	Database of dissertations. Requires a license	about.proquest.com/ products-services/dissertations/
	Study Registries	
WHO International Clinical Trials Registry Platform (ICTRP)	Openly accessible database of clinical trial registrations worldwide. Can be used to identify studies that have not (yet) been published.	www.who.int/ictrp
OSF Registries	Openly accessible interdisciplinary database of study registrations. Can be used to identify studies that have not (yet) been published.	osf.io/registries

1.4.4 Study Selection

After completing your study search, you should have been able to collect thousands of references from different sources. The next step is now to select the ones that fulfill your eligibility criteria. It is advised to follow a three-stepped procedure to do this.

In the first step, you should remove duplicate references. Especially when you search in multiple electronic bibliographical databases, it is likely that a reference will appear more than once. An easy way to do this is to first collect all your references in one place by importing them into a *reference management software*. There are several good reference management tools. Some of them, like *Zotero*[7] or *Mendeley*[8] can be downloaded for free. Other programs like *EndNote*[9] provide more functionality but usually require a license. Nearly all of those reference managers have a functionality which allows you to automatically remove duplicate articles. It is important that you write down the number of references you initially found in your study search, and how many references remained after duplicate removal. Such details should be reported later on once you make your meta-analysis public.

After duplicate removal, it is time to eliminate references that do not fit your purpose, based on their *title and abstract*. It is very likely that your study search will yield hundreds of results that are not even remotely linked to your research question[10]. Such references can be safely removed by looking at their title and abstract only. A reference manager will be helpful for this step too. You can go through each reference one after another and simply remove it when you are sure that the article is not relevant for you[11]. If you think that a study *might* contain interesting information based on the title and abstract, do *not* remove it–even if it seems unlikely that the study is important. It would be unfortunate if you put considerable time and effort into a comprehensive study search just to erroneously delete relevant references in the next step. The title and abstract-based screening of references does not require you to give a specific reason why you excluded the study. In the end, you must only document how many studies remained for the next step.

Based on title and abstract screening, it is likely that more than 90% of your initial references could be removed. In the next step, you should now retrieve the *full article* for each reference. Based on everything reported in the article, you then make a final decision if the study fulfills your eligibility criteria or not. You should be particularly thorough here because this is the final step determining if a study will be included in your meta-analysis or not. Furthermore, it is not simply sufficient to say that you removed a study because it did not fit your purpose. You have to give a *reason* here. For each study you decide to remove, you should document why exactly it was not eligible as per your defined criteria. Besides your eligibility criteria, there is one other reason why you might not be able to include a study. When going through the full article, you might discover that not enough information is provided to decide whether the study is eligible or not. It is possible that a study simply does not provide enough information on the research design. Another frequent scenario is that the results of a study are not reported in a format that would allow to calculate the effect size metric used in your meta-analysis. If this happens, you should try to contact

---

[7] `https://www.zotero.org/`

[8] `https://www.mendeley.com/`

[9] `https://endnote.com/`

[10] Lipsey and Wilson (2001) tell the amusing anecdote that, when searching articles for a meta-analysis on the relationship between alcohol consumption and aggression, they had to exclude a surprisingly large number of studies in which alcohol was given to fish to examine territorial fighting behavior.

[11] When exporting references from an electronic database, the abstract is usually added to the reference file, and can be displayed in the reference management tool. If no abstract is found for the reference, it usually only takes a quick Google search of the study title to find it.

the corresponding author of the study at least two times, and ask for the needed information. Only if the author does not respond, and if the information lacking in the published article is essential, you can exclude the study.

Once we have arrived at the final selection of studies to include, we write down all the details of the inclusion process in a *flow diagram*. A commonly used template for such a flow chart is the one provided by the PRISMA guidelines[12]. This flow chart documents all the necessary information we covered above: (1) how many references we could identify by searching electronic databases; (2) how many additional references we found through other sources; (3) the number of references that remained after duplicate removal; (4) the number of references we removed based on title and abstract; (5) the number of articles we removed based on the full manuscript, including how many articles where excluded due to which specific reason; and (6) the number of studies we included in our qualitative synthesis (systematic review) and quantitative synthesis (meta-analysis). Please note that the number of articles that were not excluded at (5) and the number of studies included in (6) are usually identical, but they do not have to be. For example, it is possible that one article reports results of two or more independent studies, all of which are suitable for meta-analysis. The number of *studies* would then be higher than the number of included *articles*.

!

Double-Screening

Nearly all relevant guidelines and consensus statements emphasize that *double screening* should be used during the study selection process (Tacconelli, 2009; Higgins et al., 2019; Campbell Collaboration, 2016). This means that at least two people should perform each of the study selection steps independently to avoid errors. Reference removal based on the title and abstract should be conducted independently by two or more researchers, and the combination of all records that have not been removed by the assessors should be forwarded to the next step. Using two or more assessors is even more important in the final step, in which full articles are screened. In this step, each person should independently assess if a study is eligible, and if it is not, give reasons why.

The assessors should then meet and compare their results. It is common that assessors disagree on the eligibility of some studies, and such disagreements can usually be resolved through discussion. If assessors fail to find an agreement, it can be helpful to determine a senior researcher beforehand who can make a final decision in such cases.

Using two or more assessors is not only advisable in the study selection

[12]http://prisma-statement.org/PRISMAStatement/FlowDiagram

process. This approach is also beneficial when extracting and coding data (see Chapter 1.4.5).

1.4.5 Data Extraction & Coding

When the selection of studies to be included in the meta-analysis is finalized, data can be extracted. There are three major types of information we should extract from the selected articles (Cuijpers, 2016):

1. Characteristics of the studies.
2. Data needed to calculate effect sizes.
3. Study quality or risk of bias characteristics.

It is conventional for high-quality meta-analyses to provide a table in which characteristics of the included studies are reported. The exact details reported in this table can vary depending on the research field and research question. However, you should always extract and report the first author of a study, and when it was published. The sample size of each study should also be reported. Apart from that, you may include some information on characteristics specified in the PICO of your meta-analysis; such as the country of origin, the mean or median age, the proportion of female and male participants, the type of intervention or exposure, the type of control group or comparison (if applicable), as well as the assessed outcomes of each study. If one or several studies have not assessed one of the characteristics, you should indicate that this detail has not been specified in the table.

It is also necessary to extract and collect the data needed to calculate the effect sizes or outcome measures we plan to pool. In Chapter 2, we will discuss in greater detail how you can structure your effect size data in a spreadsheet so that it can easily be used for calculations in R. If your analysis plan (see Chapter 1.4.2) also includes planned subgroup analyses and meta-regressions, you should also extract the data you need for these analyses from the articles.

It is common in meta-analysis to also rate and report the quality of the primary studies. The information you need to extract from each study depends on the type of rating system you are using. Countless tools to assess the quality of primary studies have been developed in the last decades (Sanderson et al., 2007). When only randomized controlled trials are eligible for your study, one of the best ways to code the study quality is to use the *Risk of Bias Tool* developed by Cochrane (Higgins et al., 2011; Sterne et al., 2019)[13]. As it says in the title, this tool does not assess the quality of studies *per se*, but their risk of bias. Study quality and risk of bias are related, but

[13]https://methods.cochrane.org/bias/resources/rob-2-revised-cochrane-risk-bias-tool-randomized-trials

not identical concepts. "Bias" refers to systematic errors in the results of a study or their interpretation. Risks of bias are aspects of the way a study was conducted, or its results, that may increase the likelihood of such systematic errors. Even when a study only applies methods that are considered the "state of the art", it is still possible that biases exist. A study can fulfill all quality standards that are perceived as important in a particular research field, but sometimes even these best practices may not be enough to shield the study from distortions. The "risk of bias" concept thus has a slightly different focus compared to study quality assessments. It primarily cares if the output of an intervention study is *believable* and focuses on criteria which are conducive to this goal (see also Higgins et al., 2019, chapter 7).

On several domains, the risk of bias tool lets you classify the risk of bias of a study as "high" or "low", or it can be determined that there are "some concerns". There are also conventions on how the risk of bias can be summarized visually (see Chapter 15, where we describe how this can be done in R). A similar resource to assess the risk of bias in non-randomized studies is the *Risk of Bias in Non-randomized Studies of Interventions*, or ROBINS-I, tool (Sterne et al., 2016)[14].

The Cochrane Risk of Bias tools have become the standard approach to assess the risk of bias in (non-)randomized clinical trials (Jørgensen et al., 2016). In other areas, current practices unfortunately still rather resemble the Wild West. In psychological research, for example, study quality assessments are often inconsistent, nontransparent, or not conducted at all (Hohn et al., 2019). If you plan to meta-analyze studies other than clinical trials, there are two things you can do. First, you can check if the Risk of Bias or ROBINS-I tool may still be applicable, for example, if your studies focus on another type of intervention that simply has no health-related focus. Another–admittedly suboptimal–way may be to search for previous high-quality meta-analyses on similar topics, and check how these studies have determined the quality of primary studies.

This ends our dive into the history of meta-analysis, its problems, and how we can avoid some of them when collecting and encoding our data. The next chapter is the beginning of the "hands-on" part of this guide. In it, we will do our own first steps in R.

□

[14]https://www.riskofbias.info/welcome/home

1.5 Questions & Answers

Test your knowledge!

1. How can meta-analyses be defined? What differentiates a meta-analysis from other types of literature reviews?
2. Can you name one of the founding mothers and fathers of meta-analysis? What achievement can be attributed to her or him?
3. Name three common problems of meta-analyses and describe them in one or two sentences.
4. Name qualities that define a good research question for a meta-analysis.
5. Have a look at the eligibility criteria of the meta-analysis on sleep interventions in college students (end of Chapter 1.4.1). Can you extract the PICO from the inclusion and exclusion criteria of this study?
6. Name a few important sources that can be used to search studies.
7. Describe the difference between "study quality" and "risk of bias" in one or two sentences.

Answers to these questions are listed in Appendix A at the end of this book.

1.6 Summary

- More and more scientific research is published each year, making it harder to keep track of available evidence. However, more research output does not automatically result in scientific progress.
- Meta-analysis aims to combine the results of previous studies in a quantitative way. It synthesizes all available evidence pertaining to a research question and can be used for decision-making.

- Meta-analytic methods trace back to the beginning of the 20th century. Modern meta-analytic approaches, however, have been developed in the second half of the 20th century, and meta-analysis has become a common research tool since then.
- There are several problems that are relevant for each meta-analysis: the "Apples and Oranges" problem, the "Garbage In, Garbage Out" problem, the "File Drawer" problem, and the "Researcher Agenda" problem.
- Many of these problems can be mitigated by defining a clear research question and eligibility criteria, writing an analysis plan, pre-registering the meta-analysis, and conducting the study search and data extraction in a systematic and reproducible way.

2

Discovering R

In this chapter, we begin our journey into the R universe. Possibly, this is your first-ever exposure to programming, and you might feel a little anxious. That is understandable, but there is no reason to worry. Over the last two decades, thousands of intelligent people all around the world have contributed ways to make working with R easier and more convenient for its users. We will also get to know an extremely powerful computer program that we can use to make writing and running R code much less cumbersome.

Nevertheless, it is still harder to work with R compared to other data analysis programs you may have used before. Hadley Wickham, one of the most important figures in the R community, once stressed that R is fundamentally different to *Graphical User Interface* (GUI)-based software for statistics (Grolemund, 2014). GUIs allow you to perform data analyses just by clicking a few buttons, but you are ultimately limited to the functionality that the developers have deemed important. R, on the other hand, has none of these limitations but can require more background knowledge. Like any language, R has to be learned, and it needs practice to become a proficient user. Frustration is an unpleasant, but natural part of this process. In the preface, you can find an entire section in which we describe a few things you can do if you get stuck.

We want to assure you that learning R *is* worth it. R is the most versatile, comprehensive and most frequently used statistical programming language. The R community is rapidly expanding each and every year, and R's appeal is so large that even the New York Times found it newsworthy to report on it (Vance, 2009). Whether you are working in academia or at a company, the things you can do in R will often seem like a superpower to others. But it is a superpower that everyone can learn, provided some time and effort. That being said, it is time to get started.

2.1 Installing R & R Studio

Before we can begin, we have to download and prepare a computer program which allows us to use R in a convenient way for our statistical analyses. Probably the best option for this at the moment is *R Studio*[1]. This program gives us a user interface

[1] `https://rstudio.com/`

DOI: 10.1201/9781003107347-2

which makes it easier to handle our data, packages and output. The best part is that R Studio is completely free and can be downloaded anytime on the Internet. Recently, an online version of R Studio has been released[2], which provides you with largely the same interface and functionality through your web browser. In this book, however, we will focus on the R Studio version we install directly on our computer.

In this chapter, we will focus on how you can install R and R Studio on your computer. If you have installed R Studio on your computer already, and if you are an experienced R user, none of this might be new for you. You may skip this chapter then. If you never used R before, bear with us.

Let us go through the necessary steps to set up R and R Studio for our first coding endeavors.

1. R Studio is an interface which allows us to write R code and run it in an easy way. But R Studio is not identical with R; it requires that R software is already installed on your computer. First, we therefore have to install the latest R version. Like R Studio, R is completely free. It can be downloaded from the *Comprehensive* R *Archive Network*, or CRAN, website. The type of R you have to download depends on whether you are using a Windows PC[3] or a Mac[4]. An important detail about R is its *version*. R is regularly updated, meaning that new versions become available. When your R version becomes too outdated, it may happen that some things will not work anymore. It is therefore helpful to update the R version regularly, maybe roughly every year, by re-installing R. For this book, we are using R version 4.0.3. By the time you install R, there may already be a higher version available, and it is advised to always install the latest version.
2. After you have downloaded and installed R, you can download "R Studio Desktop" from the R Studio website[5]. There are also versions of R Studio for which you have to buy a license, but this is definitely not required for our purposes. Simply download and install the free version of "R Studio Desktop".
3. The first time you open R Studio, it will probably look a lot like in Figure 2.1. There are three panes in R Studio. In the upper right corner, we have the *Environment* pane, which displays objects that we defined (i.e. saved) internally in R. In the bottom right corner, you can find the *Files, Plots, Packages and Help* pane. This pane serves several functions; for example, it

[2] `https://rstudio.cloud/`
[3] `https://cran.r-project.org/bin/windows/base/`
[4] `https://cran.r-project.org/bin/macosx/`
[5] `https://rstudio.com/products/rstudio/download/`

is used to display files on your computer, show plots and installed packages and to access the help pages. The heart of R Studio, however, is the left side, where you can find the *Console*. The console is where all the R code is entered and then run.

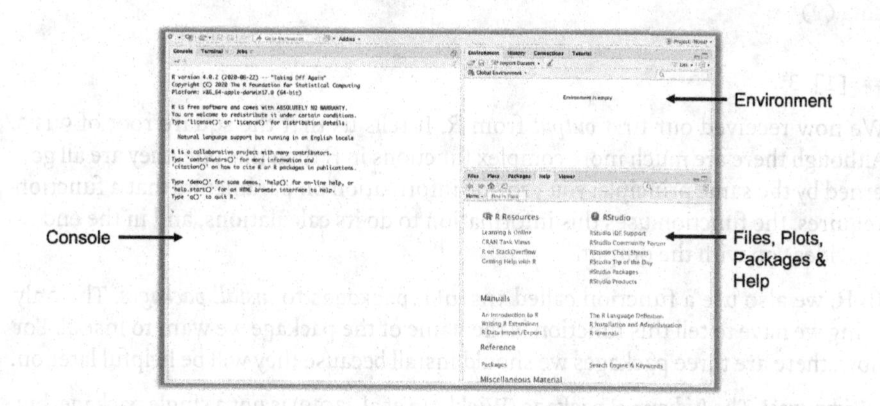

FIGURE 2.1: Panes in R Studio.

4. There is a fourth pane in R Studio that is usually not shown in the beginning, the *source* pane. You can open the source pane by clicking on *File > New File > R Script* in the menu. This will open a new pane in the upper left corner, containing an empty R *script*. R scripts are a great way to collect your code in one place; you can also save them as a file with the extension ".R" (e.g. *myscript.R*) on your computer. To run code in an R script, select it by dragging your cursor across all relevant lines, and then click on the "Run" button to the right. This will send the code to the console, where it is evaluated. A shortcut for this is Ctrl + R (Windows) or Cmd + R (Mac).

2.2 Packages

We will now install a few *packages* using R code. Packages are one of the main reasons why R is so powerful. They allow experts all around the world to develop sets of *functions* that others can download and then use in R. Functions are the core elements of R; they allow us to perform a predefined type of operation, usually on our own data. There is a parallel between the mathematical formulation of a function, $f(x)$, and the way functions are defined in R. In R, a function is coded by first writing down its name, followed by a bracket which contains the input and/or specifications for the function (so-called *arguments*). Say that we want to know what the square root of 9 is. In R, we can use the `sqrt` function for this. We simply have to provide 9 as the input to the function to get the result. You can try this yourself. Next to the little arrow (>) in your console, write down `sqrt(9)` and then hit Enter. Let us see what happens.

```
sqrt(9)
```

```
## [1] 3
```

We now received our first *output* from R. It tells us that the square root of 9 is 3. Although there are much more complex functions in R than this one, they are all governed by the same principle: you provide information on parameters that a function requires, the function uses this information to do its calculations, and in the end, it provides you with the output.

In R, we also use a function called `install.packages` to *install packages*. The only thing we have to tell this function is the name of the package we want to install. For now, there are three packages we should install because they will be helpful later on.

- ***{tidyverse}***. The *{tidyverse}* package (Wickham et al., 2019) is not a single package, but actually a bundle of packages which make it easy to manipulate and visualize data in R. When we install the *{tidyverse}* package, this provides us with the *{ggplot2}*, *{dplyr}*, *{tidyr}*, *{readr}*, *{purrr}*, *{stringr}* and *{forcats}* package at the same time. Functions included in the tidyverse have become very popular in the R community in recent years, and are used by many researchers, programmers and data scientists. If you want to learn more about the tidyverse, you can visit its website[6].
- ***{meta}***. This package contains functions which make it easy to run different types of meta-analyses (Balduzzi et al., 2019). We will primarily focus on this package in the guide, because it is easy to use, well documented and very versatile. More info on the *{meta}* package can be found on its website[7].
- ***{metafor}***. The *{metafor}* package (Viechtbauer, 2010) is also dedicated to conducting meta-analyses, and a true powerhouse in terms of functionality. Since we will use this package at times in later chapters, and because *{metafor}* is used by the *{meta}* package for many applications, it is best to have it installed. The *{metafor}* package also has an excellent documentation[8] for various meta-analysis-related topics.

The `install.packages` function only requires the name of the package we want to install as input. One package after another, our code should look like this:

```
install.packages("tidyverse")
install.packages("meta")
install.packages("metafor")
```

[6] https://www.tidyverse.org/
[7] http://www.imbi.uni-freiburg.de/lehre/lehrbuecher/meta-analysis-with-r
[8] http://www.metafor-project.org/doku.php/metafor

Simply type the code above into the console; then hit Enter to start the installation (see Figure 2.1).

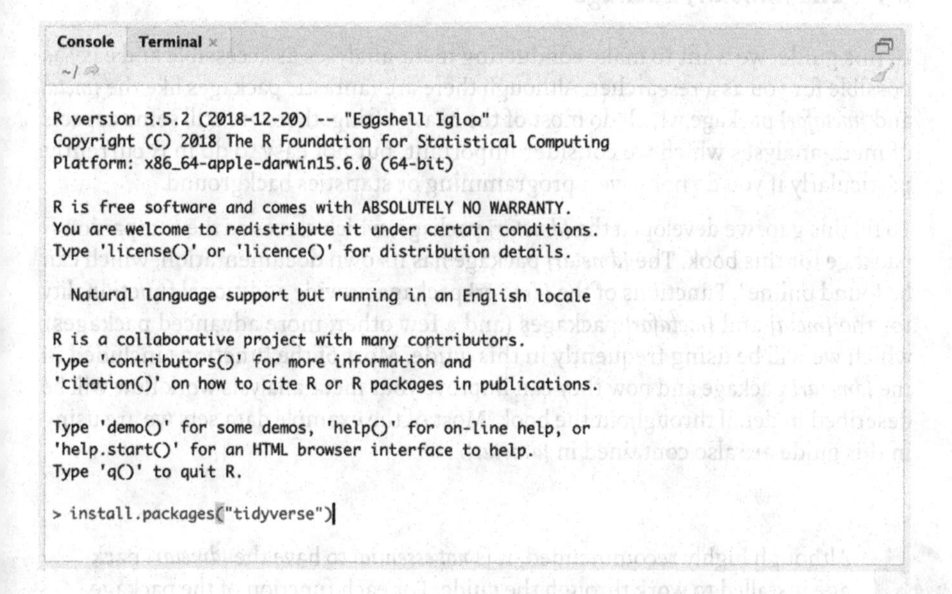

Console
Terminal
~/
R version 3.5.2 (2018-12-20) -- "Eggshell Igloo"
Copyright (C) 2018 The R Foundation for Statistical Computing
Platform: x86_64-apple-darwin15.6.0 (64-bit)
R is free software and comes with ABSOLUTELY NO WARRANTY.
You are welcome to redistribute it under certain conditions.
Type 'license()' or 'licence()' for distribution details.
Natural language support but running in an English locale
R is a collaborative project with many contributors.
Type 'contributors()' for more information and
'citation()' on how to cite R or R packages in publications.
Type 'demo()' for some demos, 'help()' for on-line help, or
'help.start()' for an HTML browser interface to help.
Type 'q()' to quit R.
> install.packages("tidyverse")

FIGURE 2.2: Installing a package.

> Do not forget to put the package names into quotation marks (""). Otherwise, you will get an error message.

After you hit Enter, R will start to install the package and print some information on the progress of the installation. When the `install.packages` function is finished, the package is ready to be used. Installed packages are added to R's *system library*. This system library can be accessed in the *Packages* pane in the bottom left corner of your R Studio screen. Whenever we want to use the installed package, we can load it from our library using the `library` function. Let us try this out and load the *{tidyverse}* package.

```
library(tidyverse)
```

2.3 The *{dmetar}* Package

In this guide, we want to make conducting meta-analyses as accessible and easy as possible for you as a researcher. Although there are fantastic packages like the *{meta}* and *{metafor}* package which do most of the heavy lifting, there are still some aspects of meta-analyses which we consider important, but not easy to do in R currently, particularly if you do not have a programming or statistics background.

To fill this gap, we developed the *{dmetar}* package, which serves as the companion R package for this book. The *{dmetar}* package has its own documentation, which can be found online[9]. Functions of the *{dmetar}* package provide additional functionality for the *{meta}* and *{metafor}* packages (and a few other, more advanced packages), which we will be using frequently in this guide. Most of the functions included in the *{dmetar}* package and how they can improve your meta-analysis work flow will be described in detail throughout the book. Most of the example data sets we are using in this guide are also contained in *{dmetar}*.

> Although highly recommended, it is *not essential* to have the *{dmetar}* package installed to work through the guide. For each function of the package, we will also provide the source code, which can be used to save the function locally on your computer, and the additional R packages those functions rely on. We will also provide supplementary download links for data sets included in the package.
>
> However, installing the *{dmetar}* package beforehand is much more convenient because this pre-installs all the functions and data sets on your computer.

To install the *{dmetar}* package, the R version on your computer must be 3.6 or higher. If you have (re-)installed R recently, this will probably be the case. To check if your R version is new enough, you can paste this line of code into the Console and then hit Enter.

```
R.Version()$version.string
```

This will display the current R version you have. If the R version is below 3.6, you will have to update it. There are good blog posts[10] on the Internet providing guidance on how to do this.

[9] https://dmetar.protectlab.org/

[10] https://www.linkedin.com/pulse/3-methods-update-r-rstudio-windows-mac-woratana-ngarmtrakulchol/

If you want to install *{dmetar}*, one package already needs to be installed on your computer already. This package is called *{devtools}*. So, if *{devtools}* is not in your system library yet, you can install it like we did before.

```
install.packages("devtools")
```

You can then install *{dmetar}* using this line of code:

```
devtools::install_github("MathiasHarrer/dmetar")
```

This will initiate the installation process. It is likely that the installation will take some time because several other packages have to be installed along with the *{dmetar}* package for it to function properly. During the installation process, the installation manager may ask you if you want to update existing R packages on your computer. The output may look something like this:

```
## These packages have more recent versions available.
## Which would you like to update?
##
## 1: All
## 2: CRAN packages only
## 3: None
## 4: ggpubr (0.2.2 -> 0.2.3) [CRAN]
## 5: zip    (2.0.3 -> 2.0.4) [CRAN]
##
## Enter one or more numbers, or an empty line to skip updates:
```

When you get this message, it is best to tell the installation manager that no packages should be updated. In our example, this means pasting 3 into the console and then hitting Enter. In the same vein, when the installation manager asks this question:

```
## There are binary versions available but the source versions are
## later:
##
##  [...]
##
##   Do you want to install from sources the package which needs
##   compilation?
##   y/n:
```

It is best to choose `n` (no). If the installation fails with this strategy (meaning that you get an `Error`), run the installation again, but update all packages this time.

When writing this book and developing the package, we made sure that everyone can install it without errors. Nevertheless, there is still a chance that installing the package does not work at the first try. If the installation problem persists, you can have a look at the "Contact Us" section in the preface of this book.

2.4 Data Preparation & Import

This chapter will tell you how to import data into R using R Studio. Data preparation can be tedious and exhausting at times, but it is the backbone of all later steps. Therefore, we have to pay close attention to bringing the data into the correct format before we can proceed.

Usually, data imported into R is stored in *Microsoft Excel* spreadsheets first. We recommend to store your data there because this makes it very easy to do the import. There are a few "Dos and Don'ts" when preparing the data in *Excel*.

- It is very important how you name the columns of your spreadsheet. If you already named the columns of your sheet adequately in *Excel*, you can save a lot of time later because your data does not have to be transformed using R. "Naming" the columns of the spreadsheet simply means to write the name of the variable into the first line of the column; R will automatically detect that this is the name of the column then.
- Column names should not contain any spaces. To separate two words in a column name, you can use underscores or dots (e.g. "column_name").
- It does *not* matter how columns are ordered in your *Excel* spreadsheet. They just have to be labeled correctly.
- There is also *no* need to format the columns in any way. If you type the column name in the first line of your spreadsheet, R will automatically detect it as a column name.
- It is also important to know that the import may distort special characters like ä, ü, ö, á, é, ê, etc. You might want to transform them into "normal" letters before you proceed.
- Make sure that your *Excel* file only contains one sheet.
- If you have one or several empty rows or columns which used to contain data, make sure to delete those columns/rows completely, because R could think that these columns contain (missing) data and import them also.

Let us start with an example data set. Imagine that you plan to conduct a meta-analysis of suicide prevention programs. The outcome you want to focus on in your study is the severity of suicidal ideation (i.e. to what degree individuals think about, consider or plan to end their life), assessed by questionnaires. You already completed the study search and data extraction and now want to import your meta-analysis data in R. The next task is therefore to prepare an *Excel* sheet containing all the relevant data. Table 2.1 presents all the data we want to import. In the first row, this table also shows how we can name our columns in the *Excel* file based on the rules we listed above. We can see that the spreadsheet lists each study in one row. For each study, the sample size (n), mean and standard deviation (SD) are included for both the intervention and control group. This is the outcome data needed to

TABLE 2.1: The suicide prevention data set.

'author'	'n.e'	'mean.e'	'sd.e'	'n.c'	'mean.c'	'sd.c'	'pubyear'	'age_group'	'control'
	Intervention Group			Control Group			Subgroups		
Author	N	Mean	SD	N	Mean	SD	Year	Age Group	Control Group
Berry et al.	90	14.98	3.29	95	15.54	4.41	2006	general	WLC
DeVries et al.	77	16.21	5.35	69	20.13	7.43	2019	older adult	no intervention
Fleming et al.	30	3.01	0.87	30	3.13	1.23	2006	general	no intervention
Hunt & Burke	64	19.32	6.41	65	20.22	7.62	2011	general	WLC
McCarthy et al.	50	4.54	2.75	50	5.61	2.66	1997	general	WLC
Meijer et al.	109	15.11	4.63	111	16.46	5.39	2000	general	no intervention
Rivera et al.	60	3.44	1.26	60	3.42	1.88	2013	general	no intervention
Watkins et al.	40	7.1	0.76	40	7.38	1.41	2015	older adult	no intervention
Zaytsev et al.	51	23.74	7.24	56	24.91	10.65	2014	older adult	no intervention

calculate effect sizes, which is something we will cover in detail in Chapter 3. The following three columns contain variables we want to analyze later on as part of the meta-analysis. We have prepared an *Excel* file for you called *"SuicidePrevention.xlsx"*, containing exactly this data. The file can be downloaded from the Internet[11].

To import our *Excel* file in R Studio, we have to set a *working directory* first. The working directory is a folder on your computer from which R can use data, and in which outputs are saved. To set a working directory, you first have to create a folder on your computer in which you want all your meta-analysis data and results to be saved. You should also save the *"SuicidePrevention.xlsx"* file we want to import in this folder. Then start R Studio and open your newly created folder in the bottom right *Files* pane. Once you have opened your folder, the *Excel* file you just saved there should be displayed. Then set this folder as the working directory by clicking on the little gear wheel on top of the pane, and then on *Set as working directory* in the drop-down menu. This will make the currently opened folder the working directory.

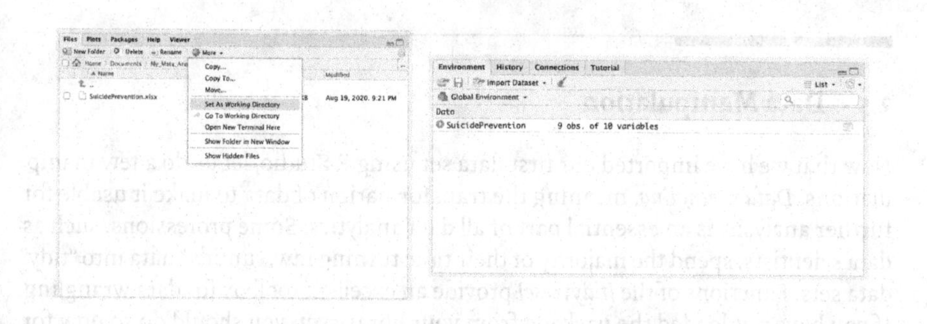

FIGURE 2.3: Setting the working directory; data set loaded in the R environment.

We can now proceed and import the data into R. In the *Files* pane, simply click on the

[11]`http://protectlab.org/meta-analysis-in-r/data/SuicidePrevention.xlsx`

"SuicidePrevention.xlsx" file. Then click on *Import Dataset*.... An import assistant should now pop up, which is also loading a preview of your data. This can be time-consuming sometimes, so you can skip this step if you want to, and click straight on *Import*.

Your data set should then be listed with the name `SuicidePrevention` in the top right *Environment* pane. This means that your data is now loaded and can be used by R code. Tabular data sets like the one we imported here are called *data frames* (`data.frame`) in R. Data frames are data sets with columns and rows, just like the *Excel* spreadsheet we just imported.

{openxlsx}

It is also possible to import data files directly using code. A good package we can use to do this is called *{openxslx}* (Schauberger and Walker, 2020). As with all R packages, you have to install it first. You can then use the `read.xlsx` function to import an *Excel* sheet.

If the file is saved in your working directory, you only have to provide the function with the file's name and assign the imported data to an object in R. If we want our data set to have the name `data` in R, for example, we can use this code:

```
library(openxlsx)
data <- read.xlsx("SuicidePrevention.xlsx")
```

2.5 Data Manipulation

Now that we have imported our first data set using R Studio, let us do a few manipulations. *Data wrangling*, meaning the transformation of data to make it usable for further analysis, is an essential part of all data analytics. Some professions, such as data scientists, spend the majority of their time turning raw, "untidy" data into "tidy" data sets. Functions of the *{tidyverse}* provide an excellent toolbox for data wrangling. If you have not loaded the package from your library yet, you should do so now for the following examples.

```
library(tidyverse)
```

2.5.1 Class Conversion

First, we should take a peek at the `SuicidePrevention` data set we imported in the last chapter. To do this, we can use the `glimpse` function provided by the *{tidyverse}*.

```
glimpse(SuicidePrevention)
```

```
## Rows: 9
## Columns: 10
## $ author    <chr> "Berry et al.", "DeVries et al."...
## $ n.e       <chr> "90", "77", "30", "64", "50", "1...
## $ mean.e    <chr> "14.98", "16.21", "3.01", "19.32...
## $ sd.e      <chr> "3.29", "5.35", "0.87", "6.41", ...
## $ n.c       <dbl> 95, 69, 30, 65, 50, 111, 60, 40, 56
## $ mean.c    <chr> "15.54", "20.13", "3.13", "20.22...
## $ sd.c      <chr> "4.41", "7.43", "1.23", "7.62", ...
## $ pubyear   <dbl> 2006, 2019, 2006, 2011, 1997, 20...
## $ age_group <chr> "general", "older adult", "gener...
## $ control   <chr> "WLC", "no intervention", "no in...
```

We see that this gives us details on the type of data we have stored in each column of our data set. There are different abbreviations signifying different types of data. In R, they are called *classes*.

- <num> stands for *numeric*. This is all data stored as numbers (e.g. 1.02).
- <chr> stands for *character*. This is all data stored as words.
- <log> stands for *logical*. These are variables which are binary, meaning that they signify that a condition is either `TRUE` or `FALSE`.
- <factor> stands for *factor*. Factors are stored as numbers, with each number signifying a different level of a variable. Possible factor levels of a variable might be 1 = "low", 2 = "medium", 3 = "high".

We can also check the class of a column using the `class` function. We can access a column in a data frame directly by adding the `$` operator to its name and then the name of the column. Let us try this out. First, we let R provide us with the data contained in the column `n.e`. After that, we check the class of the column.

```
SuicidePrevention$n.e
```

```
## [1] "90"  "77"  "30"  "64"  "50"  "109" "60"  "40"
## [9] "51"
```

```
class(SuicidePrevention$n.e)
```

```
## [1] "character"
```

We see that column `n.e` containing the sample sizes in the intervention group has the class `character`. But wait, that is the wrong class! During the import, this column was wrongly classified as a `character` variable, while it should actually have the class `numeric`. This has implications on further analysis steps. For example, if we want to calculate the mean sample size, we get this warning:

```
mean(SuicidePrevention$n.e)
```

```
## Warning in mean.default(SuicidePrevention$n.e):
## argument is not numeric or logical: returning NA

## [1] NA
```

To make our data set usable, we often have to convert our columns to the right classes first. To do this, we can use a set of functions which all begin with "`as.`": `as.numeric`, `as.character`, `as.logical` and `as.factor`. Let us go through a few examples.

In the output of the `glimpse` function from before, we see that several columns have been given the `character` class, while they should be `numeric`. This concerns columns `n.e`, `mean.e`, `sd.e`, `mean.c` and `sd.c`. We see that the publication year `pubyear` has the class `<dbl>`. This stands for *double* and means that the column is a numeric vector. It is a historical anomaly that both `double` and `numeric` are used in R to refer to numeric data types. Usually, however, this has no actual practical implications.

The fact that some numerical values are coded as *characters* in our data set, however, will lead to problems downstream, so we should change the class using the `as.numeric` function. We provide the function with the column we want to change and then save the output back to its original place using the *assignment operator* (<-). This leads to the following code.

```
SuicidePrevention$n.e <- as.numeric(SuicidePrevention$n.e)
SuicidePrevention$mean.e <- as.numeric(SuicidePrevention$mean.e)
SuicidePrevention$sd.e <- as.numeric(SuicidePrevention$sd.e)
SuicidePrevention$n.c <- as.numeric(SuicidePrevention$n.c)
SuicidePrevention$mean.c <- as.numeric(SuicidePrevention$mean.c)
SuicidePrevention$sd.c <- as.numeric(SuicidePrevention$sd.c)
SuicidePrevention$n.c <- as.numeric(SuicidePrevention$n.c)
```

We also see in the `glimpse` output that `age_group` and `control`, the subgroups in our data, are coded as characters. In reality, however, it is more adequate to encode them as factors, with two factor levels each. We can use the `as.factor` function to change the class.

```
SuicidePrevention$age_group <- as.factor(SuicidePrevention$age_group)
SuicidePrevention$control <- as.factor(SuicidePrevention$control)
```

Using the `levels` and `nlevels` function, we can also have a look at the factor labels and number of levels in a factor.

```
levels(SuicidePrevention$age_group)
```

```
## [1] "general"     "older adult"
```

```
nlevels(SuicidePrevention$age_group)
```

```
## [1] 2
```

We can also use the `levels` function to change the name of the factor labels. We simply have to assign new names to the original labels. To do this in R, we have to use the *concatenate*, or `c` function. This function can tie two or more words or numbers together and create one element. Let us try this out.

```
new.factor.levels <- c("gen", "older")
new.factor.levels
```

```
## [1] "gen"   "older"
```

Perfect. We can now use the newly created `new.factor.levels` object and assign it to the factor labels of our `age_group` column.

```
levels(SuicidePrevention$age_group) <- new.factor.levels
```

Let us check if the renaming has worked.

```
SuicidePrevention$age_group
```

```
## [1] gen   older gen   gen   gen   gen   gen   older
## [9] older
## Levels: gen older
```

It is also possible to create logicals using `as.logical`. Let us say we want to recode the column `pubyear`, so that it only displays if a study was published after 2009. To do this, we have to define a yes/no rule via code. We can do this using the "greater or equal than" operator >=, and then use this as the input for the `as.logical` function.

```
SuicidePrevention$pubyear
```

```
## [1] 2006 2019 2006 2011 1997 2000 2013 2015 2014
```

```
as.logical(SuicidePrevention$pubyear >= 2010)
```

```
## [1] FALSE  TRUE FALSE  TRUE FALSE FALSE  TRUE  TRUE
## [9]  TRUE
```

We can see that this encodes every element in pubyear as TRUE or FALSE, depending on whether the publication year was greater than or equal to 2010, or not.

2.5.2 Data Slicing

In R, there are several ways to extract a subset of a data frame. We have already covered one way, the $ operator, which can be used to extract columns. A more generic approach to extract slices from a data set is to use square brackets. The general form we have to follow when using square brackets is data.frame[rows, columns]. It is always possible to extract rows and columns by using the number in which they appear in the data set. For example, we can use the following code to extract the data in the second row of the data frame.

```
SuicidePrevention[2,]
```

```
##            author n.e mean.e sd.e n.c mean.c sd.c
## 2 DeVries et al.   77  16.21 5.35  69  20.13 7.43
##   pubyear age_group         control
## 2    2019     older no intervention
```

We can be even more specific and tell R that we want only the information in the first column of the second row.

```
SuicidePrevention[2, 1]
```

```
## [1] "DeVries et al."
```

To select specific slices, we have to use the concatenate (c) function again. For example, if we want to extract rows 2 and 3 as well as columns 4 and 6, we can use this code.

```
SuicidePrevention[c(2,3), c(4,6)]
```

```
##   sd.e mean.c
## 2 5.35  20.13
## 3 0.87   3.13
```

It is usually only possible to select rows by their number. For columns, however, it is also possible to provide the column *name* instead of the number.

```
SuicidePrevention[, c("author", "control")]
```

```
##             author         control
## 1     Berry et al.             WLC
## 2   DeVries et al. no intervention
## 3   Fleming et al. no intervention
```

```
## 4     Hunt & Burke              WLC
## 5 McCarthy et al.               WLC
## 6   Meijer et al. no intervention
## 7   Rivera et al. no intervention
## 8  Watkins et al. no intervention
## 9  Zaytsev et al. no intervention
```

Another possibility is to *filter* a data set based on row values. We can use the `filter` function to do this. In the function, we need to specify our data set name, as well as a filtering logic. A relatively straightforward example is to filter all studies in which `n.e` is equal to or smaller than 50.

```
filter(SuicidePrevention, n.e <= 50)
```

```
##            author n.e mean.e sd.e n.c mean.c sd.c
## 1  Fleming et al.  30   3.01 0.87  30   3.13 1.23
## 2 McCarthy et al.  50   4.54 2.75  50   5.61 2.66
## 3  Watkins et al.  40   7.10 0.76  40   7.38 1.41
##   pubyear age_group         control
## 1    2006       gen no intervention
## 2    1997       gen             WLC
## 3    2015     older no intervention
```

But it is also possible to filter by names. Imagine we want to extract the studies by authors *Meijer* and *Zaytsev*. To do this, we have to define our filtering logic using the `%in%` operator and the concatenate function.

```
filter(SuicidePrevention, author %in% c("Meijer et al.",
                                        "Zaytsev et al."))
```

```
##           author n.e mean.e sd.e n.c mean.c  sd.c
## 1  Meijer et al. 109  15.11 4.63 111  16.46  5.39
## 2 Zaytsev et al.  51  23.74 7.24  56  24.91 10.65
##   pubyear age_group         control
## 1    2000       gen no intervention
## 2    2014     older no intervention
```

Conversely, we can also extract all studies *except* the one by *Meijer* and *Zaytsev* by putting an exclamation mark in front of the filtering logic.

```
filter(SuicidePrevention, !author %in% c("Meijer et al.",
                                         "Zaytsev et al."))
```

2.5.3 Data Transformation

Of course it is also possible to change specific values in an R data frame, or to expand it. To change data we saved internally in R, we have to use the *assignment operator*. Let us reuse what we previously learned about data slicing to change a specific value in our data set. Imagine that we made a mistake and that the publication year of the study by *DeVries et al.* is wrongly reported as 2019 when it should be 2018. We can change the value by slicing our data set accordingly, and then assigning the new value. Remember that the results of *DeVries et al.* are reported in the second row of the data set.

```
SuicidePrevention[2, "pubyear"] <- 2018

SuicidePrevention[2, "pubyear"]
```

```
## [1] 2018
```

It is also possible to change more than one value at once. For example, if we want to add 5 to every intervention group mean in our data set, we can do that using this code.

```
SuicidePrevention$mean.e + 5
```

```
## [1] 19.98 21.21  8.01 24.32  9.54 20.11  8.44 12.10
## [9] 28.74
```

We can also use two or more columns to do calculations. A practically relevant example is that we might be interested in calculating the *mean difference* between the intervention and control group means for each study. Compared to other programming languages, this is spectacularly easy in R.

```
SuicidePrevention$mean.e - SuicidePrevention$mean.c
```

```
## [1] -0.56 -3.92 -0.12 -0.90 -1.07 -1.35  0.02 -0.28
## [9] -1.17
```

As you can see, this takes the intervention group mean of each study, and then subtracts the control group mean, each time using the value of the same row. Suppose that we want to use this mean difference later on. Therefore, we want to save it as an extra object called `md` and add it as a new column to our `SuicidePrevention` data frame. Both is easy using the assignment operator.

```
md <- SuicidePrevention$mean.e - SuicidePrevention$mean.c

SuicidePrevention$md <- SuicidePrevention$mean.e -
                        SuicidePrevention$mean.c
```

The last thing we want to show you are *pipe operators*. In R, pipes are written using the `%>%` symbol. Pipes essentially allow us to apply a function to an object without having to specify the object name directly in the function call. We simply connect the object and the function using the pipe operator. Let us give you an easy example. If we want to calculate the mean sample size in the control group, we can use the `mean` function and the pipe operator like this:

```
SuicidePrevention$n.c %>% mean()
```

```
## [1] 64
```

Admittedly, in this example, it is hard to see the added values of such pipes. The special strength of pipes stems from the fact that they allow us to *chain* many functions together. Imagine we want to know the square root of the mean control group sample size, but only of studies published after 2009. Pipes let us calculate this conveniently in one step.

```
SuicidePrevention %>%
  filter(pubyear > 2009) %>%
  pull(n.c) %>%
  mean() %>%
  sqrt()
```

```
## [1] 7.616
```

In the pipe, we used one function we have not covered before, the `pull` function. This function can be seen as an equivalent of the `$` operator that we can use in pipes. It simply "pulls out" the variable we specify in the function, so it can be fed forward to the next part of the pipe.

Accessing the R Documentation

Many functions in R require several arguments, and it is impossible to remember how all functions are used correctly. Thankfully, it is not necessary to know how each function is used by heart. R Studio makes it easy for us to access the R documentation, where every function has a detailed description page. There are two ways to search for a function documentation page. The first is to access the *Help* pane in the lower left corner of R Studio, and then use the search bar to find information on a specific function. A more convenient way is to simply run `?` followed by the name of the function in the console, e.g. `?mean`. This will open the documentation entry for this function automatically.

The R documentation of a function usually at least contains a *Usage*, *Arguments* and *Examples* section. The *Arguments* and *Examples* section is often particularly helpful to understand how a function is used.

2.5.4 Saving Data

Once we have done transformations with our data and saved them internally in R, we have to *export* them at some point. There are two types of file formats which we advise you to use when saving R data frames: *.rda* and *.csv*.

The file ending *.rda* stands for R *Data*. It is a file type specifically for R, with all advantages and disadvantages. An advantage of *.rda* files is that they can easily be re-opened in R, and that there is no risk that your data may be distorted during the export. They are also very versatile and can save data that does not fit into a spreadsheet format. The disadvantage is that they can only be opened in R; but for some projects, this is sufficient. To save an object as an *.rda* data file, you can use the save function. In the function, you have to provide (1) the name of the object and (2) the *exact* name you want the file to have, including the file ending. Running the function will then save the file to your working directory.

```
save(SuicidePrevention, file = "suicideprevention.rda")
```

The file ending *.csv* stands for *comma-separated values*. This format is one of the most commonly used ones for data in general. It can be opened by many programs, including *Excel*. You can use the `write.csv` function to save your data as a *.csv*. The code structure and behavior are nearly identical to the one of save, but the supplied object *needs* to be a data frame or other tabular data object. And of course, you need to specify the file type ".csv".

```
write.csv(SuicidePrevention, file = "suicideprevention.csv")
```

This has only been a quick overview of data manipulation strategies in R. Learning R from scratch can be exhausting at times, especially when we deal with something as supposedly easy as manipulating data. However, the best way to get accustomed to the way R works is to practice. After some time, common R commands will become second nature to you.

A good way to continue learning is to have a look at Hadley Wickham and Garrett Grolemund's book *R for Data Science* (2016). Like this guide, the book can be completely read online for free[12]. Additionally, we also collected a few exercises on the next page, which you can use for practicing what we learned here.

□

[12] `https://r4ds.had.co.nz/transform.html`

2.6 Questions & Answers

Data Manipulation Exercises

For these exercises, we will use a new data set called `data`. You can create this data set directly in R using this code:

```
data <- data.frame("Author" = c("Jones", "Goldman",
                               "Townsend", "Martin",
                               "Rose"),
                   "TE" = c(0.23, 0.56,
                            0.78, 0.23,
                            0.33),
                   "seTE" = c(0.324, 0.235,
                              0.394, 0.275,
                              0.348),
                   "subgroup" = c("one", "one",
                                  "two", "two",
                                  "three"))
```

Here are the exercises for this data set.

1. Show the variable `Author`.
2. Convert `subgroup` to a factor.
3. Select all the data of the "Jones" and "Martin" study.
4. Change the name of the study "Rose" to "Bloom".
5. Create a new variable `TE_seTE_diff` by subtracting `seTE` from `TE`. Save the results in `data`.
6. Use a pipe to (1) filter all studies in `subgroup` "one" or "two", (2) select the variable `TE_seTE_diff`, (3) take the mean of the variable, and then apply the `exp` function to it. Access the R documentation to find out what the `exp` function does.

Answers to these questions are listed in Appendix A at the end of this book.

2.7 Summary

- R has become one of the most powerful and frequently used statistical programming languages in the world.
- R is not a computer program with a graphical user interface and predefined functionality. It is a full programming language to which people all around the world can contribute freely available add-ons, so-called *packages*.
- R Studio is a computer program which allows us to use R for statistical analyses in a convenient way.
- The fundamental building blocks of R are functions. Many of these functions can be imported through packages which we can install from the Internet.
- Functions can be used to import, manipulate, analyze and save data using R.

Part II

Meta-Analysis in R

3

Effect Sizes

In the last chapter, we were able to familiarize ourselves with the R universe and learned a few helpful tools to import and manipulate data. In this second part of the book, we can now apply and expand our R knowledge while learning about core statistical techniques that are used in meta-analyses.

In Chapter 1.1, we defined meta-analysis as a technique which summarizes quantitative outcomes from several studies. In meta-analyses studies instead of individuals become the fundamental units of our analysis. This introduces new problems. In a primary study, it is usually quite easy to calculate *summary statistics* through which we can describe the data we collected. For example, it is conventional to calculate the *arithmetic mean* $\bar{x}$ and *standard deviation* s of a continuous outcome in primary studies. However, this is only possible because an essential prerequisite is usually met: we know that the outcome variable has been *measured in exactly the same way* across all study subjects. In meta-analyses, this assumption is typically not met. Suppose that we want to conduct a meta-analysis in which our outcome of interest are the math skills of grade eight students. Even if we apply strict inclusion criteria (see Chapter 1.4.1), it is likely that not every study used exactly the same test to measure math skills; some may have even only reported the proportion of students who passed or failed the test. This makes it virtually impossible to quantitatively synthesize the results.

To perform a meta-analysis, we have to find an *effect size* which can be summarized across all studies. Sometimes, such effect sizes can be directly extracted from the publication; more often, we have to calculate them from other data reported in an article. The selected effect size metric can have a nontrivial impact on the results of a meta-analysis, and their interpretability. They should therefore fulfill a few important criteria (Lipsey and Wilson, 2001; Higgins et al., 2019). In particular, the selected effect size measure for a meta-analysis should be:

- ***Comparable***. It is important that the effect size measure has the same meaning across all studies. Let us take math skills as an example again. It makes no sense to pool differences between experimental and control groups in the number of points achieved on a math test when studies used different tests. Tests may, for example, vary in their level of difficulty, or in the maximum number of points that can be achieved.
- ***Computable***. We can only use an effect size metric for our meta-analysis if it is possible to derive its numerical value from the primary study. It must be possible to calculate the effect size for all of the included studies based on their data.

DOI: 10.1201/9781003107347-3

- ***Reliable***. Even if it is possible to calculate an effect size for all included studies, we must also be able to *pool* them statistically. To use some metric in meta-analyses, it must be at least possible to calculate the *standard error* (see next chapter). It is also important that the format of the effect size is suited for the meta-analytic technique we want to apply and does not lead to errors or biases in our estimate.
- ***Interpretable***. The type of effect size we choose should be appropriate to answer our research question. For example, if we are interested in the strength of an association between two continuous variables, it is conventional to use correlations to express the size of the effect. It is relatively straightforward to interpret the magnitude of a correlation, and many researchers can understand them. In the following chapters, we will learn that it is sometimes not possible to use outcome measures which are both easy to interpret *and* ideal for our statistical computations. In such cases, it is necessary to transform effect sizes to a format with better mathematical properties before we pool them.

It is very likely that you have already stumbled upon the term "effect size" before. We also used the word here, without paying too much attention to what it precisely stands for. In the next section, we should therefore explore what we actually mean when we talk about an "effect size".

3.1 What Is an Effect Size?

In the terminology we use in this book, an effect size is defined as a metric quantifying the relationship between two entities. It captures the *direction* and *magnitude* of this relationship. If relationships are expressed as the same effect size measure, it is possible to compare them.

We want to stress here that this is just *one* way to define what the word "effect size" means. Definitions can be wider and narrower, and the term is used differently by different people (Borenstein et al., 2011, chapter 3). Some researchers only talk of effect sizes when referring to the results of intervention studies, which are usually expressed as differences between the treatment and control group (e.g. *standardized mean differences*, see Chapter 3.3.1). Using this conceptualization, "effect size" refers to the effect of a treatment, and how large this effect is. In our opinion, this is quite a narrow definition. Not only treatments can have an effect on some variable; effects can also appear *naturally* without any direct human intervention. For example, it is possible that socio-demographic variables, such as the income and education of parents, may have an effect on the educational attainment of their children. Correlations describe how well we can predict the values of a variable through the values of another and can also be seen as a form of effect size.

On the other hand, it might go too far to say that everything we can pool as part of a meta-analysis is automatically an effect size. As we will learn, there are measures of the *central tendency*, such as the sample mean, which can also be used in meta-analyses. But a sample mean alone does not quantify a relationship between two phenomena, and there is no "effect". Nevertheless, in this book, we will often use the word "effect size" as a *pars pro toto*, representing both estimates of an actual effect, as well as "one-variable" and central tendency measures. We do not do this because it is accurate, but because it is more convenient.

Others disapprove of the term "effect size" altogether. They stress that the word "effect" in "effect size" suggests that there is a *causal* relationship. However, we all know that *correlation is not causation*, and a difference between an intervention and control group must not automatically be caused by the treatment itself. In the end, it is up to you to decide which definition you prefer, but be aware that people may have different conceptualizations in mind when they talk about effect sizes.

In mathematical notation, it is common to use the greek letter *theta* (θ) as the symbol for a *true* effect size[1]. More precisely, θ_k represents the true effect size of a study k. Importantly, the true effect size is *not* identical with the *observed effect size* that we find in the published results of the study. The observed effect size is only an *estimate* of the true effect size. It is common to use a *hat* (^) symbol to clarify that the entity we refer to is only an estimate. The observed effect size in study k, our estimate of the true effect size, can therefore be written as $\hat\theta_k$. But why does $\hat\theta_k$ differ from θ_k? It differs because of the *sampling error*, which can be symbolized as ϵ_k. In every primary study, researchers can only draw a small sample from the whole population. For example, when we want to examine the benefits of regular exercise on the cardiovascular health of primary care patients, we will only be able to include a small selection of patients, not *all* primary care patients in the world. The fact that a study can only take small samples from an infinitely large population means that the observed effect will differ from the true population effect. Put simply, $\hat\theta_k$ is, therefore, the same as θ_k plus some sampling error ϵ_k[2].

$$\hat\theta_k = \theta_k + \epsilon_k \tag{3.1}$$

It is obviously desirable that the effect size estimate $\hat\theta_k$ of study k is as close as possible to the true effect size, and that ϵ_k is minimal. All things being equal, we can assume that studies with smaller ϵ will deliver a more *precise* estimate of the true effect size. Meta-analysis methods take into account how precise an effect size estimate is (see Chapter 4). When pooling the results of different studies, they give effects with a greater precision (i.e. less sampling error) a higher weight, because they are better estimators of the true effect (Hedges and Olkin, 2014).

---

[1]In this book, we will largely follow the notation used by Schwarzer et al. (2015) and Borenstein et al. (2011) when discussing effect sizes.

[2]It should be noted that there are often more reasons for our observed effect sizes to differ from the true effect size than sampling error alone; for example, biases in a study's research methodology, or measurement error. In Chapter 3.4, we will discuss this in greater detail.

But how can we know how big the sampling error is? Unsurprisingly, the true effect of a study θ_k is unknown, and so ϵ_k is also unknown. Often, however, we can use statistical theory to approximate the expected sampling error. A common way to represent ϵ is through the *standard error* (SE). The standard error is defined as the standard deviation of the *sampling distribution*. A sampling distribution is the distribution of a metric we get when we draw random samples with the same sample size n from our population *many, many times*.

We can make this more concrete by simulating data in R. We can pretend that we are drawing random samples from a larger population using the `rnorm` function. This function allows us to draw random samples from a normal distribution, therefore the name. The `rnorm` function simulates a "perfect world" in which we *know* how values are distributed in the true population and lets us take samples. The function takes three arguments: `n`, the number of observations we want to have in our sample; `mean`, the *true* mean of the population; and `sd`, the *true* standard deviation. The `rnorm` function has a random component, so to make results reproducible, we have to set a *seed* first. This can be done using the `set.seed` function, which we have to supply with a number. For our example, we chose to set a seed of 123. Furthermore, we want to simulate that the true mean of our population is $\mu = 10$, that the true standard deviation is $\sigma = 2$, and that our sample consists of $n = 50$ randomly selected observations, which we save under the name `sample`. This is what our code looks like:

```
set.seed(123)
sample <- rnorm(n = 50, mean = 10, sd = 2)
```

Now, we can calculate the mean of our sample.

```
mean(sample)
```

```
## [1] 10.07
```

We see that the mean is $\bar{x} = 10.07$, which is already very close to the true value in our population. The sampling distribution can now be created by repeating what we did here–taking a random sample and calculating its mean–*countless times*. To simulate this process for you, we conducted the steps from before 1000 times. The histogram in Figure 3.1 displays the results. We can see that the means of the samples closely resemble a normal distribution with a mean of 10. If we were to draw even more samples, the distribution of the means would get even closer to a normal distribution. This idea is expressed in one of the most fundamental tenets of statistics, the *central limit theorem* (Aronow and Miller, 2019, chapter 3.2.4).

The standard error is defined as the standard deviation of this sampling distribution. Therefore, we calculated the standard deviation of the 1000 simulated means to get an approximation of the standard error. The result is $SE = 0.267$.

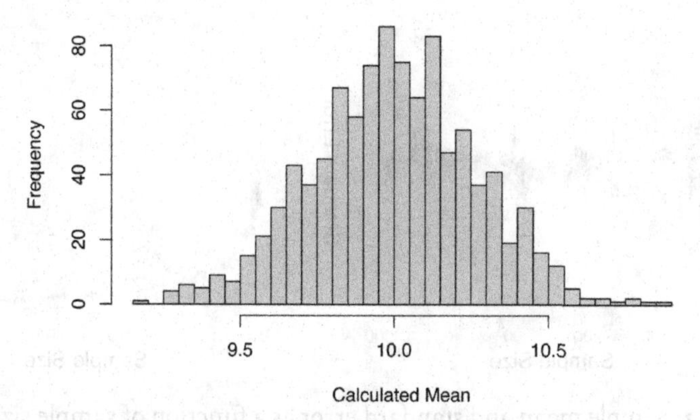

FIGURE 3.1: "Sampling distribution" of means (1000 samples).

As we mentioned before, we cannot simply calculate the standard error in real life by simulating the true sampling distribution. However, there are formulas based on statistical theory which allow us to calculate an estimate of the standard error, even when we are limited to only one observed sample–which we usually are. The formula to calculate the standard error of the *mean* is defined like this:

$$SE = \frac{s}{\sqrt{n}} \tag{3.2}$$

It defines the standard error as the standard deviation of our sample s, divided by the square root of the sample size n. Using this formula, we can easily calculate the standard error of our `sample` object from before using R. Remember that the size of our random sample was $n = 50$.

```
sd(sample)/sqrt(50)
```

```
## [1] 0.2619
```

If we compare this value to the one we found in our simulation of the sampling distribution, we see that they are nearly identical. Using the formula, we could accurately calculate the standard error using only the sample we have at hand.

In formula 3.2, we can see that the standard error of the mean depends on the sample size of a study. As n becomes larger, the standard error becomes smaller, meaning that a study's estimate of the true population mean is getting more precise. To exemplify this relationship, we conducted another simulation. Again, we used the `rnorm` function, and assumed a true population mean of $\mu = 10$ and that $\sigma = 2$. But this time, we varied the sample size, from $n = 2$ to $n = 500$. For each simulation, we calculated both the mean and the standard error using formula 3.2.

Figure 3.2 shows the results. We can see that the means look like a *funnel*: as the sample size increases, the mean estimates become more and more precise, and converge towards 10. This increase in precision is represented by the standard error: with increasing sample size, the standard error becomes smaller and smaller.

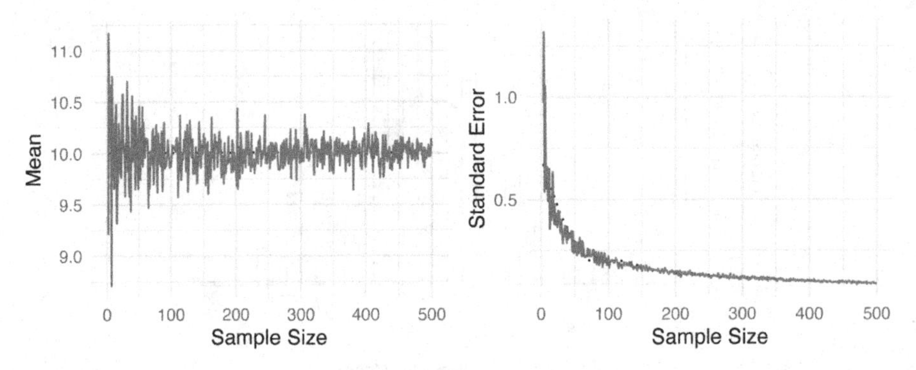

FIGURE 3.2: Sample mean and standard error as a function of sample size.

We have now explored the quintessential elements we need to conduct a meta-analysis: an (1) observed effect size or outcome measure and (2) its precision, expressed as the standard error. If these two types of information can be calculated from a published study, it is usually also possible to perform a meta-analytic synthesis (see Chapter 4).

In our simulations, we used the mean of a variable as an example. It is important to understand that the properties we saw above can also be found in other outcome measures, including commonly used effect sizes. If we would have calculated a mean *difference* in our sample instead of a mean, this mean difference would have exhibited a similarly shaped sampling distribution, and the standard error of the mean difference would have also decreased as the sample size increases (provided the standard deviation remains the identical). The same is also true, for example, for (Fisher's z transformed) correlations. In the following sections, we will go through the most commonly used effect sizes and outcome measures in meta-analyses. One reason why these effect size metrics are used so often is because they fulfill two of the criteria we defined at the beginning of this chapter: they are *reliable* and *computable*. In formula 3.2, we described how the standard error of a mean can be calculated, but this formula can *only* be readily applied to *means*. Different formulas to calculate the standard error are needed for other effect sizes and outcome measures. For the effect size metrics we cover here, these formulas luckily exist, and we will show you all of them. A collection of the formulas can be also found in the Appendix. Some of these equations are somewhat complicated, but the good news is that we hardly ever have to calculate the standard error manually. There are various functions in R which do the heavy lifting for us.

In the following section, we not only want to provide a theoretical discussion of different effect size metrics. We also show you which kind of information you have to prepare in your data set so that the R meta-analysis functions we are using later can calculate the effect sizes for us. We grouped effect sizes based on the type of research design in which they usually appear: *single group designs* (e.g. naturalistic studies, surveys, or uncontrolled trials) and *control group designs* (e.g. experimental studies or controlled clinical trials). Please note that this is just a rough classification, not a strict rule. Many of the effect sizes we present are technically applicable to any type of research design, as long as the type of outcome data is suited.

3.2 Measures & Effect Sizes in Single Group Designs

3.2.1 Means

The *arithmetic mean* is probably the most commonly used central tendency measure. Although means are rather infrequently used as outcome measures, they can easily be pooled in a meta-analysis. For example, one could investigate the mean height of males, expressed as centimeters or inches, by pooling several representative studies. The arithmetic mean $\bar{x}$ is calculated by summing all individual values x_i in a sample and then dividing the sum by the sample size.

$$\bar{x} = \frac{\sum_{i=1}^{n} x_i}{n} \tag{3.3}$$

We already covered how the standard error of the mean is calculated (see Chapter 3.1). We simply have to divide the sample standard deviation s through the square root of the sample size.

$$SE_{\bar{x}} = \frac{s}{\sqrt{n}} \tag{3.4}$$

As we have seen before, the mean and its standard error are easy to calculate in R.

```
# Set seed of 123 for reproducibility and take a random sample (n=50).
set.seed(123)
sample <- rnorm(n = 50, mean = 20, sd = 5)

# Calculate the mean
mean(sample)
```

```
## [1] 20.17
```

```
# Calculate the standard error
sd(sample)/sqrt(50)
```

```
## [1] 0.6547
```

To conduct a meta-analysis of means, our data set should at least contain the following columns:

- `n`. The number of observations (sample size) in a study.
- `mean`. The mean reported in the study.
- `sd`. The standard deviation of the variable reported in the study.

3.2.2 Proportions

A *proportion* is another type of one-variable measure. It specifies how many units of a sample fall into a certain subgroup. Proportions can take values between zero and one, which can be transformed into *percentages* by multiplying with 100. Proportions may, for example, be used as an outcome measure when we want to examine the prevalence of a disease at a given point in time. To calculate a proportion p, we have to divide the number of individuals k falling into a specific subgroup by the total sample size n.

$$p = \frac{k}{n} \tag{3.5}$$

The standard error of a proportion can be calculated this way:

$$SE_p = \sqrt{\frac{p(1-p)}{n}} \tag{3.6}$$

We can calculate the proportion and its standard error in R using this code:

```
# We define the following values for k and n:
k <- 25
n <- 125

# Calculate the proportion
p <- k/n
p
```

```
## [1] 0.2
```

```
# Calculate the standard error
sqrt((p*(1-p))/n)
```

```
## [1] 0.03578
```

The fact that the range of proportions is restricted between 0 and 1 is problematic from a mathematical standpoint (Lipsey and Wilson, 2001, chapter 3). When p is close to 0 or close to 1, the standard error is artificially compressed, which leads us to overestimate the precision of the proportion estimate. This has something to do with the sampling distribution. When values of p are very low or very high, the sampling distribution will not be approximately normal like in Figure 3.1. The distribution will be *right-skewed* or *left-skewed* because it is impossible for a random sample to have a calculated proportion outside the 0-1 range. To avoid this, proportions are commonly *logit*-transformed before they are pooled. A logit-transformation first involves calculating the *odds* (see Chapter 3.3.2.2). Odds are defined as the proportion of units which fall into a specific category, divided by the proportion of units which do not fall into that category. The natural logarithm function $\log_e$ is then used to transform the odds into a format where $p = 0.5$ equals a value of 0, and where there is no range restriction. This ensures that the sampling distribution is approximately normal and that standard errors are not biased. The calculation of logit-transformed proportions and their standard errors can be done using these formulas (Lipsey and Wilson, 2001, chapter 3):

$$p_{\text{logit}} = \log_e \left(\frac{p}{1-p} \right) \tag{3.7}$$

$$SE_{p_{\text{logit}}} = \sqrt{\frac{1}{np} + \frac{1}{n(1-p)}} \tag{3.8}$$

Luckily, the meta-analysis function we can use in R performs this logit-transformation automatically for us. Therefore, we only have to prepare the following columns in our data set:

- `event`. The number of observations which are part of a specific subgroup (k).
- `n`. The total sample size n.

3.2.3 Correlations

3.2.3.1 Pearson Product-Moment Correlation

A correlation is an effect size which expresses the amount of *co-variation* between two variables. The most common form is the *Pearson product-moment correlation*[3], which can be calculated for two continuous variables. Product-moment correlations can be used as the effect size, for example, when a meta-analyst wants to examine the relationship between relationship quality and well-being. A correlation r_{xy} between a variable x and a variable y is defined as the *covariance* $\text{Cov}(x,y) = \sigma^2_{xy}$ of x and y, divided by the *product* of their standard deviations, σ_x and σ_y.

$$r_{xy} = \frac{\sigma^2_{xy}}{\sigma_x \sigma_y} \tag{3.9}$$

Using the sample size n, the standard error of r_{xy} can be calculated like this:

$$SE_{r_{xy}} = \frac{1-r^2_{xy}}{\sqrt{n-2}} \tag{3.10}$$

[3] This type of correlation was named after Karl Pearson, the famous statistician who also played a part in the history of meta-analysis (see Chapter 1.2).

When calculating the product-moment correlation, we standardize the covariance between two variables by their standard deviations. This means that it becomes less relevant if two or more studies measured a construct on the same scale; once we calculate a correlation, it is automatically possible to compare the effects. Correlations can take values between −1 and 1. The magnitude of a correlation is often interpreted using Cohen's (1988) conventions:

- $r \approx 0.10$: small effect.
- $r \approx 0.30$: moderate effect.
- $r \approx 0.50$: large effect.

It should be noted, however, that these conventions may be at best seen as rules of thumb. It is often much better to quantify a correlation as small or large depending on the topic and previous research.

Unfortunately, like proportions (Chapter 3.2.2), correlations are restricted in their range, and it can introduce bias when we estimate the standard error for studies with a small sample size (Alexander et al., 1989). In meta-analyses, correlations are therefore usually transformed into *Fisher's* z[4]. Like the logit-transformation, this also entails the use of the natural logarithm function to remove the range restriction and make sure that the sampling distribution is approximately normal (see Chapter 3.3.2 for a more detailed explanation). The formula looks like this:

$$z = 0.5 \log_e \left(\frac{1+r}{1-r} \right) \tag{3.11}$$

If we know the sample size n, the approximate standard error of Fisher's z can be obtained through this formula (Olkin and Finn, 1995):

$$SE_z = \frac{1}{\sqrt{n-3}} \tag{3.12}$$

We can also calculate r_{xy} and z directly in R, using the `cor` and `log` function.

```
# Simulate two continuous variables x and y
set.seed(12345)
x <- rnorm(20, 50, 10)
y <- rnorm(20, 10, 3)

# Calculate the correlation between x and y
r <- cor(x,y)
r
```

```
## [1] 0.2841
```

[4]Fisher's z was named after yet another famous statistician we mentioned in Chapter 1.2, Ronald A. Fisher.

```
# Calculate Fisher's z
z <- 0.5*log((1+r)/(1-r))
z
```

```
## [1] 0.2921
```

Thankfully, we do not have to perform Fisher's z transformation manually when conducting a meta-analysis of correlations in R. The only columns we need in our data set are:

- `cor`. The (nontransformed) correlation coefficient of a study.
- `n`. The sample size of the study.

3.2.3.2 Point-Biserial Correlation

The Pearson product-moment correlation describes the relationship between two continuous variables. In cases where only one variable y is continuous, while the other variable x is dichotomous (i.e. only takes two values), a *point-biserial correlation* can be calculated, which expresses how well y can be predicted from the group membership in x. Point-biserial correlations can be calculated using this formula:

$$r_{pb} = \frac{(\bar{y_1} - \bar{y_2})\sqrt{p_1(1-p_1)}}{s_y} \tag{3.13}$$

In this formula, $\bar{y_1}$ is the mean of the continuous variable when only the first group of the dichotomous variable x is considered, and $\bar{y_2}$ is the mean when only the second group of x is considered; p_1 is the proportion of cases that fall into group 1 in x, and s_y is the standard deviation of y. The point-biserial correlation can be calculated in R using the `cor` function (see previous section). If one of the supplied variables only assumes two values while the other is continuous, the (approximate) point-biserial correlation is automatically calculated.

The point-biserial correlation bears a close resemblance to the *standardized mean difference*, which we will cover later (Chapter 3.3.1.2). Both effect size metrics quantify how much values of a continuous variable differ between two groups. However, it is less common that point-biserial correlations are pooled in meta-analyses. Like the product-moment correlation, the point-biserial correlation has undesirable statistical properties for meta-analyses, such as range restriction when the group proportions are unequal (Bonett, 2020). When we are interested in group differences on a continuous outcome variable, it is therefore advised to convert point-biserial correlations to standardized mean differences for meta-analyses (Lipsey and Wilson, 2001, chapter 3). A formula to convert a point-biserial correlation to a standardized mean difference can be found in Chapter 17.3 in the "Helpful Tools" section of this book.

3.3 Effect Sizes in Control Group Designs

3.3.1 (Standardized) Mean Differences

3.3.1.1 Between-Group Mean Difference

The *between-group mean difference* $\text{MD}_{\text{between}}$ is defined as the raw, unstandardized difference of the means of two *independent* groups. Between-group mean differences can be calculated when a study contained at least two groups, as is usually the case in controlled trials or other types of experimental studies. In meta-analyses, mean differences can only be used when *all* the studies measured the outcome of interest on *exactly* the same scale. Weight, for example, is nearly always measured in kilograms in scientific research; and in diabetology, the HbA_{1c} value is commonly used to measure the blood sugar. The mean difference is defined as the mean of group 1, $\bar{x}_1$, minus the mean of group 2, $\bar{x}_2$:

$$\text{MD}_{\text{between}} = \bar{x}_1 - \bar{x}_2 \tag{3.14}$$

The standard error can be obtained using this formula:

$$SE_{\text{MD}_{\text{between}}} = s_{\text{pooled}}\sqrt{\frac{1}{n_1}+\frac{1}{n_2}} \tag{3.15}$$

In the formula, n_1 represents the sample size in group 1, n_2 the sample size in group 2, and s_{pooled} the *pooled standard deviation* of both groups. Using the standard deviation of group 1 (s_1) and group 2 (s_2), the value of s_{pooled} can be calculated this way:

$$s_{\text{pooled}} = \sqrt{\frac{(n_1-1)s_1^2+(n_2-1)s_2^2}{(n_1-1)+(n_2-1)}} \tag{3.16}$$

Here is an example showing how we can calculate a mean difference and its standard error in R.

```
# Generate two random variables with different population means
set.seed(123)
x1 <- rnorm(n = 20, mean = 10, sd = 3)
x2 <- rnorm(n = 20, mean = 15, sd = 3)

# Calculate values we need for the formulas
s1 <- sd(x1)
s2 <- sd(x2)
n1 <- 20
```

```
n2 <- 20

# Calculate the mean difference
MD <- mean(x1) - mean(x2)
MD
```

```
## [1] -4.421
```

```
# Calculate s_pooled
s_pooled <- sqrt(
  (((n1-1)*s1^2) + ((n2-1)*s2^2))/
    ((n1-1)+(n2-1))
)

# Calculate the standard error
se <- s_pooled*sqrt((1/n1)+(1/n2))
se
```

```
## [1] 0.8577
```

It is usually not necessary to do these calculations manually. For a meta-analysis of mean differences, we only have to prepare the following columns in our data set:

- `n.e`. The number of observations in the intervention/experimental group.
- `mean.e`. The mean of the intervention/experimental group.
- `sd.e`. The standard deviation in the intervention/experimental group.
- `n.c`. The number of observations in the control group.
- `mean.c`. The mean of the control group.
- `sd.c`. The standard deviation in the control group.

3.3.1.2 Between-Group Standardized Mean Difference

The standardized between-group mean difference $\text{SMD}_{\text{between}}$ is defined as the difference in means between two independent groups, standardized by the pooled standard deviation s_{pooled}. In the literature, the standardized mean difference is also often called *Cohen's d*, named after the psychologist and statistician Jacob Cohen. In contrast to unstandardized mean differences, $\text{SMD}_{\text{between}}$ expresses the difference between two groups in *units of standard deviations*. This can be achieved by dividing the raw mean difference of two groups, $\bar{x_1}$ and $\bar{x_2}$, through the pooled standard deviation s_{pooled} of both groups:

$$\text{SMD}_{\text{between}} = \frac{\bar{x}_1 - \bar{x}_2}{s_{\text{pooled}}} \tag{3.17}$$

Where s_{pooled} is calculated using the same formula (3.3) we already covered in Chapter 3.3.1.2. Standardized mean differences are much more often used in meta-analyses than unstandardized mean differences. This is because $\text{SMD}_{\text{between}}$ can be compared between studies, even if those studies did not measure the outcome of interest using the same instruments. The standardization has the effect that $\text{SMD}_{\text{between}} = 1$ always means that the two group means are one sample standard deviation away from each other (see Figure 3.3); $\text{SMD}_{\text{between}} = 2$ then represents a difference of 2 standard deviations, and so forth[5].

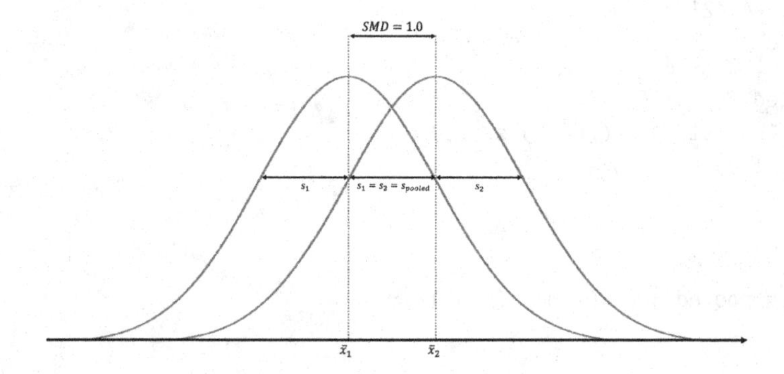

FIGURE 3.3: Standardized mean difference of 1 (assuming normality and equal standard deviations in both groups).

The standardization makes it much easier to evaluate the magnitude of the mean difference. Standardized mean differences are often interpreted using the conventions by Cohen (1988):

- $\text{SMD} \approx 0.20$: small effect.
- $\text{SMD} \approx 0.50$: moderate effect.
- $\text{SMD} \approx 0.80$: large effect.

Like the guideline for Pearson product-moment correlations (Chapter 3.2.3.1), these are rules of thumb at best. It is usually much better to interpret standardized mean differences based on their "real-life" implications. An effect size may be small according to the criteria by Cohen, but it can still be extremely important. For many serious diseases, for example, even a very small statistical effect can still have a huge impact on the population level, and potentially save millions of lives. One study showed that, for depression treatment, even effects as small as $\text{SMD}_{\text{between}} = 0.24$ can have a clinically important impact on the lives of patients (Cuijpers et al., 2014).

The standard error of $\text{SMD}_{\text{between}}$ can be calculated using this formula (Borenstein et al., 2011):

[5] Kristoffer Magnusson developed a great interactive tool visualizing the distribution of two groups for varying values of the standardized mean difference. The tool can be found online: `https://www.rpsychologist.com/d3/cohend/`

$$SE_{\text{SMD}_{\text{between}}} = \sqrt{\frac{n_1 + n_2}{n_1 n_2} + \frac{\text{SMD}^2_{\text{between}}}{2(n_1 + n_2)}} \tag{3.18}$$

where n_1 and n_2 are the sample sizes of group 1 and group 2, and with $\text{SMD}_{\text{between}}$ being the calculated between-group standardized mean difference.

There are several functions in R which allow us to calculate $\text{SMD}_{\text{between}}$/Cohen's d in one step. Here, we use the esc_mean_sd function, which is part of the *{esc}* package (Lüdecke, 2019). We have not used this package before, so it is necessary to install it first (see Chapter 2.2).

```
# Load esc package
library(esc)

# Define the data we need to calculate SMD/d
grp1m <- 50   # mean of group 1
grp2m <- 60   # mean of group 2
grp1sd <- 10  # sd of group 1
grp2sd <- 10  # sd of group 2
grp1n <- 100  # n of group1
grp2n <- 100  # n of group2

# Calculate effect size
esc_mean_sd(grp1m = grp1m, grp2m = grp2m,
            grp1sd = grp1sd, grp2sd = grp2sd,
            grp1n = grp1n, grp2n = grp2n)
```

```
## Effect Size Calculation for Meta Analysis
##
##      Conversion: mean and sd to effect size d
##     Effect Size:  -1.0000
##  Standard Error:   0.1500
##           [...]
```

The output contains two important pieces of information. First, we see that the calculated standardized mean difference is exactly 1. This makes sense because the difference between the two means we defined is equal to the (pooled) standard deviation. Second, we see that the effect size is *negative*. This is because the mean of group 2 is larger than the mean of group 1. While this is mathematically correct, we sometimes have to change the sign of calculated effect sizes so that others can interpret them more easily. Imagine that the data in this example came from a study measuring the mean number of cigarettes people were smoking per week after receiving an intervention (group 1) or no intervention (group 2). In this context, the results of the study were *positive* because the mean number of smoked cigarettes was lower in the intervention group. Therefore, it makes sense to report the effect size as 1.0 instead of −1.0, so that other people can intuitively understand that the intervention had a positive effect.

The sign of effect sizes becomes particularly important when some studies used measures for which *higher* values mean better outcomes, while others used a measure for which *lower* values indicate better outcomes. In this case, it is essential that all effect sizes are consistently coded in the same direction (we have to ensure that, for example, higher effect sizes mean better outcomes in the intervention group in all studies in our meta-analysis).

Often, a small-sample correction is applied to standardized mean differences, which leads to an effect size called *Hedges' g*. We will cover this correction in Chapter 3.4.1.

To conduct a meta-analysis of standardized mean differences, our data set should at least contain the following columns:

- `n.e`. The number of observations in the intervention/experimental group.
- `mean.e`. The mean of the intervention/experimental group.
- `sd.e`. The standard deviation in the intervention/experimental group.
- `n.c`. The number of observations in the control group.
- `mean.c`. The mean of the control group.
- `sd.c`. The standard deviation in the control group.

Standardizing by External Estimates of the Standard Deviation

When calculating SMDs, we use s_{pooled} because it serves as a proxy of the true standard deviation in our population. Especially when a study is small, however, the standard deviation calculated based on its sample may be a bad estimator of the population standard deviation. In this case, a possible solution is to use *external* estimates of s_{pooled} to standardize the mean difference (Higgins et al., 2019, chapter 6.5.1.2). Such external estimates may be extracted from larger cross-sectional studies which used the same instrument as the study in a similar population.

3.3.1.3 Within-Group (Standardized) Mean Difference

Within-group unstandardized or standardized mean differences can be calculated when a difference *within* one group is examined. This is usually the case when the same group of people is measured at two different time points (e.g. before an intervention and after an intervention). In contrast to between-group mean differences, $\text{(S)MD}_{\text{within}}$ is calculated using data that is *not independent*. For example, it is likely that the value of person i at measurement point t_1 has influenced the value of the same person at measurement point t_2. Due to the fact that within-group mean differences are usually based on data measured at different time points, they are also known as the *(standardized) mean gain*. The within-group mean difference $\text{MD}_{\text{within}}$ is calculated the same way as $\text{MD}_{\text{between}}$ (see Chapter 3.3.1.1), except that we now compare the values of the same group at two different time points, t_1 and t_2.

$$\text{MD}_{\text{within}} = \bar{x}_{\text{t}_2} - \bar{x}_{\text{t}_1} \tag{3.19}$$

Things become more complicated when we want to calculate a standardized version of the within-group mean difference. There is no full consensus on how $\text{SMD}_{\text{within}}$ should be computed. In a blog post[6], Jake Westfall points out that there are at least five distinct ways to calculate it. However, probably the most frequently used formula is the one by Becker (1988). The formula is quite complicated; it requires us to know the standard deviation at both assessment points, s_{t_1} and s_{t_2}, as well as the correlation between both assessment points, $r_{\text{t}_1\text{t}_2}$:

$$\text{SMD}_{\text{within}} = \frac{\bar{x}_{\text{t}_2} - \bar{x}_{\text{t}_1}}{\left(\frac{\sqrt{(s^2_{\text{t}_1} + s^2_{\text{t}_2})/2}}{\sqrt{2(1 - r_{\text{t}_1\text{t}_2})}}\right)} \tag{3.20}$$

The standard errors of $\text{MD}_{\text{within}}$ and $\text{SMD}_{\text{within}}$ can be calculated using these formulas (Borenstein et al., 2011, chapter 24; Becker, 1988):

$$SE_{\text{MD}_{\text{within}}} = \sqrt{\frac{s^2_{\text{t}_1} + s^2_{\text{t}_2} - (2r_{\text{t}_1\text{t}_2} s_{\text{t}_1} s_{\text{t}_2})}{n}} \tag{3.21}$$

$$SE_{\text{SMD}_{\text{within}}} = \sqrt{\frac{2(1 - r_{\text{t}_1\text{t}_2})}{n} + \frac{\text{SMD}^2_{\text{within}}}{2n}} \tag{3.22}$$

The fact that we need two know the correlation $r_{\text{t}_1\text{t}_2}$ between the assessment points when calculating within-group (standardized) mean differences is often problematic in practice. The pre-post correlation of a variable is hardly ever reported in published research, which forces us to assume a value of $r_{\text{t}_1\text{t}_2}$ based on previous research. However, if we do not get the correlation exactly right, this can lead to substantial biases in our results. It should therefore at best be avoided to calculate within-group effect sizes for a meta-analysis (Cuijpers et al., 2017). Especially when we have data from both an experimental *and* control group, it is much better to calculate the *between-group* (standardized) mean differences at t_2 to measure the effect of a treatment, instead of pre-post comparisons. Within-group mean difference may be calculated, however, when our meta-analysis focuses solely on studies which did not include a control group.

The within-group standardized mean difference (also known as within-group Cohen's *d*) can be calculated like this in R:

[6] `http://jakewestfall.org/blog/index.php/2016/03/25/five-different-cohens-d-statistics-for-within-subject-designs/`

```
# Define data needed for effect size calculation
x1 <- 20    # mean at t1
x2 <- 30    # mean at t2
sd1 <- 13   # sd at t1
sd2 <- 16   # sd at t2
n <- 80     # sample size
r <- 0.5    # correlation between t1 and t2

# Caclulate the raw mean difference
md_within <- x2 - x1

# Calculate the smd
smd_within <- md_within/
                  ((sqrt((sd1^2+sd2^2)/2))/
                   (sqrt(2*(1-r))))
smd_within
```

```
## [1] 0.686
```

```
# Calculate standard error
se_within <- sqrt(
               ((2*(1-r))/n) +
               (smd_within^2/(2*n))
               )
se_within
```

```
## [1] 0.1243
```

Meta-analyses of within-group (standardized) mean differences can only be performed in R using *pre-calculated effect sizes* (see Chapter 3.5.1). The following columns are required in our data set:

- `TE`: The calculated within-group effect size.
- `seTE`: The standard error of the within-group effect size.

3.3.2 Risk & Odds Ratios

3.3.2.1 Risk Ratio

As it says in the name, a *risk ratio* (also known as the *relative risk*) is a ratio of two *risks*. Risks are essentially *proportions* (see Chapter 3.2.2). They can be calculated when we are dealing with binary, or *dichotomous*, outcome data. We use the term "risk" instead of "proportion" because this type of outcome data is frequently found in medical research, where one examines the *risk* of developing a disease or dying. Such occurrences are known as *events*. Imagine we are conducting a controlled clinical trial comprising a treatment group and a control group. We are interested in how many patients experienced some event E during the study period. The results we get from such a study can be categorized in a 2×2 table (Schwarzer et al., 2015, chapter 3.1):

TABLE 3.1: Results of controlled studies using binary outcome data.

	Event	No Event	
Treatment	a	b	n_{treat}
Control	c	d	n_{control}
	n_E	$n_{\neg E}$	

Based on this data, we can calculate the risk of experiencing event E during the study period for both the treatment group and control group. We simply divide the number of people experiencing E in one group by the total sample size of that group. The risk in the treatment group, $p_{E_{\text{treat}}}$, is therefore calculated like this:

$$p_{E_{\text{treat}}} = \frac{a}{a+b} = \frac{a}{n_{\text{treat}}} \tag{3.23}$$

And the risk in the control group, $p_{E_{\text{control}}}$, like this:

$$p_{E_{\text{control}}} = \frac{c}{c+d} = \frac{c}{n_{\text{control}}} \tag{3.24}$$

The risk ratio is then defined as the risk in the treatment/intervention group divided by the risk in the control group:

$$\text{RR} = \frac{p_{E_{\text{treat}}}}{p_{E_{\text{control}}}} \tag{3.25}$$

Because both $p_{E_{\text{treat}}}$ and $p_{E_{\text{control}}}$ can only have values between 0 and 1, the RR has a few interesting properties. First of all, a risk ratio can never be negative. Second, if there is no difference between the treatment group and the control group, RR has a value of 1 (instead of 0, like SMDs). If an RR is larger than 1, this means that the treatment group increases the risk of event E; if RR is smaller than 1, the intervention reduces the risk. A peculiarity of the RR is that same-sized effects are not *equidistant*. For example, RR = 0.5 means that the risks are halved in the intervention group. However, the direct opposite of this effect, the risk being doubled due to the intervention, is not expressed by RR = 1.5, but by RR = 2. This means that risk ratios do not follow a normal distribution, which can be problematic in meta-analyses. To avoid this issue, risk ratios are often transformed into the *log-risk ratio* before pooling. This ensures normality, that effect sizes can assume any value, and that values are centered around 0 (meaning no effect). The transformation is performed by taking the natural logarithm of RR:

$$\log \text{RR} = \log_e(\text{RR}) \tag{3.26}$$

The standard error of the log-risk ratio can then be calculated using this formula:

$$SE_{\log \text{RR}} = \sqrt{\frac{1}{a} + \frac{1}{c} - \frac{1}{a+b} - \frac{1}{c+d}} \tag{3.27}$$

We can calculate the (log-)risk ratio in R like this:

```
# Define data
a <- 46           # events in the treatment group
c <- 77           # events in the control group
n_treat <- 248  # sample size treatment group
n_contr <- 251  # sample size control group

# Calculate the risks
p_treat <- a/n_treat
p_contr <- c/n_contr

# Calculate the risk ratio
rr <- p_treat/p_contr
rr
```

```
## [1] 0.6046
```

```
# Calculate the log-risk ratio and its standard error
log_rr <- log(rr)
log_rr
```

```
## [1] -0.5031
```

```
se_log_rr <- sqrt((1/a) + (1/c) - (1/n_treat) - (1/n_contr))
se_log_rr
```

```
## [1] 0.1634
```

The calculation of risk ratios becomes difficult when there are *zero cells*. It is possible in practice that a or c (or both) are zero, meaning that no event was recorded in the treatment or control group. If you have a look at the formula used to calculate RRs, it is easy to see why this is problematic. If a (events in the treatment group) is zero, $p_{E_{\text{treat}}}$ is also zero, and the RR will be zero. The case of c being zero is even more problematic: it means that $p_{E_{\text{control}}}$ is zero, and we all know that we *cannot divide by zero*. This issue is often dealt with using a *continuity correction*. The most common continuity correction method is to add an increment of 0.5 in all cells that are zero (Gart and Zweifel, 1967). When the sample sizes of the control group and treatment group are very uneven, we can also use the *treatment arm continuity correction* (Sweeting et al., 2004). However, there is evidence that such corrections can lead to biased results (Efthimiou, 2018). The (fixed-effect) *Mantel-Haenszel* method, a meta-analytic pooling technique we will discover in Chapter 4.2.3.1.1, can handle zero cells *without* correction, unless they exist in *every study* in our meta-analysis. It may therefore be advisable to avoid continuity corrections unless the latter scenario applies.

A special form of the *zero cell* problem are *double-zero studies*. These are studies in which both a and c are zero. Intuitively, one might think that the results of such studies simply mean that the risk in the intervention and control group are similar, and that RR = 1. Unfortunately, it is not that easy. It is very much possible that there is a true effect between the two groups, but that the sample size was too small to detect this difference. This is particularly likely when the probability that E occurs is very low. Imagine that a crazy scientist conducts a randomized controlled trial in which he assesses the effect of *Fulguridone*, a medication that allegedly reduces the risk of getting struck by lightning. He allocates 100 people evenly to either a medication group or a control group, and observes them for three years. The results of the trial are disappointing, because no one was struck by lightning, neither in the treatment group nor in the control group. However, we know how unlikely it is, *in general*, to get struck by lightning. Observing only 100 people is simply not enough to detect differences in such a rare event, even if we accept the somewhat bizarre idea that the treatment works. For this reason, double-zero studies are often discarded completely when pooling the effects.

This leads us to one last caveat pertaining to risk ratios: they give us no information on how common an event is *in general*. If a meta-analysis reports a risk ratio of 0.5, for example, we know that an intervention reduced the risk by half. But we do not know if it reduced the risk from 40% to 20%, or from 0.004% to 0.002%. Whether a risk ratio is practically relevant depends on the context. If a risk ratio of 0.5 represents a risk reduction of 0.002%, this may not have a large impact on a population level, but it may still be important if the event of interest is, for example, a severe and debilitating disease.

When we conduct a meta-analysis in R, it is usually not necessary to calculate the log-risk ratio of a study by hand. We also do not have to worry about zero cells when importing the data. The following columns should be included in our data set:

- `event.e`. The number of events in the treatment or experimental group.
- `n.e`. The sample size of the treatment or experimental group.
- `event.c`. The number of events in the control group.
- `n.c`. The sample size of the control group.

3.3.2.2 Odds Ratio

Like the risk ratio (Chapter 3.3.2.1), *odds ratios* can also be calculated when we have binary outcome data of two groups. In the previous chapter on proportions (Chapter 3.2.2), we already defined the odds as the number of cases which fall into a specific category, divided by the number of units which do not fall into that category. Using the notation in Table 3.1, the formula for the odds in the treatment and control group looks like this:

$$\text{Odds}_{\text{treat}} = \frac{a}{b} \tag{3.28}$$

$$\text{Odds}_{\text{control}} = \frac{c}{d} \tag{3.29}$$

It can be difficult to correctly interpret what odds actually mean. They describe the ratio of events to non-events, not the *probability* of the event. Imagine that we studied three individuals. Two experienced the event of interest, while one person did not. Based on this data, the probability (or risk) of the event would be $p = 2/3 \approx 66\%$. However, the odds of the event would be Odds = $\frac{2}{1}$ = 2, meaning that there are two events for one non-event.

The odds ratio (OR) is then defined as the odds in the treatment group, divided by the odds in the control group:

$$\text{OR} = \frac{a/b}{c/d} \tag{3.30}$$

Like the risk ratio (see Chapter 3.3.2.1), the odds ratio has undesirable statistical properties for meta-analyses. It is therefore also common to transform the odds ratio to the *log-odds ratio* using the natural logarithm:

$$\log \text{OR} = \log_e(\text{OR}) \tag{3.31}$$

The standard error of the log-odds ratio can be calculated using this formula (we use the notation in Table 3.1):

$$SE_{\log \text{OR}} = \sqrt{\frac{1}{a} + \frac{1}{b} + \frac{1}{c} + \frac{1}{d}} \tag{3.32}$$

The esc_2x2 function in the *{esc}* package provides an easy way to calculate the (log) odds ratio in R.

```
library(esc)

# Define data
grp1yes <- 45  # events in the treatment group
grp1no <- 98   # non-events in the control group
grp2yes <- 67  # events in the control group
grp2no <- 76   # non-events in the control group

# Calculate OR by setting es.type to "or"
```

```
esc_2x2(grp1yes = grp1yes, grp1no = grp1no,
        grp2yes = grp2yes, grp2no = grp2no,
        es.type = "or")
```

```
##
## Effect Size Calculation for Meta Analysis
##
##     Conversion: 2x2 table (OR) coefficient to effect size odds ratio
##    Effect Size:  0.5209
## Standard Error:  0.2460
##       Variance:  0.0605
##       Lower CI:  0.3216
##       Upper CI:  0.8435
##         Weight: 16.5263
```

```
# Calculate logOR by setting es.type to "logit"
esc_2x2(grp1yes = grp1yes, grp1no = grp1no,
              grp2yes = grp2yes, grp2no = grp2no,
              es.type = "logit")
```

```
##
## Effect Size Calculation for Meta Analysis
##
##     Conversion: 2x2 table (OR) to effect size logits
##    Effect Size: -0.6523
## Standard Error:  0.2460
##       Variance:  0.0605
##       Lower CI: -1.1344
##       Upper CI: -0.1701
##         Weight: 16.5263
```

The same problems pertaining to risk ratios, *zero cells* and *double-zero studies* (see Chapter 3.3.2.1), are also relevant when calculating odds ratios. However, the odds ratio has one additional disadvantage compared to RRs: many people find it harder to understand, and ORs are often erroneously interpreted as RRs. It is, therefore, often preferable to either only use risk ratios in a meta-analysis or to convert odds ratios to risk ratios when reporting the results (Higgins et al., 2019, chapter 6.4.1.2). The conversion can be performed using this formula (Zhang and Yu, 1998):

$$\text{RR} = \frac{\text{OR}}{\left(1 - \frac{c}{n_{\text{control}}}\right) + \left(\frac{c}{n_{\text{control}}} \times \text{OR}\right)} \tag{3.33}$$

To conduct a meta-analysis of odds ratios in R, the following columns should be included in our data set:

- `event.e`. The number of events in the treatment or experimental group.
- `n.e`. The sample size of the treatment or experimental group.
- `event.c`. The number of events in the control group.
- `n.c`. The sample size of the control group.

3.3.3 Incidence Rate Ratios

The effect sizes for binary outcome data we examined previously, risk ratios and odds ratios, are ways to compare the number of events in two groups. However, they do not encode the *time* during which these events occurred. When calculating a risk or odds ratio, we tacitly assume that the observation periods in both groups are comparable. Furthermore, risk and odds ratios do not provide us with any information on *how long* it took until the events occurred. In some cases, this is just fine, because the time frame is not overly important for our research question. It is also possible that our binary data is cross-sectional and has no time dimension at all[7]. In these cases, the risk or odds ratio is an appropriate effect size metric.

But now, imagine a study in which we examine the mortality of individuals in two groups over 10 years. It might be possible that the number of events over these 10 years (i.e. death) is roughly the same in both groups. However, once we have a closer look at *when* the deaths occurred, we see that more events in one group occurred before the end of the first year, while in the other group, more events occurred at the end of our 10-year observation period. The calculated odds or risk ratio for our data would be 1, indicating no group difference. But this misses something important: that participants in one group survived *somewhat longer*, even if they died eventually.

To incorporate time into our effect size estimate, we can calculate *incidence rate ratios*, which are sometimes simply called *rate ratios*. Incidence rate ratios consist of two *incidence rates*. To calculate these incidence rates, we have to first understand the concept of *person-time*.

The person-time expresses the total time in which participants in a study were at risk of having an event. To calculate the person-time, we sum up the time at risk (expressed as days, weeks, or years) of all study subjects. However, the time at risk differs from person to person. To exemplify this, imagine we are conducting a study with 6 participants. The study lasts for exactly 10 years. After each year, we interview the participants to examine if they experienced the event of interest. Whenever we observe that the event has occurred, the study ends for the affected participant, and we do not examine her or him until the study ends. The results of our study are visualized in Figure 3.4.

We see that only two of our participants, Victoria and Lea, remained in the study until the end. This is because they did not experience the event during the entire 10-year observation period. Therefore, both were *at risk* for 10 years. All other participants

[7]For example, we can also use risk or odds ratios to express differences in the proportion of smokers between females and males, based on cross-sectional survey data.

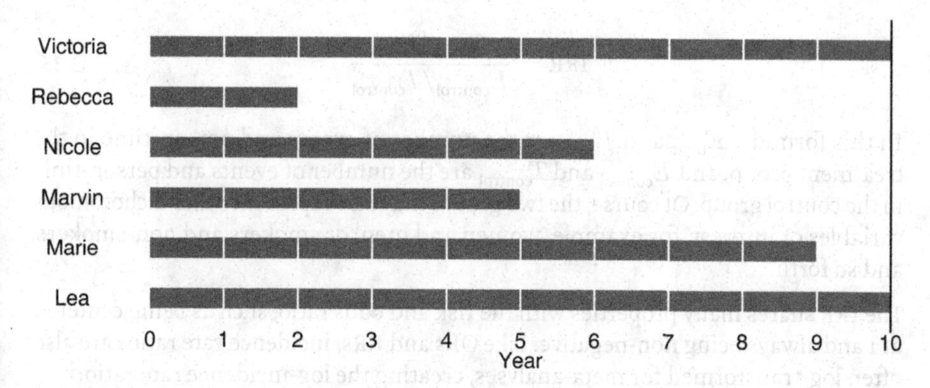

FIGURE 3.4: Example of time-to-event data.

experienced the event during the study period. When Rebecca was examined at year 2, for example, we found out that she experienced the event during the last year. However, we only know *that* the event occurred during year 2, not when exactly. Research data like this is called *interval censored* data, and very frequently found in clinical trials which conduct a so-called *survival analysis*. Data being censored means that we only partially know how long Rebecca was at risk before she finally experienced the event. We know that she had the event after year 1 and before the end of year 2, but not more. Lacking other information, we may therefore assume that the event occurred somewhere in the middle, and settle with a time at risk of 1.5 years. If we apply the same scheme for all our censored data, we can calculate the *person-years* at risk in our study:

$$10 + 1.5 + 5.5 + 4.5 + 8.5 + 10 = 40$$

So the estimated total person-years at risk in our study is 40. Knowing that a year has 52 weeks, we can also calculate the *person-weeks* of our study: $40 \times 52 = 2080$.

Now that we know the person-years in our experiment, which we will denote as T, we can also calculate the incidence rate within one year. We know that four participants experienced the event during the study period, so the number of events is $E = 4$. We can then calculate the incidence rate IR using this formula:

$$\text{IR} = \frac{E}{T} \tag{3.34}$$

In our example, this gives an incidence rate of $3/40 = 0.075$. This incidence rate means that, if we would follow 1000 people for one year, 75 would experience the event during that time.

To calculate the incidence rate ratio IRR, we have to divide the incidence rate of one group by the incidence rate of another group:

$$\text{IRR} = \frac{E_{\text{treat}}/T_{\text{treat}}}{E_{\text{control}}/T_{\text{control}}} \tag{3.35}$$

In this formula, E_{treat} and T_{treat} are the number of events and person-time in the treatment group, and E_{control} and T_{control} are the number of events and person-time in the control group. Of course, the two groups may also represent other dichotomous variables of interest, for example, women and men, or smokers and non-smokers, and so forth.

The IRR shares many properties with the risk and odds ratio, such as being centered at 1 and always being non-negative. Like ORs and RRs, incidence rate ratios are also often log-transformed for meta-analyses, creating the log-incidence rate ratio:

$$\log \text{IRR} = \log_e(\text{IRR}) \tag{3.36}$$

For which we can calculate the standard error like this (Rothman et al., 2008, chapter 14):

$$SE_{\log \text{IRR}} = \sqrt{\frac{1}{E_{\text{treat}}} + \frac{1}{E_{\text{control}}}} \tag{3.37}$$

We can calculate the (log-)incidence ratio and the standard error in R like this:

```
# Define Data
e_treat <- 28    # Number of events in the treatment group
e_contr <- 28    # Number of events in the control group
t_treat <- 3025  # Person-time in the treatment group
t_contr <- 2380  # Person-time in the control group

# Calculate IRR
irr <- (e_treat/t_treat)/(e_contr/t_contr)
irr
```

```
## [1] 0.7868
```

```
# Calculate log-IRR
log_irr <- log(irr)

# Calculate standard error
se_log_irr <- sqrt((1/e_treat)+(1/e_contr))
```

In this example, we simulated a case in which the number of events E_{treat} and E_{control} is exactly equal, but where the treatment group has a longer person-time at risk. This time difference is accounted for when we calculate IRRs. Therefore, the result we get is not 1, but IRR $\approx$ 0.79, indicating that the incidence rate is smaller in the treatment group.

Incidence rate ratios are commonly used in epidemiology and prevention research. They can be used when participants are followed for a longer period of time, and when there are regular assessments in between. In practice, however, there is one caveat we should consider when calculating IRRs as part of a meta-analysis: it is important that the incidence data reported in the included articles is fine-grained enough. Sometimes, papers only report the total number of events during the entire study period and not the number of events recorded at each assessment point in between. It is also possible that no interim assessments were made to begin with. In our example above (see Figure 3.4), we simply took the *midpoint* between the last "event-free" assessment point and the assessment point in which the event was recorded to estimate the time at risk of a participant. It is important to keep in mind that this is only a *best guess* of when the event happened exactly. Even when taking the midpoint, our estimates can still be off by about half a year in our example.

Our estimate of the person-time will be best if the time between assessment points is as small as possible. If assessment intervals in a study are too coarse depends on the context of the meta-analysis, but it is always advisable to conduct sensitivity analyses (Panageas et al., 2007). This means to recalculate the IRR of studies based on different estimates of the person-time: (1) using the midpoint of the interval, (2) using the last "event-free" assessment point, and (3) using the assessment point in which the event was detected. If the results of all three of these meta-analyses point in the same direction, we can be more confident in our findings. We should also make sure that the assessment periods do not differ too much between studies (e.g. one study examining events daily, and the other only each year). When there are doubts about the applicability of IRRs for a meta-analysis, there is always the possibility to calculate risk or odds ratios instead (or in addition). However, when we do this, we should make sure that the assessment point was similar in each study (e.g. after one year).

To calculate a meta-analysis based on incidence rate ratios in R, the following columns need to be prepared in our data set:

- `event.e`: The total number of events in the treatment or experimental group.
- `time.e`: The person-time in the treatment or experimental group. The person-time has to be expressed in the same units (person-days, person-weeks, or person-years) in all studies.
- `event.c`: The total number of events in the control group.
- `time.c`: The person-time in the control group. The person-time has to be expressed in the same units (person-days, person-weeks, or person-years) in all studies.

3.4 Effect Size Correction

In Chapter 3.1, we covered that the effect size $\hat{\theta}_k$ we calculate for a study k is an estimate of the study's true effect size θ_k, and that $\hat{\theta}_k$ deviates from θ_k due to the sampling error ϵ_k. Unfortunately, in many cases, this is an oversimplification. In the equation we discussed before, the only thing that separates the estimated effect size from the true effect is the sampling error. Following the formula, as the sampling error decreases, the effect size estimate "naturally" converges with the true effect size in the population. This is not the case, however, when our effect size estimate is additionally burdened by systematic error, or *bias*. Such biases can have different reasons. Some are caused by the mathematical properties of an effect size metric itself, while other biases are created by the way a study was conducted.

We can deal with biases arising from the way a study was conducted by evaluating its risk of bias (see Chapter 1.4.5 for an introduction to risk of bias assessment tools and Chapter 15 for ways to visualize the risk of bias). This judgment can then also be used to determine if the risk of bias is associated with differences in the pooled effect, for example, in subgroup analyses (Chapter 7). To deal with biases arising from the statistical properties of an effect size metric, we can use specific *effect size correction* methods to adjust our data before we begin with the meta-analysis.

In this chapter, we will cover three commonly used effect size correction procedures, and how we can implement them in R.

3.4.1 Small Sample Bias

In Chapter 3.3.1, we covered standardized mean differences (SMDs), an effect size we can calculate when we have continuous outcome data of two groups. The standardized mean difference, however, has been found to have an *upward bias* when the sample size of a study is small, especially when $n \leq 20$ (Hedges, 1981). This small sample bias means that SMDs systematically overestimate the true effect size when the total sample size of a study is small–which is unfortunately often the case in practice. It is therefore sensible to correct the standardized mean differences of all included studies for small-sample bias, which produces an effect size called Hedges' g. Hedges' g was named after Larry Hedges, the inventor of this correction. The formula to convert uncorrected SMDs/Cohen's d to Hedges' g looks like this:

$$g = \text{SMD} \times \left(1 - \frac{3}{4n-9}\right) \tag{3.38}$$

In this formula, n represents the total sample size of the study. We can easily convert unstandardized SMDs/Cohen's d to Hedges' g using the `hedges_g` function in the *{esc}* package.

```
# Load esc package
library(esc)

# Define uncorrected SMD and sample size n
SMD <- 0.5
n <- 30

# Convert to Hedges g
g <- hedges_g(SMD, n)
g
```

```
## [1] 0.4865
```

As we can see in the output, Hedges' g is smaller than the uncorrected SMD. Hedges' g can never be larger than the uncorrected SMD, and the difference between the two metrics is larger when the sample size is smaller (see Figure 3.5).

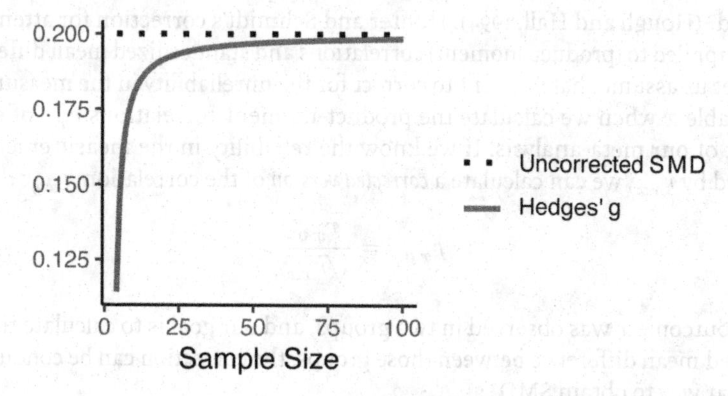

FIGURE 3.5: Corrected and uncorrected SMD of 0.2 for varying sample sizes.

It is important to note that the terms SMD and Hedges' g are sometimes used interchangeably in research reports. When a study reports results as the SMD, it is, therefore, relevant to check if the authors indeed refer to the uncorrected standardized mean difference, or if the small-sample bias correction has been applied (meaning that Hedges' g was used).

3.4.2 Unreliability

It is also possible that effect size estimates are biased due to *measurement error*. Most questionnaires or tests do not measure an outcome of interest perfectly. The less prone an instrument is to produce measurement errors, the more *reliable* it is. The reliability of an instrument measuring some variable x can be expressed through a reliability coefficient r_{xx}, which can take values between 0 and 1. Reliability is often defined as the *test-retest-reliability*, and can be calculated by taking two or more measurements of the same person under similar circumstances within a short period of time, and then calculating the correlation between the values[8].

When we examine the relationship of two continuous variables, a lack of reliability in one or both of the instruments used to assess these variables can lead to a phenomenon called *attenuation*. This problem has been described as early as 1904 by the famous psychologist Charles Spearman (1904). When we calculate a correlation, for example, and one or both variables are measured with error, this causes us to *underestimate* the true correlation. The correlation is *diluted*. But there are good news. If we have an estimate of the (un)reliability of a measurement, it is possible to correct for this attenuation in order to get a better estimate of the true effect size.

John Hunter and Frank Schmidt, two important contributors to the field of meta-analysis, have developed and promoted a method through which a correction for attenuation can be conducted as part of meta-analyses (Hunter and Schmidt, 2004, chapters 3 and 7). This correction is one of several other procedures, which together are sometimes called "Hunter and Schmidt techniques" or the "Hunter and Schmidt method" (Hough and Hall, 1994). Hunter and Schmidt's correction for attenuation can be applied to (product-moment) correlations and standardized mean differences. First, let us assume that we want to correct for the unreliability in the measurement of variable x when we calculate the product-moment correlations r_{xy} of studies as part of our meta-analysis. If we know the reliability in the measurement of x, denoted by r_{xx}, we can calculate a *corrected version* of the correlation, r_{xy_c}:

$$r_{xy_c} = \frac{r_{xy}}{\sqrt{r_{xx}}} \tag{3.39}$$

When outcome x was observed in two groups, and our goal is to calculate the standardized mean difference between those groups, the correction can be conducted in a similar way to obtain SMD_c:

$$\text{SMD}_c = \frac{\text{SMD}}{\sqrt{r_{xx}}} \tag{3.40}$$

When we calculate a product-moment correlation using two continuous variables x and y, it is also possible to correct for the unreliability of both x and y, provided we also know y's reliability coefficient r_{yy}:

$$r_{xy_c} = \frac{r_{xy}}{\sqrt{r_{xx}}\sqrt{r_{yy}}} \tag{3.41}$$

Lastly, we also have to correct the standard error. The standard error is corrected in the same way as the effect size itself. If we want to correct one variable x, we can use this formula:

[8] An accessible and more detailed discussion of various methods to estimate the reliability of an instrument can be found in Hunter and Schmidt (2004), chapter 3.

$$SE_c = \frac{SE}{\sqrt{r_{xx}}} \tag{3.42}$$

If we want to correct (a product-moment correlation) for both x and y, we can use this formula.

$$SE_c = \frac{SE}{\sqrt{r_{xx}}\sqrt{r_{yy}}} \tag{3.43}$$

After the correlation or SMD has been corrected, it is possible to apply other common transformations, such as converting r_{xy_c} to Fisher's z (Chapter 3.2.3) or SMD_c to Hedges' g (Chapter 3.4.1).

Let us try out the correction procedure in an example using R.

```
# Define uncorrected correlation and SMD with their standard error
r_xy <- 0.34
se_r_xy <- 0.09
smd <- 0.65
se_smd <- 0.18

# Define reliabilities of x and y
r_xx <- 0.8
r_yy <- 0.7

# Correct SMD for unreliability in x
smd_c <- smd/sqrt(r_xx)
smd_c
```

```
## [1] 0.7267
```

```
se_c <- se_smd/sqrt(r_xx)
se_c
```

```
## [1] 0.2012
```

```
# Correct correlation for unreliability in x and y
r_xy_c <- r_xy/(sqrt(r_xx)*sqrt(r_yy))
r_xy_c
```

```
## [1] 0.4543
```

```
se_c <- se_r_xy/(sqrt(r_xx)*sqrt(r_yy))
se_c
```

```
## [1] 0.1203
```

Take a close look at the results in this example. We see that due to the correction, the correlation and SMD are larger than the initial uncorrected value. However, we also see that the standard errors increase. This result is intended; we correct the standard error so that it also incorporates the measurement error we assume for our data.

It is common in some fields, for example, in organizational psychology, to apply attenuation corrections. However, in other disciplines, including the biomedical field, this procedure is rarely used. In meta-analyses, we can only perform a correction for unreliability if the reliability coefficient r_{xx} (and r_{yy}) is reported in each study. Very often, this is not the case. In this scenario, we may assume a value for the reliability of the instrument based on previous research. However, given that the correction has a large impact on the value of the effect size, taking an inappropriate estimate of r_{xx} can distort the results considerably. Also, it is not possible to only correct *some* effect sizes in our meta-analysis, while leaving others uncorrected. Due to these reasons, the applicability of the reliability correction is unfortunately often limited in practice.

3.4.3 Range Restriction

Another effect size adjustment proposed by Hunter and Schmidt (2004, chapters 3 and 7) deals with the problem of range restriction. Range restriction is a phenomenon which occurs when the variation in some variable x is smaller in a study than in the actual population of interest. This often happens when a study recruited a very selective sample of individuals which may not represent the population as a whole.

For example, consider the case where a study reports the correlation between the age of a participant and her or his cognitive functioning. Intuitively, one may assume that there is indeed an association between these variables. However, if the study only included participants which were 65 to 69 years old, it is very unlikely that a (high) correlation will be found between the two variables. This is because age in the study sample is highly range restricted. There is no real variation in age, which means that this variable cannot be a good predictor of cognitive abilities. Like unreliability of our measurement instruments (see previous chapter), this leads to an artificial attenuation in the effects we calculate for a study: even when there is in fact an important association, we are not able to detect it.

It is possible to correct for range restriction in SMDs or correlations r_{xy}. However, this requires that we know (or estimate) the unrestricted standard deviation $s_{\text{unrestricted}}$ of our population of interest. The population of interest is determined by the research question of our meta-analysis. For example, if we want to examine the relationship between age and cognitive functioning in *older age*, we might want to search for an estimate of the standard deviation in large representative samples of individuals who are older than 65 (this is commonly how "older person" is defined in research). Of course, this is still a range restriction, but it restricts age to a range that *matters*, because it reflects the study population we are dealing with in our meta-analysis.

To correct for range restriction, we have to calculate U, the ratio between the unrestricted population standard deviation $s_{\text{unrestricted}}$, and the standard deviation of the restricted variable in our study, $s_{\text{restricted}}$.

$$U = \frac{s_{\text{unrestricted}}}{s_{\text{restricted}}} \tag{3.44}$$

The value of $s_{\text{unrestricted}}$ can be obtained, for example, from previous representative studies which assessed the variable of interest. We can then use U to correct the value of a correlation r_{xy} using this formula:

$$r_{xy_c} = \frac{U \times r_{xy}}{\sqrt{(U^2-1)r_{xy}^2+1}} \tag{3.45}$$

This lets us obtain the corrected correlation r_{xy_c}. The same formula can also be used to calculate a corrected version of the SMD:

$$\text{SMD}_c = \frac{U \times \text{SMD}}{\sqrt{(U^2-1)\text{SMD}^2+1}} \tag{3.46}$$

The standard errors of r_{xy} and SMD, respectively, must also be corrected using these formulas:

$$SE_{r_{xy_c}} = \frac{r_{xy_c}}{r_{xy}} SE_{r_{xy}} \tag{3.47}$$

$$SE_{\text{SMD}_c} = \frac{\text{SMD}_c}{\text{SMD}} SE_{\text{SMD}} \tag{3.48}$$

After the correlation or SMD has been corrected, it is possible to apply other common transformations, such as converting r_{xy_c} to Fisher's z (Chapter 3.2.3.1) or SMD_c to Hedges' g (see Chapter 3.4.1). Let us now try out the correction using R.

```
# Define correlation to correct
r_xy <- 0.34
se_r_xy <- 0.09

# Define restricted and unrestricted SD
sd_restricted <- 11
sd_unrestricted <- 18

# Calculate U
U <- sd_unrestricted/sd_restricted

# Correct the correlation
```

```
r_xy_c <- (U*r_xy)/sqrt((U^2-1)*r_xy^2+1)
r_xy_c
```

```
## [1] 0.5092
```

```
# Correct the standard error
se_r_xy_c <- (r_xy_c/r_xy)*se_r_xy
se_r_xy_c
```

```
## [1] 0.1348
```

Like other Hunter and Schmidt adjustments, corrections of range restriction are more commonly found in some research areas than in others. When we decide to apply a correction for range restriction, it is important that the correction is performed for *all* effect sizes in our meta-analysis. It is technically possible to correct for range restriction in every meta-analysis, but often, this is not necessary.

In practice, it is hardly ever the case that each study perfectly represents the scope of our meta-analysis. In fact, the purpose of meta-analysis is to go *beyond* the results of individual studies. A correction of range restriction may therefore only be necessary when the range of several studies is heavily restricted.

Further Reading

In this guide, we only cover corrections for unreliability and range restriction, because these problems are most commonly found in practice. However, Hunter and Schmidt have proposed various other kinds of artifact corrections. Along with a few additional methods, these techniques are sometimes called *psychometric meta-analysis*. If you want to learn more about the Hunter and Schmidt methods, you can have a look at their book *Methods of Meta-Analysis* (Hunter and Schmidt, 2004), which provides an !accessible and comprehensive overview. A shorter introduction can also be found in Borenstein et al. (2011), chapter 38. Many of the techniques of Hunter and Schmidt are also implemented in an R package called *{psychmeta}*[a] (Dahlke and Wiernik, 2019).

[a] https://psychmeta.com/

3.5 Common Problems

In this chapter, we want to devote a little more time to problems that we are often faced with in practice when calculating effect sizes. First, we will discuss what we can do when effect size data is reported in different formats. After that, we examine the unit-of-analysis problem, which has implications on the meta-analytic pooling in later steps.

3.5.1 Different Effect Size Data Formats

When we described effect size metrics in the last chapters, we also mentioned the type of variables we need as columns in our data set. These variables are needed so that R functions can calculate the effect sizes and perform a meta-analysis for us. To calculate a meta-analysis of between-group standardized mean differences, for example, we have to prepare the mean, standard deviation, and sample size of both groups. If we can extract this information from all studies, everything is fine. In practice, however, one may soon find that not all studies report their results in a suitable format. Some studies, for example, may not report the raw data of two groups, but only a calculated standardized mean difference, and its confidence interval. Others may only report the results of a t-test or *analysis of variance* (ANOVA) examining the difference between two groups.

If this is the case, it often becomes impossible to use raw effect size data for our meta-analysis. Instead, we have to *pre-calculate* the effect size of each study before we can pool them. In Chapter 3.1, we already found out that the minimum information we need to do a meta-analysis is the effect size and standard error of a study. Therefore, as long as we can transform the results into an estimate of the effect size and its standard error, a study can be included. In Chapter 17 in the "Helpful Tools" section, we present several effect size converters which can help you to derive an effect size from other types of reported data. However, it is still possible that there are studies for which effect sizes cannot be calculated, even with these tools. As mentioned in Chapter 1.4.4, one remaining possibility under such circumstances is to contact the authors of the respective publication several times, and ask them if they can provide the data you need to calculate the effect size. If this also fails, the study has to be excluded.

In Chapter 4.2.1, we will learn about a special function in R called `metagen`. This function allows us to perform a meta-analysis of effect size data that had to be pre-calculated. To use the function, we have to prepare the following columns in our data set:

- `TE`. The calculated effect size of each study.
- `seTE`. The standard error of each effect size.

3.5.2 The Unit-of-Analysis Problem

It is not uncommon that a study contributes more than one effect size to our meta-analysis. In particular, it may be that (1) a study included more than two groups or that (2) a study measured an outcome using two or more instruments. Both cases cause problems. If studies contribute more than one effect size in a meta-analysis, we violate one of its core assumptions: that each effect size in a meta-analysis is *independent* (Higgins et al., 2019, chapters 6.2 and 23; Borenstein et al., 2011, chapter 25). If this assumption is not met, we are dealing with a *unit-of-analysis* problem.

Let us begin with the first case, where a study has more than two groups; for example, one group examining treatment A, another one in which treatment B is administered, and a control group C. We can calculate *two* effect sizes for this study. Depending on the outcome data, these can be risk, odds or incidence rate ratios, or standardized mean differences. We have one effect size $\hat{\theta}_{\text{A-C}}$ comparing treatment A to control, and another effect size $\hat{\theta}_{\text{B-C}}$, which expresses the effect of treatment B compared to control. When both $\hat{\theta}_{\text{A-C}}$ and $\hat{\theta}_{\text{B-C}}$ are included in the same meta-analysis, these effect sizes are not independent, because the information in C is included twice. This issue is also known as *double-counting*.

Due to the double-counting of C, the two effect sizes are *correlated*. If the sample size is equal in all groups, we know that this correlation is $r = 0.5$ (Borenstein et al., 2011, chapter 25). This is because A and B are independent groups, and therefore uncorrelated. However, the control group in both effect sizes is identical, which leads to a perfect correlation of 1; the midpoint is 0.5. Double-counting of a group leads us to overestimate the precision (i.e. the standard error) of the affected effect sizes. This inflates the weight we give these effects in our meta-analysis, and ultimately distorts our results. There are three ways to deal with this issue:

1. *Split the sample size of the shared group.* This would mean to split the sample size of group C (e.g. $n = 200$) evenly between the comparison with A, and the comparison with C, when we calculate the effect size. If we are dealing with binary outcome data, the number of events is also split evenly. In our example, we would calculate the two effect sizes like we have before, but now we pretend that C only consisted of 100 individuals in both calculations. This approach solves the problem that the precision of the effect sizes is artificially inflated due to double-counting. However, it is still sub-optimal, because the effect sizes will remain correlated (Higgins et al., 2019, chapter 23.3.4).

2. *Remove groups.* A brute force approach is to simply remove one comparison, e.g. $\hat{\theta}_{\text{B-C}}$, entirely from the meta-analysis. This solves the unit-of-analysis problem, but causes new issues. If we simply discard one effect size, we lose potentially relevant information.

3. *Combine groups.* This approach involves combining the results of two groups so that only one comparison remains. In our example, this would

mean to combine the data of A and B and then compare the pooled results with C. This is relatively easy for binary outcome data, where we only have to sum up the number of participants and the number of events in both groups. If we have continuous outcome data, i.e. means and standard deviations, things are a little more complicated. In Chapter 17.9 in the "Helpful Tools" section, you can find an R function which allows us to combine such data. By combining groups, we avoid both double-counting and correlated effect sizes. This is why this approach is also recommended by Cochrane (Higgins et al., 2019, chapter 23.3.4). Nevertheless, the method also has its drawbacks. It is possible that two groups are so different that we lump something together which can not actually be compared. Imagine that the treatments in groups A and B were completely different, with A being a state-of-the-art intervention and B being an outdated approach with a limited evidence base. If we combine these two treatments and find no effect, it is nearly impossible to disentangle if this is true for both types of interventions, or if the ineffectiveness of B simply diluted the effects of A. Approaches (1) and (2) may therefore be used when two groups are too dissimilar.

The unit-of-analysis problem also arises when a study measured an outcome using multiple instruments. This is commonly the case when there is no clear "gold standard" determining how a variable of interest should be measured. If we calculate an effect size for each of these measurements and include them into our meta-analysis, this also results in double-counting. Furthermore, the effect sizes will be correlated, because the same sample was used to measure the effects. There are two approaches to deal with this situation. First, we can simply select one instrument per study. It is important that this selection is done in a systematic and reproducible way. At best, our analysis plan (Chapter 1.4.2) should already define a hierarchy of instruments for our meta-analysis. This hierarchy can be based on previous evidence on the reliability of certain instruments, or based on which type of measurement reflects the content of our research question best. The hierarchy then clearly determines which instrument we select when more than one is available. The second approach is to include data from all available instruments, and use meta-analytic models which can account for the fact that studies in our meta-analysis contribute more than one effect size. This can be achieved by "three-level" meta-analysis models, which we will examine in Chapter 10.

□

3.6 Questions & Answers

Test your knowledge!

1. Is there a clear definition of the term "effect size"? What do people refer to when they speak of effect sizes?
2. Name a primary reason why observed effect sizes deviate from the true effect size of the population. How can it be quantified?
3. Why are large studies better estimators of the true effect than small ones?
4. What criteria does an effect size metric have to fulfill to be usable for meta-analyses?
5. What does a standardized mean difference of 1 represent?
6. What kind of transformation is necessary to pool effect sizes based on ratios (e.g. an odds ratio) using the inverse-variance method?
7. Name three types of effect size corrections.
8. When does the unit-of-analysis problem occur? How can it be avoided?

Answers to these questions are listed in Appendix A at the end of this book.

3.7 Summary

- Effect sizes are the building blocks of meta-analyses. To perform a meta-analysis, we need at least an estimate of the effect sizes and their standard error.
- The standard error of an effect size represents how *precise* the study's estimate of the effect is. Meta-analysis gives effect sizes with a greater precision a higher weight because they are better estimators of the true effect.

- There are various effect sizes we can use in meta-analyses. Common ones are "one-variable" relationship measures (such as means and proportions), correlations, (standardized) mean differences, as well as risk, odds, and incidence rate ratios.
- Effect sizes can also be biased, for example, by measurement error and range restriction. There are formulas to correct for some biases, including the small sample bias of standardized mean differences, attenuation due to unreliability, as well as range restriction problems.
- Other common problems are that studies report the data needed to calculate effect sizes in different formats, as well as the unit-of-analysis problem, which arises when studies contribute more than one effect size.

4

Pooling Effect Sizes

A long and winding road already lies behind us. Fortunately, we have now reached the core part of every meta-analysis: the pooling of effect sizes. We hope that you were able to resist the temptation of starting directly with this chapter. We have already discussed various topics in this book, including the definition of research questions, guidelines for searching, selecting, and extracting study data, as well as how to prepare our effect sizes. Thorough preparation is a key ingredient of a good meta-analysis and will be immensely helpful in the steps that are about to follow. We can assure you that the time you spent working through the previous chapters was well invested.

There are many packages which allow us to pool effect sizes in R. Here, we will focus on functions of the *{meta}* package, which we already installed in Chapter 2.2. This package is very user-friendly and provides us with nearly all important meta-analysis results using just a few lines of code. In the previous chapter, we covered that effect sizes come in different "flavors", depending on the outcome of interest. The *{meta}* package contains specialized meta-analysis functions for each of these effect size metrics. All of the functions also follow nearly the same structure. Thus, once we have a basic understanding of how *{meta}* works, coding meta-analyses becomes straightforward, no matter which effect size we are focusing on. In this chapter, we will cover the general structure of the *{meta}* package. And of course, we will also explore the meta-analysis functions of the package in greater detail using hands-on examples.

The *{meta}* package allows us to tweak many details about the way effect sizes are pooled. As we previously mentioned, meta-analysis comes with many "researcher degrees of freedom". There are a myriad of choices concerning the statistical techniques and approaches we can apply, and if one method is better than the other often depends on the context.

Before we begin with our analyses in R, we therefore have to get a basic understanding of the statistical assumptions of meta-analyses, and the maths behind it. Importantly, we will also discuss the "idea" behind meta-analyses. In statistics, this "idea" translates to a *model*, and we will have a look at what the meta-analytic model looks like. As we will see, the nature of the meta-analysis requires us to make a fundamental decision right away: we have to assume either a *fixed-effect model* or a *random-effects model*. Knowledge of the concept behind meta-analytic pooling is needed to make an informed decision which of these two models, along with other analytic specifications, is more appropriate in which context.

DOI: 10.1201/9781003107347-4

4.1 The Fixed-Effect & Random-Effects Model

Before we specify the meta-analytic model, we should first clarify what a statistical model actually is. Statistics is full of "models", and it is likely that you have heard the term in this context before. There are "linear models", "generalized linear models", "mixture models", "gaussian additive models", "structural equation models", and so on. The ubiquity of models in statistics indicates how important this concept is. In one way or the other, models build the basis of virtually all parts of our statistical toolbox. There is a model behind t-tests, ANOVAs, and regression. Every hypothesis test has its corresponding statistical model.

When defining a statistical model, we start with the information that is already given to us. This is, quite literally, our *data*[1]. In meta-analyses, the data are effect sizes that were observed in the included studies. Our model is used to describe the process through which these observed data were generated. The data are seen as the product of a *black box*, and our model aims to illuminate what is going on inside that black box.

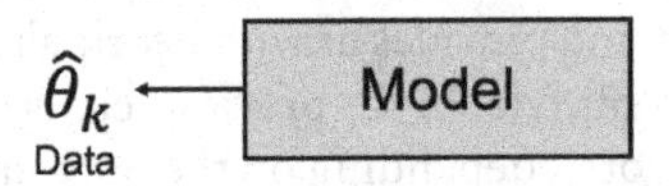

Typically, a statistical model is like a special type of *theory*. Models try to explain the mechanisms that generated our observed data, especially when those mechanisms themselves cannot be directly observed. They are an *imitation of life*, using a mathematical formula to describe processes in the world around us in an idealized way. This explanatory character of models is deeply ingrained in modern statistics, and meta-analysis is no exception. The conceptualization of models as a vehicle for explanation is the hallmark of a statistical "culture" to which, as Breiman (2001) famously estimated, 98% of all statisticians adhere.

By specifying a statistical model, we try to find an approximate representation of the "reality" behind our data. We want a mathematical formula that explains how we can find the *true* effect size underlying all of our studies, based on their observed results. As we learned in Chapter 1.1, one of the ultimate goals of meta-analysis is to find one numerical value that characterizes our studies *as a whole*, even though the observed effect sizes vary from study to study. A meta-analysis model must therefore explain the reasons why and how much observed study results differ, even though there is only one overall effect. There are two models which try to answer exactly this question, the *fixed-effect model* and the *random-effects model*. Although both are based

[1]"Data" is derived from the Latin word *datum*, meaning "a thing that is given".

on different assumptions, there is still a strong link between them, as we will soon see.

4.1.1 The Fixed-Effect Model

The fixed-effect model assumes that all effect sizes stem from a single, homogeneous population. It states that all studies share the *same* true effect size. This true effect is the overall effect size we want to calculate in our meta-analysis, denoted with θ. According to the fixed-effect model, the only reason why a study k's observed effect size $\hat{\theta}_k$ deviates from θ is because of its sampling error ϵ_k. The fixed-effect model tells us that the process generating studies' different effect sizes, the content of the "black box", is simple: all studies are estimators of the same true effect size. Yet, because every study can only draw somewhat bigger or smaller samples of the infinitely large study population, results are burdened by sampling error. This sampling error causes the observed effect to deviate from the overall, true effect. We can describe the relationship like this (Borenstein et al., 2011, chapter 11):

$$\hat{\theta}_k = \theta + \epsilon_k \tag{4.1}$$

To the alert reader, this formula may seem oddly similar to the one in Chapter 3.1. You are not mistaken. In the previous formula, we defined that an observed effect size $\hat{\theta}_k$ of some study k is an estimator of that study's true effect size θ_k, burdened by the study's sampling error ϵ_k. There is only a tiny, but insightful difference between the previous formula, and the one of the fixed-effect model. In the formula of the fixed-effect model, the true effect size is not symbolized by θ_k, but by θ; the subscript k is dropped. Previously, we only made statements about the true effect size of *one* individual study k. The fixed-effect model goes one step further. It tells us that if we find the true effect size of study k, this effect size is not only true for k specifically, but for *all* studies in our meta-analysis. A *study's* true effect size θ_k, and the *overall*, pooled effect size θ, are *identical*.

The **idea behind the fixed-effect model** is that observed effect sizes may vary from study to study, but this is only because of the sampling error. In reality, their true effect sizes are *all the same*: they are *fixed*.

The formula of the fixed-effect models tells us that there is only one reason why observed effect sizes θ_k deviate from the true overall effect: because of the sampling error ϵ_k. In Chapter 3.1, we already discussed that there is a link between the sampling error and the sample size of a study. All things being equal, as the sample size becomes larger, the sampling error becomes smaller. We also learned that the sampling error can be represented numerically by the *standard error*, which also grows smaller when the sample size increases.

Although we do not know the true overall effect size of our studies, we can exploit this relationship to arrive at the best possible estimate of the true overall effect, $\hat{\theta}$. We know that a smaller standard error corresponds with a smaller sampling error; therefore, studies with a small standard error should be better estimators of the true overall effect than studies with a large standard error. We can illustrate this with a simulation. Using the `rnorm` function we already used before, we simulated a selection of studies in which the true overall effect is $\theta = 0$. We took several samples but varied the sample size so that the standard error differs between the "observed" effects. The results of the simulation can be found in Figure 4.1.

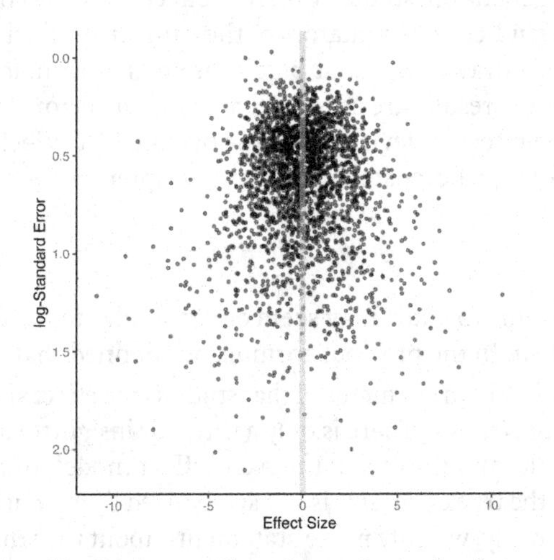

FIGURE 4.1: Relationship between effect size and standard error.

The results of the simulation show an interesting pattern. We see that effect sizes with a small sampling error are tightly packed around the true effect size $\theta = 0$. As the standard error on the y-axis[2] increases, the *dispersion* of effect sizes becomes larger and larger, and the observed effects deviate more and more from the true effect. This behavior can be predicted by the formula of the fixed-effect model. We know that studies with a smaller standard error have a smaller sampling error, and their estimate of the overall effect size is therefore more likely to be closer to the truth.

We have seen that, while all observed effect sizes are estimators of the true effect, some are better than others. When we pool the effects in our meta-analysis, we should therefore give effect sizes with a higher *precision* (i.e. a smaller standard error) a greater *weight*. If we want to calculate the pooled effect size under the fixed-effect model, we therefore simply use a *weighted average* of all studies. To calculate the weight w_k for each study k, we can use the standard error, which we square to obtain the

[2]We log-transformed the standard error before plotting so that the pattern can be more easily seen.

variance s_k^2 of each effect size. Since a *lower* variance indicates higher precision, the *inverse* of the variance is used to determine the weight of each study.

$$w_k = \frac{1}{s^2_k} \tag{4.2}$$

Once we know the weights, we can calculate the weighted average, our estimate of the true pooled effect $\hat\theta$. We only have to multiply each study's effect size $\hat\theta_k$ with its corresponding weight w_k, sum the results across all studies K in our meta-analysis, and then divide by the sum of all the individual weights.

$$\hat\theta = \frac{\sum^{K}_{k=1} \hat\theta_k w_k}{\sum^{K}_{k=1} w_k} \tag{4.3}$$

This method is the most common approach to calculate average effects in meta-analyses. Because we use the inverse of the variance, it is often called *inverse-variance weighting* or simply *inverse-variance meta-analysis*. For binary effect size data, there are alternative methods to calculate the weighted average, including the *Mantel-Haenszel*, *Peto*, or the sample size weighting method by Bakbergenuly (2020). We will discuss these methods in Chapter 4.2.3.1.

The *{meta}* package makes it very easy to perform a fixed-effect meta-analysis. Before, however, let us try out the inverse-variance pooling "manually" in R. In our example, we will use the `SuicidePrevention` data set, which we already imported in Chapter 2.4.

The "SuicidePrevention" Data Set

The `SuicidePrevention` data set is also included directly in the *{dmetar}* package. If you have installed *{dmetar}*, and loaded it from your library, running `data(SuicidePrevention)` automatically saves the data set in your R environment. The data set is then ready to be used. If you do not have *{dmetar}* installed, you can download the data set as an *.rda* file from the Internet[a], save it in your working directory, and then click on it in your R Studio window to import it.

[a] `https://www.protectlab.org/meta-analysis-in-r/data/suicideprevention.rda`

The `SuicidePrevention` data set contains raw effect size data, meaning that we have to calculate the effect sizes first. In this example, we calculate the small-sample adjusted standardized mean difference (Hedges' g). To do this, we use the `esc_mean_sd` function in the *{esc}* package (Chapter 3.3.1.2). The function has an additional argument, `es.type`, through which we can specify that the small-sample correction should be performed (by setting `es.type = "g"`; Chapter 3.4.1).

```
# Load dmetar, esc and tidyverse (for pipe)
library(dmetar)
library(esc)
library(tidyverse)

# Load data set from dmetar
data(SuicidePrevention)

# Calculate Hedges' g and the Standard Error
# - We save the study names in "study".
# - After that, we use the pipe operator to directly transform
#   the results to a data frame.
SP_calc <- esc_mean_sd(grp1m = SuicidePrevention$mean.e,
                       grp1sd = SuicidePrevention$sd.e,
                       grp1n = SuicidePrevention$n.e,
                       grp2m = SuicidePrevention$mean.c,
                       grp2sd = SuicidePrevention$sd.c,
                       grp2n = SuicidePrevention$n.c,
                       study = SuicidePrevention$author,
                       es.type = "g") %>%
                     as.data.frame()

# Let us catch a glimpse of the data
# The data set contains Hedges' g ("es") and standard error ("se")
glimpse(SP_calc)
```

```
## Rows: 9
## Columns: 9
## $ study       <chr> "Berry et al.", "DeVries et al." …
## $ es          <dbl> -0.14279447, -0.60770928, -0.11117965 …
## $ weight      <dbl> 46.09784, 34.77314, 14.97625, 32.18243, 24.52054 …
## $ sample.size <dbl> 185, 146, 60, 129, 100, 220, 120, 80, 107
## $ se          <dbl> 0.1472854, 0.1695813, 0.2584036, 0.1762749 …
## $ var         <dbl> 0.02169299, 0.02875783, 0.06677240, 0.03107286 …
## $ ci.lo       <dbl> -0.4314686, -0.9400826, -0.6176413, -0.4724727 …
## $ ci.hi       <dbl> 0.145879624, -0.275335960, 0.395282029 …
## $ measure     <chr> "g", "g", "g", "g", "g", "g", "g", "g", "g"
```

```
# We now calculate the inverse variance-weights for each study
SP_calc$w <- 1/SP_calc$se^2

# Then, we use the weights to calculate the pooled effect
pooled_effect <- sum(SP_calc$w*SP_calc$es)/
                 sum(SP_calc$w)
pooled_effect
```

```
## [1] -0.2311121
```

The results of our calculations reveal that the pooled effect size, assuming a fixed-effect model, is $g \approx -0.23$.

4.1.2 The Random-Effects Model

As we have seen, the fixed-effect model is one way to conceptualize the genesis of our meta-analysis data, and how effects can be pooled. However, the important question is: does this approach adequately reflect reality? The fixed-effect model assumes that all our studies are part of a homogeneous population and that the only cause for differences in observed effects is the sampling error of studies. If we were to calculate the effect size of each study without sampling error, all true effect sizes would be absolutely the same.

Subjecting this notion to a quick reality check, we see that the assumptions of the fixed-effect model might be too simplistic in many real-world applications. It is simply unrealistic that studies in a meta-analysis are always completely homogeneous. Studies will very often differ, even if only in subtle ways. The outcome of interest may have been measured in different ways. Maybe the type of treatment was not exactly the same or the intensity and length of the treatment. The target population of the studies may not have been exactly identical, or maybe there were differences in the control groups that were used. It is likely that the studies in your meta-analysis will not only vary on one of these aspects but several ones at the same time. If this is true, we can anticipate considerable between-study *heterogeneity* in the true effects.

All of this casts the validity of the fixed-effect model into doubt. If some studies used different types of a treatment, for example, it seems perfectly normal that one format is more effective than the other. It would be far-fetched to assume that these differences are only noise, produced by the studies' sampling error. Quite the opposite, there may be countless reasons why *real* differences exist in the *true* effect sizes of studies. The random-effects model addresses this concern. It provides us with a model that often reflects the reality behind our data much better.

In the random-effects model, we want to account for the fact that effect sizes show more variance than when drawn from a single homogeneous population (Hedges and Vevea, 1998). Therefore, we assume that effects of individual studies do not only deviate due to sampling error alone but that there is *another* source of variance. This additional variance component is introduced by the fact that studies do not stem from one single population. Instead, each study is seen as an independent draw from a "universe" of populations.

The random-effects model assumes that there is not only one true effect size but a *distribution* of true effect sizes. The goal of the random-effects model is therefore not to estimate the one true effect size of all studies, but the *mean* of the *distribution* of true effects.

Let us see how the random-effects model can be expressed in a formula. Similar to the fixed-effect model, the random-effects model starts by assuming that an observed effect size $\hat\theta_k$ is an estimator of the study's true effect size θ_k, burdened by sampling error ϵ_k:

$$\hat\theta_k = \theta_k + \epsilon_k \tag{4.4}$$

The fact that we use θ_k instead of θ already points to an important difference. The random-effects model only assumes that θ_k is the true effect size of *one* single study k. It stipulates that there is a second source of error, denoted by ζ_k. This second source of error is introduced by the fact that even the true effect size θ_k of study k is only part of an over-arching distribution of true effect sizes with mean μ.

$$\theta_k = \mu + \zeta_k \tag{4.5}$$

The random-effects model tells us that there is a hierarchy of two processes happening inside our black box (Thompson et al., 2001): the observed effect sizes of a study deviate from their true value because of the sampling error. But even the true effect sizes are only a draw from a universe of true effects, whose mean μ we want to estimate as the pooled effect of our meta-analysis. By plugging the second formula into the first one (i.e. replacing θ_k with its definition in the second formula), we can express the random-effects model in one line (Borenstein et al., 2011, chapter 12):

$$\hat\theta_k = \mu + \zeta_k + \epsilon_k \tag{4.6}$$

This formula makes it clear that our observed effect size deviates from the pooled effect μ because of two error terms, ζ_k and ϵ_k. This relationship is visualized in Figure 4.2.

A crucial assumption of the random-effects model is that the size of ζ_k is *independent* of k. Put differently, we assume that there is nothing which indicates *a priori* that ζ_k in one study is higher than in another. We presuppose that the size of ζ_k is a product of chance, and chance alone. This is known as the *exchangeability* assumption of the random-effects model (Higgins et al., 2009). All true effect sizes are assumed to be exchangeable in so far as we have nothing that could tell us how big ζ_k will be in some study k before seeing the data.

Which Model Should I Use?

In practice, is it very uncommon to find a selection of studies that is perfectly homogeneous. This is true even when we follow best practices, and try to make the scope of our analysis as precise as possible through our PICO (Chapter 1.4.1).

In many fields, including medicine and the social sciences, it is therefore conventional to *always* use a random-effects model, since some degree of between-study heterogeneity can virtually always be anticipated. A fixed-effect model may only be used when we could not detect any between-study heterogeneity (we will discuss how this is done in Chapter 5) *and* when we have very good reasons to assume that the true effect is fixed. This may be the case when, for example, only exact replications of a study are considered, or when we meta-analyze subsets of one big study. Needless to say, this is seldom the case, and applications of the fixed-effect model "in the wild" are rather rare.

Even though it is conventional to use the random-effects model *a priori*, this approach is not undisputed. The random-effects model pays more attention to small studies when calculating the overall effect of a meta-analysis (Schwarzer et al., 2015, chapter 2.3). Yet, small studies in particular are often fraught with biases (see Chapter 9.2.1). This is why some have argued that the fixed-effect model is (sometimes) preferable (Poole and Greenland, 1999; Furukawa et al., 2003).

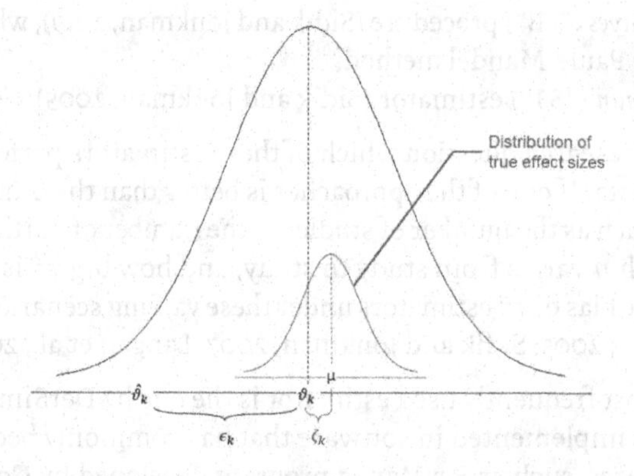

FIGURE 4.2: Illustration of parameters of the random-effects model.

4.1.2.1 Estimators of the Between-Study Heterogeneity

The challenge associated with the random-effects model is that we have to take the error ζ_k into account. To do this, we have to estimate the *variance* of the distribution of true effect sizes. This variance is known as τ^2, or *tau-squared*. Once we know the value of τ^2, we can include the between-study heterogeneity when determining the inverse-variance weight of each effect size. In the random-effects model, we therefore calculate an adjusted *random-effects weight* w^*_k for each observation. The formula looks like this:

$$w^*_k = \frac{1}{s^2_k + \tau^2} \tag{4.7}$$

Using the adjusted random-effects weights, we then calculate the pooled effect size using the inverse variance method, just like we did using the fixed-effect model:

$$\hat{\theta} = \frac{\sum^K_{k=1} \hat{\theta}_k w^*_k}{\sum^K_{k=1} w^*_k} \tag{4.8}$$

There are several methods to estimate τ^2, most of which are too complicated to do by hand. Luckily, however, these estimators are implemented in the functions of the *{meta}* package, which does the calculations automatically for us. Here is a list of the most common estimators, and the code by which they are referenced in *{meta}*:

- The *DerSimonian-Laird* (`"DL"`) estimator (DerSimonian and Laird, 1986).
- The *Restricted Maximum Likelihood* (`"REML"`) or *Maximum Likelihood* (`"ML"`) procedures (Viechtbauer, 2005).
- The *Paule-Mandel* (`"PM"`) procedure (Paule and Mandel, 1982).
- The *Empirical Bayes* (`"EB"`) procedure (Sidik and Jonkman, 2019), which is practically identical to the Paule-Mandel method.
- The *Sidik-Jonkman* (`"SJ"`) estimator (Sidik and Jonkman, 2005).

It is an ongoing research question which of these estimators performs best for different kinds of data. If one of the approaches is better than the other often depends on parameters such as the number of studies k, the number of participants n in each study, how much n varies from study to study, and how big τ^2 is. Several studies have analyzed the bias of τ^2 estimators under these varying scenarios (Veroniki et al., 2016; Viechtbauer, 2005; Sidik and Jonkman, 2007; Langan et al., 2019).

Arguably, the most frequently used estimator is the one by DerSimonian and Laird. The estimator is implemented in software that has commonly been used by meta-analysts in the past, such as *RevMan* (a program developed by Cochrane) or *Comprehensive Meta-Analysis*. It is also the default estimator used in *{meta}*. Due to this historic legacy, one often finds research papers in which "using a random-effects model" is used synonymously with employing the DerSimonian-Laird estimator. However, it has been found that this estimator can be biased, particularly when the number of studies is small and heterogeneity is high (Hartung, 1999; Hartung and Knapp, 2001b,a; Follmann and Proschan, 1999; Makambi, 2004). This is quite problematic because it is very common to find meta-analyses with few studies and high heterogeneity.

In an overview paper, Veroniki and colleagues (2016) reviewed evidence on the robustness of various τ^2 estimators. They recommended the Paule-Mandel method for both binary and continuous effect size data, and the restricted maximum likelihood estimator for continuous outcomes. The restricted maximum-likelihood estimator is also the default method used by the *{metafor}* package. A more recent simulation study by Langan and colleagues (2019) came to a similar result but found that the Paule-Mandel estimator may be suboptimal when the sample size of studies varies drastically. Another study by Bakbergenuly and colleagues (2020) found that the Paule-Mandel estimator is well suited especially when the number of studies is small. The Sidik-Jonkman estimator, also known as the *model error variance method*, is only well suited when τ^2 is very large (Sidik and Jonkman, 2007).

!

Which Estimator Should I Use?

There are no iron-clad rules when exactly which estimator should be used. In many cases, there will only be minor differences in the results produced by various estimators, meaning that you should not worry about this issue *too* much.

When in doubt, you can always rerun your analyses using different τ^2 estimators, and see if this changes the interpretation of your results. Here are a few tentative guidelines that you may follow in your own meta-analysis:

1. For effect sizes based on continuous outcome data, the restricted maximum likelihood estimator may be used as a first start.
2. For binary effect size data, the Paule-Mandel estimator is a good first choice, provided there is no extreme variation in the sample sizes.
3. When you have very good reason to believe that the heterogeneity of effects in your sample is very large, and if avoiding false positives has a very high priority, you may use the Sidik-Jonkman estimator.
4. If you want that others can replicate your results as precisely as possible outside R, the DerSimonian-Laird estimator is the method of choice.

Overall, estimators of τ^2 fall into two categories. Some, like the DerSimonian-Laird and Sidik-Jonkman estimator, are based on *closed-form expressions*, meaning that they can be directly calculated using a formula. The (restricted) maximum likelihood,

Paule-Mandel and empirical Bayes estimator find the optimal value of τ^2 through an *iterative algorithm*. Latter estimators may therefore sometimes take a little longer to calculate the results. In most real-world cases, however, these time differences are minuscule at best.

4.1.2.2 Knapp-Hartung Adjustments

In addition to our selection of the τ^2 estimator, we also have to decide if we want to apply so-called Knapp-Hartung adjustments[3] (Knapp and Hartung, 2003; Sidik and Jonkman, 2002). These adjustments affect the way the standard error (and thus the confidence intervals) of our pooled effect size $\hat{\theta}$ is calculated. The Knapp-Hartung adjustments try to control for the uncertainty in our estimate of the between-study heterogeneity. While significance tests of the pooled effect usually assume a normal distribution (so-called *Wald-type* tests), the Knapp-Hartung method is based on a t-distribution. Knapp-Hartung adjustments can only be used in random-effects models, and usually cause the confidence intervals of the pooled effect to become slightly larger.

Reporting the Type of Model Used In Your Meta-Analysis

It is highly advised to specify the type of model you used in the methods section of your meta-analysis report. Here is an example:

> *"As we anticipated considerable between-study heterogeneity, a random-effects model was used to pool effect sizes. The restricted maximum likelihood estimator (Viechtbauer, 2005) was used to calculate the heterogeneity variance τ^2. We used Knapp-Hartung adjustments (Knapp & Hartung, 2003) to calculate the confidence interval around the pooled effect."*

Applying a Knapp-Hartung adjustment is usually sensible. Several studies (IntHout et al., 2014; Langan et al., 2019) showed that these adjustments can reduce the chance of false positives, especially when the number of studies is small. The use of the Knapp-Hartung adjustment, however, is not uncontroversial. Wiksten and colleagues (2016), for example, argued that the method can cause anti-conservative results in (seldom) cases when the effects are very homogeneous.

[3]This approach is also known as "Hartung-Knapp adjustments" or the "Hartung-Knapp-Sidik-Jonkman" (HKSJ) method.

4.2 Effect Size Pooling in R

Time to put what we learned into practice. In the rest of this chapter, we will explore how we can run meta-analyses of different effect sizes directly in R. The *{meta}* package we will use to do this has a special structure. It contains several meta-analysis functions which are each focused on one type of effect size data. There is a set of parameters which can be specified in the same way across all of these functions; for example, if we want to apply a fixed- or random-effects model, or which τ^2 estimator should be used. Apart from that, there are *function-specific* arguments which allow us to tweak details of our meta-analysis that are only relevant for a specific type of data.

Figure 4.3 provides an overview of *{meta}*'s structure. To determine which function to use, we first have to clarify what kind of effect size data we want to synthesize. The most fundamental distinction is the one between *raw* and *pre-calculated* effect size data. We speak of "raw" data when we have all the necessary information needed to calculate the desired effect size stored in our data frame but have not yet calculated the actual effect size. The `SuicidePrevention` data set we used earlier contains raw data: the mean, standard deviation and sample size of two groups, which is needed to calculate the standardized mean difference. We call effect size data "pre-calculated", on the other hand, when they already contain the final effect size of each study, as well as the standard error. If we want to use a corrected version of an effect metric (such as Hedges' g, Chapter 3.4.1), it is necessary that this correction has already been applied to pre-calculated effect size data before we start the pooling.

If possible, it is preferable to use raw data in our meta-analysis. This makes it easier for others to understand how we calculated the effect sizes and replicate the results. Yet, using raw data is often not possible in practice, because studies often report their results in a different way (Chapter 3.5.1). This leaves us no other choice than to pre-calculate the desired effect size for each study right away so that all have the same format. Chapter 17 in the "Helpful Tools" part of this book presents a few formulas which can help you to convert a reported effect size into the desired metric.

The function of choice for pre-calculated effect sizes is `metagen`. Its name stands for generic inverse variance meta-analysis. If we use `metagen` with binary data (e.g. proportions, risk ratios, odds ratios), it is important, as we covered in Chapter 3.3.2, that the effect sizes are log-transformed before the function is used.

When we can resort to raw effect size data, *{meta}* provides us with a specialized function for each effect size type. We can use the `metamean`, `metacont` and `metacor` function for means, (standardized) mean differences and correlations, respectively. We can pool (incidence) rates, proportions and incidence rate ratios using the `metarate`, `metaprop` and `metainc` functions. The `metabin` function can be employed when we are dealing with risk or odds ratios.

All meta-analysis functions in *{meta}* follow the same structure. We have to provide the functions with the (raw or pre-calculated) effect size data, as well as further

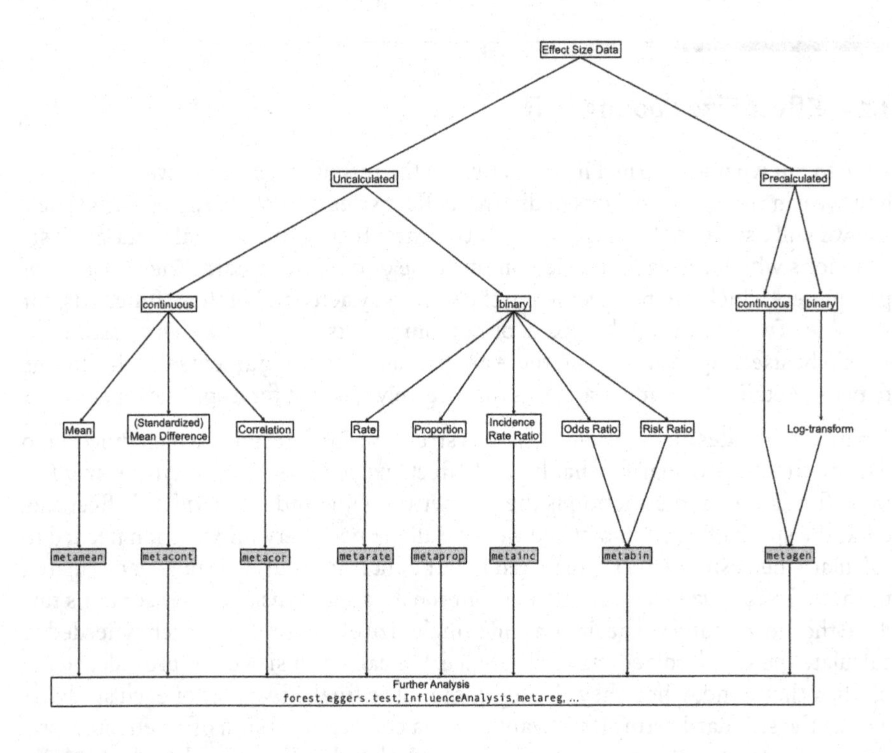

FIGURE 4.3: Conceptual overview of meta-analysis functions.

arguments, which control the specifics of the analysis. There are six core arguments which can be specified in each function:

- `studlab`. This argument associates each effect size with a *study label*. If we have the name or authors of our studies stored in our data set, we simply have to specify the name of the respective column (e.g. `studlab = author`).
- `sm`. This argument controls the *summary measure*, the effect size metric we want to use in our meta-analysis. This option is particularly important for functions using raw effect size data. The *{meta}* package uses codes for different effect size formats, for example, `"SMD"` or `"OR"`. The available summary measures are not the same in each function, and we will discuss the most common options in each case in the following sections.
- `comb.fixed`. We need to provide this argument with a logical (`TRUE` or `FALSE`), indicating if a fixed-effect model meta-analysis should be calculated.
- `comb.random`. In a similar fashion, this argument controls if a random-effects model should be used. If both `comb.fixed` and `comb.random` are set to `TRUE`, both models are calculated and displayed.
- `method.tau`. This argument defines the τ^2 estimator. All functions use the codes for different estimators that we already presented in the previous chapter (e.g. for the DerSimonian-Laird method: `method.tau = "DL"`).

- `hakn`. This is yet another logical argument, and controls if the Knapp-Hartung adjustments should be applied when using the random-effects model.
- `data`. In this argument, we provide *{meta}* with the name of our meta-analysis data set.
- `title` (*not mandatory*). This argument takes a character string with the name of the analysis. While it is not essential to provide input for this argument, it can help us to identify the analysis later on.

There are also a few additional arguments which we will get to know in later chapters. In this guide, we will not be able to discuss *all* arguments of the *{meta}* functions: there are more than 100. Thankfully, most of these arguments are rarely needed or have sensible defaults. When in doubt, you can always run the name of the function, preceded by a question mark (e.g. `?metagen`) in the R console; this will open the function documentation.

Default Arguments & Position Matching

For R beginners, it is often helpful to learn about *default arguments* and *position-based matching* in functions.

Default arguments are specified by the person who wrote the function. They set a function argument to a predefined value, which is automatically used unless we explicitly provide a different value. In *{meta}* many, but not all arguments have default values.

Default values are displayed in the "usage" section of the function documentation. If a function has defined a default value for an argument, it is not necessary to include it in our function call, unless we are not satisfied with the default behavior.

Arguments *without* default values always need to be specified in our function call. The *{meta}* package has a convenience function called `gs` which we can use to check the default value used for a specific argument. For example, try running `gs("method.tau")`. If there is no default value, `gs` will return `NULL`.

Another interesting detail about R functions is position matching. Usually, we have to write down the name of an argument and its value in a function call. Through position matching, however, we can leave out the name of the argument, and only have to type in the argument value. We can do this if we specify the argument in the same *position* in which it appears in the documentation. Take the `sqrt` function. A written out call of this function would be `sqrt(x = 4)`. However, because we know that `x`, the number, is the first argument, we can simply type in `sqrt(4)` with the same result.

4.2.1 Pre-Calculated Effect Size Data

Let us begin our tour of meta-analysis functions with `metagen`. As we learned, this function can be used for pre-calculated effect size data. In our first example, we will use the function to perform a meta-analysis of the `ThirdWave` data set.

This data set contains studies examining the effect of so-called "third wave" psychotherapies on perceived stress in college students. For each study, the standardized mean difference between a treatment and control group at post-test was calculated, and a small sample correction was applied. The effect size measure used in this meta-analysis, therefore, is Hedges' g. Let us have a look at the data.

The "ThirdWave" Data Set

The `ThirdWave` data set is included directly in the *{dmetar}* package. If you have installed *{dmetar}*, and loaded it from your library, running `data(ThirdWave)` automatically saves the data set in your R environment. The data set is then ready to be used. If you do not have *{dmetar}* installed, you can download the data set as an *.rda* file from the Internet[a], save it in your working directory, and then click on it in your R Studio window to import it.

[a] https://www.protectlab.org/meta-analysis-in-r/data/thirdwave.rda

```
library(tidyverse) # needed for 'glimpse'
library(dmetar)
library(meta)

data(ThirdWave)
glimpse(ThirdWave)
```

```
## Rows: 18
## Columns: 8
## $ Author                <chr> "Call et al.", "Cavan...
```

```
## $ TE                     <dbl> 0.7091, 0.3549, 1.791...
## $ seTE                   <dbl> 0.2608, 0.1964, 0.345...
## $ RiskOfBias             <chr> "high", "low", "high"...
## $ TypeControlGroup       <chr> "WLC", "WLC", "WLC", ...
## $ InterventionDuration <chr> "short", "short", "sh...
## $ InterventionType       <chr> "mindfulness", "mindf...
## $ ModeOfDelivery         <chr> "group", "online", "g...
```

We see that the data set has eight columns, the most important of which are `Author`, `TE` and `seTE`. The `TE` column contains the g value of each study, and `seTE` is the standard error of g. The other columns represent variables describing the subgroup categories that each study falls into. These variables are not relevant for now.

We can now start to think about the type of meta-analysis we want to perform. Looking at the subgroup columns, we see that studies vary at least with respect to their risk of bias, control group, intervention duration, intervention type, and mode of delivery. This makes it quite clear that some between-study heterogeneity can be expected, and that it makes no sense to assume that all studies have a fixed true effect. We may therefore use the random-effects model for pooling. Given its robust performance in continuous outcome data, we choose the restricted maximum likelihood (`"REML"`) estimator in this example. We will also use the Knapp-Hartung adjustments to reduce the risk of a false positive result.

Now that we have these fundamental questions settled, the specification of our call to `metagen` becomes fairly straightforward. There are two function-specific arguments which we always have to specify when using the function:

- `TE`. The name of the column in our data set which contains the calculated effect sizes.
- `seTE`. The name of the column in which the standard error of the effect size is stored.

The rest are generic *{meta}* arguments that we already covered in the last chapter. Since the analysis deals with standardized mean differences, we also specify `sm = "SMD"`. However, in this example, this has no actual effect on the results, since effect sizes are already calculated for each study. It will only tell the function to label effect sizes as SMDs in the output.

This gives us all the information we need to set up our first call to `metagen`. We will store the results of the function in an object called `m.gen`.

```
m.gen <- metagen(TE = TE,
                 seTE = seTE,
                 studlab = Author,
                 data = ThirdWave,
                 sm = "SMD",
                 comb.fixed = FALSE,
                 comb.random = TRUE,
```

```
                 method.tau = "REML",
                 hakn = TRUE,
                 title = "Third Wave Psychotherapies")
```

Our m.gen object now contains all the meta-analysis results. An easy way to get an overview is to simply call the object directly in the R console.

```
m.gen
```

```
## Review:     Third Wave Psychotherapies
##                          SMD             95%-CI %W(random)
## Call et al.           0.7091 [ 0.1979; 1.2203]        5.0
## Cavanagh et al.       0.3549 [-0.0300; 0.7397]        6.3
## DanitzOrsillo         1.7912 [ 1.1139; 2.4685]        3.8
## de Vibe et al.        0.1825 [-0.0484; 0.4133]        7.9
## Frazier et al.        0.4219 [ 0.1380; 0.7057]        7.3
## Frogeli et al.        0.6300 [ 0.2458; 1.0142]        6.3
## Gallego et al.        0.7249 [ 0.2846; 1.1652]        5.7
## Hazlett-Steve…        0.5287 [ 0.1162; 0.9412]        6.0
## Hintz et al.          0.2840 [-0.0453; 0.6133]        6.9
## Kang et al.           1.2751 [ 0.6142; 1.9360]        3.9
## Kuhlmann et al.       0.1036 [-0.2781; 0.4853]        6.3
## Lever Taylor…         0.3884 [-0.0639; 0.8407]        5.6
## Phang et al.          0.5407 [ 0.0619; 1.0196]        5.3
## Rasanen et al.        0.4262 [-0.0794; 0.9317]        5.1
## Ratanasiripong        0.5154 [-0.1731; 1.2039]        3.7
## Shapiro et al.        1.4797 [ 0.8618; 2.0977]        4.2
## Song & Lindquist      0.6126 [ 0.1683; 1.0569]        5.7
## Warnecke et al.       0.6000 [ 0.1120; 1.0880]        5.2
##
## Number of studies combined: k = 18
##
##                            SMD            95%-CI    t  p-value
## Random effects model 0.5771 [0.3782; 0.7760] 6.12 < 0.0001
##
## Quantifying heterogeneity:
##  tau^2 = 0.0820 [0.0295; 0.3533]; tau = 0.2863 [0.1717; 0.5944];
##  I^2 = 62.6% [37.9%; 77.5%]; H = 1.64 [1.27; 2.11]
##
## Test of heterogeneity:
##      Q d.f. p-value
##  45.50   17  0.0002
##
## Details on meta-analytical method:
```

```
## - Inverse variance method
## - Restricted maximum-likelihood estimator for tau^2
## - Q-profile method for confidence interval of tau^2 and tau
## - Hartung-Knapp adjustment for random effects model
```

Here we go, the results of our first meta-analysis using R. There is a lot to unpack, so let us go through the output step by step.

- The first part of the output contains the individual studies, along with their effect sizes and confidence intervals. Since the effects were pre-calculated, there is not much new to be seen here. The `%W(random)` column contains the weight (in percent) that the random-effects model attributed to each study. We can see that, with 7.3%, the greatest weight in our meta-analysis has been given to the study by de Vibe. The smallest weight has been given to the study by Ratanasiripong. Looking at the confidence interval of this study, we can see why this is the case. The CIs around the pooled effect are extremely wide, meaning that the standard error is very high, and that the study's effect size estimate is therefore not very precise.
- Furthermore, the output tells us the total number of studies in our meta-analysis. We see that $K = 18$ studies were combined.
- The next section provides us with the core result: the pooled effect size. We see that the estimate is $g \approx 0.58$ and that the 95% confidence interval ranges from $g \approx 0.38$ to 0.78. We are also presented with the results of a test determining if the effect size is significant. This is the case ($p < 0.001$). Importantly, we also see the associated test statistic, which is denoted with `t`. This is because we applied the Knapp-Hartung adjustment, which is based on a t-distribution.
- Underneath, we see results concerning the between-study heterogeneity. We will learn more about some of the results displayed here in later chapters, so let us only focus on τ^2. Next to `tau^2`, we see an estimate of the variance in true effects: $\tau^2 = 0.08$. We see that the confidence interval of `tau^2` does not include zero (0.03–0.35), meaning that τ^2 is significantly greater than zero. All of this indicates that between-study heterogeneity exists in our data and that the random-effects model was a good choice.
- The last section provides us with details about the meta-analysis. We see that effects were pooled using the inverse variance method, that the restricted maximum-likelihood estimator was used, and that the Knapp-Hartung adjustment was applied.

We can also access information stored in `m.gen` directly. Plenty of objects are stored by default in the meta-analysis results produced by *{meta}*, and a look into the "value" section of the documentation reveals what they mean. We can use the `$` operator to print specific results of our analyses. The pooled effect, for example, is stored as `TE.random`.

```
m.gen$TE.random
```

```
## [1] 0.5771
```

Even when we specify `comb.fixed = FALSE`, *{meta}*'s functions always also calculate results for the fixed-effect model internally. Thus, we can also access the pooled effect assuming a fixed-effect model.

```
m.gen$TE.fixed
```

```
## [1] 0.4805
```

We see that this estimate deviates considerably from the random-effects model result.

When we want to adapt some details of our analyses, the `update.meta` function can be helpful. This function needs the *{meta}* object as input, and the argument we want to change. Let us say that we want to check if results differ substantially if we use the Paule-Mandel instead of the restricted maximum likelihood estimator. We can do that using this code:

```
m.gen_update <- update.meta(m.gen,
                            method.tau = "PM")

# Get pooled effect
m.gen_update$TE.random
```

```
## [1] 0.5874
```

```
# Get tau^2 estimate
m.gen_update$tau2
```

```
## [1] 0.1105
```

We see that while the pooled effect does not differ much, the Paule-Mandel estimator gives us a somewhat larger approximation of τ^2.

Lastly, it is always helpful to save the results for later. Objects generated by *{meta}* can easily be saved as *.rda* (R data) files, using the `save` function.

```
save(m.gen, file = "path/to/my/meta-analysis.rda") # example path
```

4.2.2 (Standardized) Mean Differences

Raw effect size data in the form of means and standard deviations of two groups can be pooled using `metacont`. This function can be used for both standardized and unstandardized between-group mean differences. These can be obtained by either specifying `sm = "SMD"` or `sm = "MD"`. Otherwise, there are seven function-specific arguments we have to provide:

- `n.e`. The number of observations in the treatment/experimental group.
- `mean.e`. The mean in the treatment/experimental group.
- `sd.e`. The standard deviation in the treatment/experimental group.
- `n.c`. The number of observations in the control group.
- `mean.c`. The mean in the control group.
- `sd.c`. The standard deviation in the control group.
- `method.smd`. This is only relevant when `sm = "SMD"`. The `metacont` function allows us to calculate three different types of standardized mean differences. When we set `method.smd = "Cohen"`, the uncorrected standardized mean difference (Cohen's *d*) is used as the effect size metric. The two other options are `"Hedges"` (default and recommended), which calculates Hedges' g, and `"Glass"`, which will calculate Glass' Δ (*delta*). Glass' Δ uses the control group standard deviation instead of the pooled standard deviation to standardize the mean difference. This effect size is sometimes used in primary studies when there is more than one treatment group, but usually not the preferred metric for meta-analyses.

For our example analysis, we will recycle the `SuicidePrevention` data set we already worked with in Chapters 2.4 and 4.1.1. Not all studies in our sample are absolutely identical, so using a random-effects model is warranted. We will also use Knapp-Hartung adjustments again, as well as the restricted maximum likelihood estimator for τ^2. We tell `metacont` to correct for small-sample bias, producing Hedges' g as the effect size metric. Results are saved in an object that we name `m.cont`. Overall, our code looks like this:

```
# Make sure meta and dmetar are already loaded
library(meta)
library(dmetar)
library(meta)

# Load dataset from dmetar (or download and open manually)
data(SuicidePrevention)

# Use metcont to pool results.
m.cont <- metacont(n.e = n.e,
                   mean.e = mean.e,
                   sd.e = sd.e,
                   n.c = n.c,
                   mean.c = mean.c,
                   sd.c = sd.c,
                   studlab = author,
                   data = SuicidePrevention,
                   sm = "SMD",
                   method.smd = "Hedges",
```

```
                  comb.fixed = FALSE,
                  comb.random = TRUE,
                  method.tau = "REML",
                  hakn = TRUE,
                  title = "Suicide Prevention")
```

Let us see what the results are:

```
m.cont
```

```
## Review:     Suicide Prevention
##                       SMD             95%-CI %W(random)
## Berry et al.     -0.1428 [-0.4315;  0.1459]       15.6
## DeVries et al.   -0.6077 [-0.9402; -0.2752]       12.3
## Fleming et al.   -0.1112 [-0.6177;  0.3953]        5.7
## Hunt & Burke     -0.1270 [-0.4725;  0.2185]       11.5
## McCarthy et al. -0.3925 [-0.7884;  0.0034]        9.0
## Meijer et al.    -0.2676 [-0.5331; -0.0021]       17.9
## Rivera et al.     0.0124 [-0.3454;  0.3703]       10.8
## Watkins et al.   -0.2448 [-0.6848;  0.1952]        7.4
## Zaytsev et al.   -0.1265 [-0.5062;  0.2533]        9.7
##
## Number of studies combined: k = 9
##
##                           SMD             95%-CI     t
## Random effects model -0.2304 [-0.3734; -0.0874] -3.71
##                      p-value
## Random effects model  0.0059
##
## Quantifying heterogeneity:
##  tau^2 = 0.0044 [0.0000; 0.0924]; tau = 0.0661 [0.0000; 0.3040];
##  I^2 = 7.4% [0.0%; 67.4%]; H = 1.04 [1.00; 1.75]
##
## Test of heterogeneity:
##     Q d.f. p-value
##  8.64    8  0.3738
##
## Details on meta-analytical method:
## - Inverse variance method
## - Restricted maximum-likelihood estimator for tau^2
## - Q-profile method for confidence interval of tau^2 and tau
## - Hartung-Knapp adjustment for random effects model
## - Hedges' g (bias corrected standardised mean difference)
```

Looking at the output and comparing it to the one we received in Chapter 4.2.1, we already see one of *{meta}*'s greatest assets. Although `metagen` and `metacont` are different functions requiring different data types, the structure of the output looks nearly identical. This makes interpreting the results quite easy. We see that the pooled effect according to the random-effects model is $g = -0.23$, with the 95% confidence interval ranging from -0.09 to -0.37. The effect is significant ($p = 0.006$).

We see that the effect sizes have a negative sign. In the context of our meta-analysis, this represents a favorable outcome, because it means that suicidal ideation was lower in the treatment groups compared to the control groups. To make this clearer to others, we may also consistently reverse the sign of the effect sizes (e.g. write $g = 0.23$ instead), so that positive effect sizes always represent "positive" results.

The restricted maximum likelihood method estimated a between-study heterogeneity variance of $\tau^2 = 0.004$. Looking at `tau^2`, we see that the confidence interval includes zero, meaning that the variance of true effect sizes is not significantly greater than zero.

In the details section, we are informed that Hedges' g was used as the effect size metric–just as we requested.

4.2.3 Binary Outcomes

4.2.3.1 Risk & Odds Ratios

The `metabin` function can be used to pool effect sizes based on binary data, particularly risk and odds ratios. Before we start using the function, we first have to discuss a few particularities of meta-analyses based on these effect sizes.

It is possible to pool binary effect sizes using the generic inverse variance method we covered in Chapter 4.1.1 and 4.1.2.1. We need to calculate the log-odds or risk ratio, as well as the standard error of each effect, and can then use the inverse of the effect size variance to determine the pooling weights. However, this approach is suboptimal for binary outcome data (Higgins et al., 2019, chapter 10.4.1). When we are dealing with *sparse* data, meaning that the number of events or the total sample size of a study is small, the calculated standard error may not be a good estimator of the precision of the binary effect size.

4.2.3.1.1 *The Mantel-Haenszel Method*

The *Mantel-Haenszel* method (Mantel and Haenszel, 1959; Robins et al., 1986) is therefore commonly used as an alternative to calculate the weights of studies with binary outcome data. It is also the default approach used in `metabin`. This method uses the number of events and non-events in the treatment and control group to determine a study's weight. There are different formulas depending on if we want to calculate the risk or odds ratio.

Risk Ratio:

$$w_k = \frac{(a_k + b_k)c_k}{n_k} \tag{4.9}$$

Odds Ratio:

$$w_k = \frac{b_k c_k}{n_k} \tag{4.10}$$

In the formulas, we use the same notation as in Chapter 3.3.2.1, with a_k being the number of events in the treatment group, c_k the number of event in the control group, b_k the number of non-events in the treatment group, d_k the number of non-events in the control group, and n_k being the total sample size.

4.2.3.1.2 *The Peto Method*

A second approach is the *Peto* method (Yusuf et al., 1985). In its essence, this approach is based on the inverse variance principle we already know. However, it uses a special kind of effect size, the *Peto odds ratio*, which we will denote with $\hat{\psi}_k$. To calculate $\hat{\psi}_k$, we need to know O_k, the observed events in the treatment group, and calculate E_k, the *expected* number of cases in the treatment group. The difference $O_k - E_k$ is then divided by the variance V_k of the difference between O_k and E_k, resulting in a log-transformed version of $\hat{\psi}_k$. Using the same cell notation as before, the formulas to calculate E_k, O_k and V_k are the following:

$$O_k = a_k \tag{4.11}$$

$$E_k = \frac{(a_k + b_k)(a_k + c_k)}{a_k + b_k + c_k + d_k} \tag{4.12}$$

$$V_k = \frac{(a_k + b_k)(c_k + d_k)(a_k + c_k)(b_k + d_k)}{(a_k + b_k + c_k + d_k)^2(a_k + b_k + c_k + d_k - 1)} \tag{4.13}$$

$$\log \hat{\psi}_k = \frac{O_k - E_k}{V_k} \tag{4.14}$$

The inverse of the variance of $\log \hat{\psi}_k$ is then used as the weight when pooling the effect sizes.

4.2.3.1.3 *The Bakbergenuly-Sample Size Method*

Recently, Bakbergenuly and colleagues (2020) proposed another method in which the weight of effects is only determined by a study's sample size and showed that this approach may be preferable to the one by Mantel and Haenszel. We will call this the *sample size method*. The formula for this approach is fairly easy. We only need to know the sample size n_{treat_k} and n_{control_k} in the treatment and control group, respectively.

$$w_k = \frac{n_{\text{treat}_k} n_{\text{control}_k}}{n_{\text{treat}_k} + n_{\text{control}_k}} \tag{4.15}$$

When we implement this pooling method in `metabin`, the weights and overall effect using the fixed- and random-effects model will be identical. Only the p-value and confidence interval of the pooled effect will differ.

Which Pooling Method Should I Use?

In Chapter 3.3.2.1, we already talked extensively about the problem of *zero-cells* and *continuity correction*. While both the Peto and sample size method can be used without modification when there are zero cells, it is common to add 0.5 to zero cells when using the Mantel-Haenszel method. This is also the default behavior in `metabin`. Using continuity corrections, however, has been discouraged (Efthimiou, 2018), as they can lead to biased results. The Mantel-Haenszel method only *really* requires a continuity correction when one specific cell is zero in *all* included studies, which is rarely the case. Usually, it is therefore advisable to use the *exact* Mantel-Haenszel method without continuity corrections by setting `MH.exact = TRUE` in `metabin`.

The Peto method also has its limitations. First of all, it can only be used for odds ratios. Simulation studies also showed that the approach only works well when (1) the number of observations in the treatment and control group is similar, (2) when the observed event is rare (<1%), and (3) when the treatment effect is not overly large (Bradburn et al., 2007; Sweeting et al., 2004). The Bakbergenuly-sample size method, lastly, is a fairly new approach, meaning that it is not as well studied as the other two methods.

All in all, it may be advisable in most cases to follow Cochrane's general assessment (Higgins et al., 2019, chapter 10.4), and use the Mantel-Haenszel method (without continuity correction). The Peto method may be used when the odds ratio is the desired effect size metric, and when the event of interest is expected to be rare.

4.2.3.1.4 *Pooling Binary Effect Sizes in R*

There are eight important function-specific arguments in `metabin`:

- `event.e`. The number of events in the treatment/experimental group.
- `n.e`. The number of observations in the treatment/experimental group.
- `event.c`. The number of events in the control group.
- `n.c`. The number of observations in the control group.
- `method`. The pooling method to be used. This can either be `"Inverse"` (generic inverse-variance pooling), `"MH"` (Mantel-Haenszel; default and recommended), `"Peto"` (Peto method), or `"SSW"` (Bakbergenuly-sample size method).
- `sm`. The summary measure (i.e. effect size metric) to be calculated. We can use `"RR"` for the risk ratio and `"OR"` for the odds ratio.
- `incr`. The increment to be added for continuity correction of zero cells. If we specify `incr = 0.5`, an increment of 0.5 is added. If we set `incr = "TACC"`, the treatment arm continuity correction method is used (see Chapter 3.3.2.1). As mentioned before, it is usually recommended to leave out this argument and not apply continuity corrections.
- `MH.exact`. If `method = "MH"`, we can set this argument to `TRUE`, indicating that we do not want that a continuity correction is used for the Mantel-Haenszel method.

For our hands-on example, we will use the `DepressionMortality` data set. This data set is based on a meta-analysis by Cuijpers and Smit (2002), which examined the effect of suffering from depression on all-cause mortality. The data set contains the number of individuals with and without depression, and how many individuals in both groups had died after several years.

The "DepressionMortality" Data Set

The `DepressionMortality` data set is included directly in the *{dmetar}* package. If you have installed *{dmetar}*, and loaded it from your library, running `data(DepressionMortality)` automatically saves the data set in your R environment. The data set is then ready to be used. If you do not have *{dmetar}* installed, you can download the data set as an *.rda* file from the Internet[a], save it in your working directory, and then click on it in your R Studio window to import it.

[a]https://www.protectlab.org/meta-analysis-in-r/data/depressionmortality.rda

Let us have a look at the data set first:

```
library(dmetar)
library(tidyverse)
library(meta)

data(DepressionMortality)

glimpse(DepressionMortality)
```

```
## Rows: 18
## Columns: 6
## $ author  <chr> "Aaroma et al., 1994", "Black et a...
## $ event.e <dbl> 25, 65, 5, 26, 32, 1, 24, 15, 15, ...
## $ n.e     <dbl> 215, 588, 46, 67, 407, 44, 60, 61,...
## $ event.c <dbl> 171, 120, 107, 1168, 269, 87, 200,...
## $ n.c     <dbl> 3088, 1901, 2479, 3493, 6256, 1520...
## $ country <chr> "Finland", "USA", "USA", "USA", "S...
```

In this example, we will calculate the risk ratio as the effect size metric, as was done by Cuijpers and Smit. We will use a random-effects pooling model, and, since we are dealing with binary outcome data, we will use the Paule-Mandel estimator for τ^2. Looking at the data, we see that the sample sizes vary considerably from study to study, a scenario in which the Paule-Mandel method may be slightly biased (see Chapter 4.1.2.1). Keeping this in mind, we can also try out another τ^2 estimator as a sensitivity analysis to check if the results vary by a lot.

The data set contains no zero cells, so we do not have to worry about continuity correction, and can use the exact Mantel-Haenszel method right away. We save the meta-analysis results in an object called `m.bin`.

```
m.bin <- metabin(event.e = event.e,
                 n.e = n.e,
                 event.c = event.c,
                 n.c = n.c,
                 studlab = author,
                 data = DepressionMortality,
                 sm = "RR",
                 method = "MH",
                 MH.exact = TRUE,
                 comb.fixed = FALSE,
                 comb.random = TRUE,
                 method.tau = "PM",
                 hakn = TRUE,
                 title = "Depression and Mortality")

m.bin
```

```
## Review:     Depression and Mortality
##                              RR          95%-CI %W(random)
## Aaroma et al., 1994       2.09 [1.41;  3.12]        6.0
## Black et al., 1998        1.75 [1.31;  2.33]        6.6
## Bruce et al., 1989        2.51 [1.07;  5.88]        3.7
## Bruce et al., 1994        1.16 [0.85;  1.57]        6.5
## Enzell et al., 1984       1.82 [1.28;  2.60]        6.3
## Fredman et al., 1989      0.39 [0.05;  2.78]        1.2
## Murphy et al., 1987       1.76 [1.26;  2.46]        6.4
## Penninx et al., 1999      1.46 [0.93;  2.29]        5.8
## Pulska et al., 1998       1.94 [1.34;  2.81]        6.2
## Roberts et al., 1990      2.30 [1.92;  2.75]        7.0
## Saz et al., 1999          2.18 [1.55;  3.07]        6.3
## Sharma et al., 1998       2.05 [1.07;  3.91]        4.7
## Takeida et al., 1997      6.97 [4.13; 11.79]        5.3
## Takeida et al., 1999      5.81 [3.88;  8.70]        6.0
## Thomas et al., 1992       1.33 [0.77;  2.27]        5.3
## Thomas et al., 1992       1.77 [1.10;  2.83]        5.6
## Weissman et al., 1986     1.25 [0.66;  2.33]        4.8
## Zheng et al., 1997        1.98 [1.40;  2.80]        6.3
##
## Number of studies combined: k = 18
##
##                            RR           95%-CI    t  p-value
## Random effects model 2.0217 [1.5786; 2.5892] 6.00 < 0.0001
##
## Quantifying heterogeneity:
```

```
## tau^2 = 0.1865 [0.0739; 0.5568]; tau = 0.4319 [0.2718; 0.7462];
## I^2 = 77.2% [64.3%; 85.4%]; H = 2.09 [1.67; 2.62]
##
## Test of heterogeneity:
##      Q d.f.  p-value
## 74.49   17 < 0.0001
##
## Details on meta-analytical method:
## - Mantel-Haenszel method
## - Paule-Mandel estimator for tau^2
## - Q-profile method for confidence interval of tau^2 and tau
## - Hartung-Knapp adjustment for random effects model
```

We see that the pooled effect size is RR $= 2.02$. The pooled effect is significant ($p < 0.001$) and indicates that suffering from depression doubles the mortality risk. We see that our estimate of the between-study heterogeneity variance is $\tau^2 \approx 0.19$. The confidence interval of τ^2 does not include zero, indicating substantial heterogeneity between studies. Lastly, a look into the details section of the output reveals that the `metabin` function used the Mantel-Haenszel method for pooling, as intended.

As announced above, let us have a look if the method used to estimate τ^2 has an impact on the results. Using the `update.meta` function, we re-run the analysis, but use the restricted maximum likelihood estimator this time.

```
m.bin_update <- update.meta(m.bin,
                            method.tau = "REML")
```

Now, let us have a look at the pooled effect again by inspecting `TE.random`. We have to remember here that meta-analyses of binary outcomes are actually performed by using a log-transformed version of the effect size. When presenting the results, `metabin` just reconverts the effect size metrics to their original form for our convenience. This step is not performed if we inspect elements in our meta-analysis object. To retransform log-transformed effect sizes, we have to *exponentiate* the value. Exponentiation can be seen as the "antagonist" of log-transforming data, and can be performed in R using the exp function. Let us put this into practice.

```
exp(m.bin_update$TE.random)
```

```
## [1] 2.024
```

We see that the pooled effect using the restricted maximum likelihood estimator is virtually identical. Now, let us see the estimate of τ^2:

```
m.bin_update$tau2
```

```
## [1] 0.1647
```

This value deviates somewhat, but not to a degree that should make us worry about the validity of our initial results.

Our call to `metabin` would have looked exactly the same if we had decided to pool odds ratios. The only thing we need to change is the `sm` argument, which has to be set to `"OR"`. Instead of writing down the entire function call one more time, we can use the `update.meta` function again to calculate the pooled OR.

```
m.bin_or <- update.meta(m.bin,
                        sm = "OR")

m.bin_or
```

```
## Review:     Depression and Mortality
##
## [...]
##
## Number of studies combined: k = 18
##
##                           OR            95%-CI    t  p-value
## Random effects model 2.2901 [1.7512; 2.9949] 6.52 < 0.0001
##
## Quantifying heterogeneity:
##  tau^2 = 0.2032 [0.0744; 0.6314]; tau = 0.4508 [0.2728; 0.7946];
##  I^2 = 72.9% [56.7%; 83.0%]; H = 1.92 [1.52; 2.43]
##
## Test of heterogeneity:
##      Q d.f.  p-value
##  62.73   17 < 0.0001
##
## Details on meta-analytical method:
## - Mantel-Haenszel method
## - Paule-Mandel estimator for tau^2
## - Q-profile method for confidence interval of tau^2 and tau
## - Hartung-Knapp adjustment for random effects model
```

In the output, we see that the pooled effect using odds ratios is OR = 2.29.

4.2.3.1.5 *Pooling Pre-Calculated Binary Effect Sizes*

It is sometimes not possible to extract the raw effect size data needed to calculate risk or odds ratios in each study. For example, a primary study may report an odds ratio, but not the data on which this effect size is based on. If the authors do not provide us with the original data, this may require us to perform a meta-analysis based on pre-calculated effect size data. As we learned, the function we can use to do this is `metagen`.

When dealing with binary outcome data, we should be really careful if there is no other option than using pre-calculated effect size data. The `metagen` function uses the inverse-variance method to pool effect sizes, and better options such as the Mantel-Haenszel approach cannot be used. However, it is still a viable alternative if everything else fails.

Using the `DepressionMortality` data set, let us simulate that we are dealing with a pre-calculated effect size meta-analysis. We can extract the `TE` and `seTE` object in `m.bin` to get the effect size and standard error of each study. We save this information in our `DepressionMortality` data set.

```
DepressionMortality$TE <- m.bin$TE
DepressionMortality$seTE <- m.bin$seTE
```

Now, imagine that there is one effect for which we know the lower and upper bound of the confidence interval, but not the standard error. To simulate such a scenario, we will (1) define the standard error of study 7 (Murphy et al., 1987) as missing (i.e. set its value to `NA`), (2) define two new empty columns, `lower` and `upper`, in our data set, and (3) fill `lower` and `upper` with the log-transformed "reported" confidence interval in study 7.

```
# Set seTE if study 7 to NA
DepressionMortality$seTE[7] <- NA

# Create empty columns 'lower' and 'upper'
DepressionMortality[,"lower"] <- NA
DepressionMortality[,"upper"] <- NA

# Fill in values for 'lower' and 'upper' in study 7
# As always, binary effect sizes need to be log-transformed
DepressionMortality$lower[7] <- log(1.26)
DepressionMortality$upper[7] <- log(2.46)
```

Now let us have a look at the data we just created.

```
DepressionMortality[,c("author", "TE", "seTE", "lower", "upper")]
```

```
##                  author      TE    seTE  lower  upper
## 1    Aaroma et al., 1994  0.7418 0.20217     NA     NA
## 2     Black et al., 1998  0.5603 0.14659     NA     NA
## 3     Bruce et al., 1989  0.9235 0.43266     NA     NA
## 4     Bruce et al., 1994  0.1488 0.15526     NA     NA
## 5    Enzell et al., 1984  0.6035 0.17986     NA     NA
## 6   Fredman et al., 1989 -0.9236 0.99403     NA     NA
## 7    Murphy et al., 1987  0.5675      NA 0.2311 0.9001
## 8   Penninx et al., 1999  0.3816 0.22842     NA     NA
```

```
## [...]
```

It is not uncommon to find data sets like this one in practice. It may be possible to calculate the log-risk ratio for most studies, but for a few other ones, the only information we often have is the (log-transformed) risk ratio and its confidence interval.

Fortunately, `metagen` allows us to pool even such data. We only have to provide the name of the columns containing the lower and upper bound of the confidence interval to the `lower` and `upper` argument. The `metagen` function will then use this information to weight the effects when the standard error is not available. Our function call looks like this:

```
m.gen_bin <- metagen(TE = TE,
                     seTE = seTE,
                     lower = lower,
                     upper = upper,
                     studlab = author,
                     data = DepressionMortality,
                     sm = "RR",
                     method.tau = "PM",
                     comb.fixed = FALSE,
                     comb.random = TRUE,
                     title = "Depression Mortality (Pre-calculated)")

m.gen_bin
```

```
## Review:     Depression Mortality (Pre-calculated)
##
## [...]
##
## Number of studies combined: k = 18
##
##                          RR           95%-CI    z  p-value
## Random effects model 2.0218 [1.6066; 2.5442] 6.00 < 0.0001
##
## Quantifying heterogeneity:
##  tau^2 = 0.1865 [0.0739; 0.5568]; tau = 0.4319 [0.2718; 0.7462];
##  I^2 = 77.2% [64.3%; 85.4%]; H = 2.09 [1.67; 2.62]
##
## [...]
```

In the output, we see that all $K = 18$ studies could be combined in the meta-analysis, meaning that `metagen` used the information in `lower` and `upper` provided for study 7. The output also shows that the results using the inverse variance method are nearly identical to the ones of the Mantel-Haenszel method from before.

4.2.3.2 Incidence Rate Ratios

Effect sizes based on incidence rates (i.e. incidence rate ratios, Chapter 3.3.3) can be pooled using the `metainc` function. The arguments of this function are very similar to `metabin`:

- `event.e`: The number of events in the treatment/experimental group.
- `time.e`: The person-time at risk in the treatment/experimental group.
- `event.c`: The number of events in the control group.
- `time.c`: The person-time at risk in the control group.
- `method`: Like `metabin`, the default pooling method is the one by Mantel and Haenszel (`"MH"`). Alternatively, we can also use generic inverse variance pooling (`"Inverse"`).
- `sm`: The summary measure. We can choose between the incidence rate ratio (`"IRR"`) and the incidence rate difference (`"IRD"`).
- `incr`: The increment we want to add for the continuity correction of zero cells.

In contrast to `metabin`, `metainc` does not use a continuity correction by default. Specifying `MH.exact` as `TRUE` is therefore not required. A continuity correction is only performed when we choose the generic inverse variance pooling method (`method = "Inverse"`).

In our hands-on example, we will use the `EatingDisorderPrevention` data set. This data is based on a meta-analysis which examined the effects of college-based preventive interventions on the incidence of eating disorders (Harrer et al., 2020). The person-time at risk is expressed as person-years in this data set.

The "EatingDisorderPrevention" Data Set

The `EatingDisorderPrevention` data set is included in the *{dmetar}* package. If you have installed *{dmetar}*, and loaded it from your library, running `data(EatingDisorderPrevention)` automatically saves the data set in your R environment. The data set is then ready to be used. If you do not have *{dmetar}* installed, you can download the data set as an *.rda* file from the Internet[a], save it in your working directory, and then click on it in your R Studio window to import it.

[a] https://www.protectlab.org/meta-analysis-in-r/data/eatingdisorderprevention.rda

As always, let us first have a glimpse at the data:

```
library(dmetar)
library(tidyverse)
library(meta)

data(EatingDisorderPrevention)

glimpse(EatingDisorderPrevention)
```

```
## Rows: 5
## Columns: 5
## $ Author  <chr> "Stice et al., 2013", "Stice et al...
## $ event.e <dbl> 6, 22, 6, 8, 22
## $ time.e  <dbl> 362, 235, 394, 224, 160
## $ event.c <dbl> 16, 8, 9, 13, 29
## $ time.c  <dbl> 356, 74, 215, 221, 159
```

We use `metainc` to pool the effect size data, with the incidence rate ratio as the effect size metric. The Mantel-Haenszel method is used for pooling, and the Paule-Mandel estimator to calculate the between-study heterogeneity variance.

```
m.inc <- metainc(event.e = event.e,
                 time.e = time.e,
                 event.c = event.c,
                 time.c = time.c,
                 studlab = Author,
                 data = EatingDisorderPrevention,
                 sm = "IRR",
                 method = "MH",
                 comb.fixed = FALSE,
                 comb.random = TRUE,
                 method.tau = "PM",
                 hakn = TRUE,
                 title = "Eating Disorder Prevention")

m.inc
```

```
## Review:     Eating Disorder Prevention
##                        IRR           95%-CI %W(random)
## Stice et al., 2013  0.3688 [0.1443; 0.9424]       13.9
## Stice et al., 2017a 0.8660 [0.3855; 1.9450]       18.7
## Stice et al., 2017b 0.3638 [0.1295; 1.0221]       11.5
## Taylor et al., 2006 0.6071 [0.2516; 1.4648]       15.8
## Taylor et al., 2016 0.7539 [0.4332; 1.3121]       40.0
##
## Number of studies combined: k = 5
```

```
##
##                              IRR            95%-CI     t
## Random effects model 0.6223 [0.3955; 0.9791] -2.91
##                         p-value
## Random effects model  0.0439
##
## Quantifying heterogeneity:
##  tau^2 = 0 [0.0000; 1.1300]; tau = 0 [0.0000; 1.0630];
##  I^2 = 0.0% [0.0%; 75.1%]; H = 1.00 [1.00; 2.00]
##
## Test of heterogeneity:
##      Q d.f. p-value
##  3.34    4  0.5033
##
## Details on meta-analytical method:
## - Mantel-Haenszel method
## - Paule-Mandel estimator for tau^2
## - Q-profile method for confidence interval of tau^2 and tau
## - Hartung-Knapp adjustment for random effects model
```

We see that the pooled effect is IRR = 0.62. This effect is significant ($p = 0.04$), albeit being somewhat closer to the conventional significance threshold than in the previous examples. Based on the pooled effect, we can say that the preventive interventions reduced the incidence of eating disorders within one year by 38%. Lastly, we see that the estimate of the heterogeneity variance τ^2 is zero.

4.2.4 Correlations

Correlations can be pooled using the `metacor` function, which uses the generic inverse variance pooling method. In Chapter 3.2.3.1, we covered that correlations should be Fisher's z-transformed before pooling. By default, `metacor` does this transformation automatically for us. It is therefore sufficient to provide the function with the original, untransformed correlations reported in the studies. The `metacor` function has only two relevant function-specific arguments:

- `cor`. The (untransformed) correlation coefficient.
- `n`. The number of observations in the study.

To illustrate `metacor`'s functionality, we will use the `HealthWellbeing` data set. This data set is loosely based on a large meta-analysis examining the association between health and well-being (Ngamaba et al., 2017).

The "HealthWellbeing" Data Set

The `HealthWellbeing` data set is included in the *{dmetar}* package. If you have installed *{dmetar}*, and loaded it from your library, running `data(HealthWellbeing)` automatically saves the data set in your R environment. The data set is then ready to be used.

If you do not have *{dmetar}* installed, you can download the data set as an *.rda* file from the Internet[a], save it in your working directory, and then click on it in your R Studio window to import it.

[a] https://www.protectlab.org/meta-analysis-in-r/data/healthwellbeing.rda

Let us have a look at the data:

```
library(dmetar)
library(tidyverse)
library(meta)

data(HealthWellbeing)

glimpse(HealthWellbeing)
```

```
## Rows: 29
## Columns: 5
## $ author     <chr> "An, 2008", "Angner, 2013", "Ba...
## $ cor        <dbl> 0.620, 0.372, 0.290, 0.333, 0.7...
## $ n          <dbl> 121, 383, 350000, 1764, 42331, ...
## $ population <chr> "general population", "chronic ...
## $ country    <chr> "South Korea", "USA", "USA", "I...
```

We expect considerable between-study heterogeneity in this meta-analysis, so a random-effects model is employed. The restricted maximum likelihood estimator is used for τ^2.

```
m.cor <- metacor(cor = cor,
                 n = n,
                 studlab = author,
                 data = HealthWellbeing,
                 comb.fixed = FALSE,
                 comb.random = TRUE,
                 method.tau = "REML",
                 hakn = TRUE,
                 title = "Health and Wellbeing")
```

```
m.cor
```

```
## Review:     Health and Wellbeing
##                      COR            95%-CI %W(random)
## An, 2008          0.6200 [0.4964; 0.7189]        2.8
## Angner, 2013      0.3720 [0.2823; 0.4552]        3.4
## Barger, 2009      0.2900 [0.2870; 0.2930]        3.8
## Doherty, 2013     0.3330 [0.2908; 0.3739]        3.7
## Dubrovina, 2012   0.7300 [0.7255; 0.7344]        3.8
## Fisher, 2010      0.4050 [0.2373; 0.5493]        2.8
## [...]
##
## Number of studies combined: k = 29
##
##                         COR            95%-CI     t  p-value
## Random effects model 0.3632 [0.3092; 0.4148] 12.81 < 0.0001
##
## Quantifying heterogeneity:
##  tau^2 = 0.0241 [0.0141; 0.0436]; tau = 0.1554 [0.1186; 0.2088];
##  I^2 = 99.8% [99.8%; 99.8%]; H = 24.14 [23.29; 25.03]
##
## Test of heterogeneity:
##         Q d.f. p-value
##  16320.87   28       0
##
## Details on meta-analytical method:
## - Inverse variance method
## - Restricted maximum-likelihood estimator for tau^2
## - Q-profile method for confidence interval of tau^2 and tau
## - Hartung-Knapp adjustment for random effects model
## - Fisher's z transformation of correlations
```

We see that the pooled association between health and well-being is $r = 0.36$, and that this effect is significant ($p < 0.001$). Using Cohen's convention, this can be considered a moderate-sized correlation. In the output, `metacor` already reconverted the Fisher's z-transformed correlations to the original form. A look at the last line of the details section, however, tells us that z-values have indeed been used to pool the effects. Lastly, we see that the heterogeneity variance estimated for this meta-analysis is significantly larger than zero.

4.2.5 Means

A meta-analysis of means can be conducted using the `metamean` function. This function uses the generic inverse variance method to pool the data. When using `metamean`, we have to determine first if we want to perform a meta-analysis of raw or log-transformed means. In contrast to odds and risk ratios, a log-transformation of means is usually not necessary. However, it is advisable to use the transformation when dealing with means of a non-negative quantity (e.g. height), and when some means are close to zero. This is controlled via the `sm` argument. If we set `sm = "MRAW`, the raw means are pooled. The log-transformation is performed when `sm = "MLN"`. The function-specific arguments are:

- `n`: The number of observations.
- `mean`: The mean.
- `sd`: The standard deviation of the mean.
- `sm`: The type of summary measure to be used for pooling (see above).

For our hands-on example, we will use the `BdiScores` data set. This data set contains the mean score of the Beck Depression Inventory II (Beck et al., 1996), measured in samples of depression patients participating in psychotherapy and antidepressant trials (Furukawa et al., 2020).

The "BdiScores" Data Set

The `BdiScores` data set is included directly in the *{dmetar}* package. If you have installed *{dmetar}*, and loaded it from your library, running `data(BdiScores)` automatically saves the data set in your R environment. The data set is then ready to be used. If you do not have *{dmetar}* installed, you can download the data set as an *.rda* file from the Internet[a], save it in your working directory, and then click on it in your R Studio window to import it.

[a] https://www.protectlab.org/meta-analysis-in-r/data/bdiscores.rda

```
library(dmetar)
library(tidyverse)
library(meta)

data(BdiScores)

# We only need the first four columns
glimpse(BdiScores[,1:4])
```

```
## Rows: 6
## Columns: 4
## $ author <chr> "DeRubeis, 2005", "Dimidjian, 2006"...
## $ n      <dbl> 180, 145, 48, 142, 301, 104
## $ mean   <dbl> 32.6, 31.9, 28.6, 30.3, 31.9, 29.8
## $ sd     <dbl> 9.4, 7.4, 9.9, 9.1, 9.2, 8.6
```

Our goal is to calculate the overall mean depression score based on this collection of studies. We will use a random-effects model and the restricted maximum-likelihood estimator to pool the raw means in our data set. We save the results in an object called `m.mean`.

```
m.mean <- metamean(n = n,
                   mean = mean,
                   sd = sd,
                   studlab = author,
                   data = BdiScores,
                   sm = "MRAW",
                   comb.fixed = FALSE,
                   comb.random = TRUE,
                   method.tau = "REML",
                   hakn = TRUE,
                   title = "BDI-II Scores")

m.mean
```

```
## Review:     BDI-II Scores
##                    mean           95%-CI %W(random)
## DeRubeis, 2005  32.6000 [31.2268; 33.9732]       18.0
## Dimidjian, 2006 31.9000 [30.6955; 33.1045]       19.4
## Dozois, 2009    28.6000 [25.7993; 31.4007]        9.1
## Lesperance, 2007 30.3000 [28.8033; 31.7967]      17.0
## McBride, 2007   31.9000 [30.8607; 32.9393]       20.7
## Quilty, 2014    29.8000 [28.1472; 31.4528]       15.8
##
## Number of studies combined: k = 6
##
##                          mean           95%-CI
## Random effects model 31.1221 [29.6656; 32.5786]
##
## Quantifying heterogeneity:
##  tau^2 = 1.0937 [0.0603; 12.9913]; tau = 1.0458 [0.2456; 3.6043];
##  I^2 = 64.3% [13.8%; 85.2%]; H = 1.67 [1.08; 2.60]
##
## Test of heterogeneity:
##      Q d.f. p-value
```

```
## 14.00     5  0.0156
##
## Details on meta-analytical method:
## - Inverse variance method
## - Restricted maximum-likelihood estimator for tau^2
## - Q-profile method for confidence interval of tau^2 and tau
## - Hartung-Knapp adjustment for random effects model
## - Untransformed (raw) means
```

The pooled mean assuming a random-effects model is m = 31.12. We also see that the between-study heterogeneity variance τ^2 in this meta-analysis is significantly greater than zero.

4.2.6 Proportions

The `metaprop` function can be used to pool proportions. In Chapter 3.2.2, we already discussed that it is best to logit-transform proportions before the meta-analysis is performed. The `metaprop` function does this automatically for us if we specify `sm = "PLOGIT"`. If the raw proportions should be pooled, we can use `sm = "PRAW"`, but remember that this is discouraged.

The default method through which `metaprop` pools proportions is somewhat special. If we use logit-transformed values, the function does not use the inverse-variance method for pooling, but builds a *generalized linear mixed-effects model* (GLMM). Essentially, the function fits a logistic regression model to our data, which includes random-effects to account for the fact that true effect sizes vary between studies.

You may have heard of the term "mixed-effects model" before. Such models are commonly used in primary studies across many research fields. In Chapters 7 and 8, we will delve into this topic a little deeper by discussing subgroup analysis and meta-regression, which are special applications of mixed-effects models. For now, however, it is sufficient to understand the general idea of what a mixed-effects model is.

Mixed-effects models are regression models which contain both "fixed" and "random" components. The fixed elements are the β weights. A very simple regression model contains two β terms; the intercept β_0, as well as a regression term $\beta_1 x$. These are used in combination to predict observed data y through some other quantity x. This prediction will hardly ever be perfect, leaving some random error ϵ_i. Together, this gives the following formula:

$$y_i = \beta_0 + \beta_1 x_i + \epsilon_i \tag{4.16}$$

The crucial point is that the value of the β weights in this equation remains the same for each observation i. The value of x may vary from observation to observation, but β_0 and β_1 never do, since they are fixed.

This regression equation can be turned into a *mixed*-effects model when random effects are added. We denote this random-effect term with u_i. As indicated by the subscript i, the random effect term can have different values for each observation. The u_i term is centered around zero and can increase or decrease the estimate produced by the fixed effects:

$$y_i = \beta_0 + \beta_1 x_i + u_i + \epsilon_i \tag{4.17}$$

Meta-analysis can be seen as a special type of this model in which there is no $\beta_1 x_i$ term. The model only contains an intercept β_0, which corresponds with the overall effect size μ in the random-effects model. The u_i and ϵ_i parts correspond with the ζ_k and ϵ_k error terms in meta-analyses. This makes it clear that meta-analysis is equivalent to a mixed-effects regression model. This mixed-effects model, however, only contains an intercept, as well as a random effect connected to that intercept. Using a binomial logit-link, we can therefore apply a (generalized) logistic mixed-effect model to estimate the pooled effect[4].

GLMMs can be applied not only to proportions but also to other outcome measures based on binary and count data, such as odds ratios or incidence rate ratios (Stijnen et al., 2010). While GLMMs are not universally recommended for meta-analyses of binary outcome data (Bakbergenuly and Kulinskaya, 2018), their use has been advocated for proportions (Schwarzer et al., 2019).

Using GLMMs as part of `metaprop` has three implications: (1) the output will display no meta-analytic weights for each effect, (2) the τ^2 estimator can only be set to `"ML"` (since maximum-likelihood is used to estimate the GLMM), and (3) there will be no confidence intervals for our estimate of τ^2. If this information is required, you may switch to performing an inverse-variance meta-analysis. There are five function-specific arguments for `metaprop`:

- `event`. The number of events.
- `n`. The number of observations.
- `method`. The pooling method. Can be either a GLMM (`method = "GLMM"`), or inverse-variance pooling (`method = "Inverse"`).
- `incr`. The increment to be added for continuity correction in zero cells. This is only relevant when inverse-variance pooling is used.
- `sm`. The summary measure to be used. It is advised to use logit-transformed proportions by setting `sm = "PLOGIT"` (default).

[4]For study k, the logit transformation is defined as $\theta_k^{\text{LO}} = \log_e\left(\frac{p_k}{1-p_k}\right)$, leading to $\theta_k^{\text{LO}} \sim \theta + u_k$, with $u_k \sim \mathcal{N}(0, \tau^2)$. In meta-analytic GLMMs for proportions, the number of events in a study (a_k) is assumed to follow a binomial distribution: $a_k \sim \text{B}\left(n_k, \frac{\exp(\theta_k^{\text{LO}})}{1+\exp(\theta_k^{\text{LO}})}\right)$. For a more detailed description, see Schwarzer et al. (2019), A.2.2.

For our illustration of the `metaprop` function, we will use the `OpioidMisuse` data set. This data is derived from a meta-analysis which examined the 12-month prevalence of prescription opioid misuse among adolescents and young adults in the United States (Jordan et al., 2017).

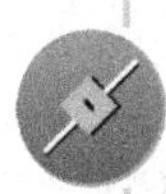

The "OpioidMisuse" Data Set

The `OpioidMisuse` data set is included directly in the *{dmetar}* package. If you have installed *{dmetar}*, and loaded it from your library, running `data(OpioidMisuse)` automatically saves the data set in your R environment. The data set is then ready to be used. If you do not have *{dmetar}* installed, you can download the data set as an *.rda* file from the Internet[a], save it in your working directory, and then click on it in your R Studio window to import it.

[a] `https://www.protectlab.org/meta-analysis-in-r/data/opioidmisuse.rda`

Let us load the data set and have a look at it:

```
library(dmetar)
library(meta)
library(tidyverse)

data(OpioidMisuse)

glimpse(OpioidMisuse)
```

```
## Rows: 15
## Columns: 3
## $ author <chr> "Becker, 2008", "Boyd, 2009", "Boyd...
## $ event  <dbl> 2186, 91, 126, 543, 6496, 10850, 86...
## $ n      <dbl> 21826, 912, 1084, 7646, 55215, 1147...
```

We pool the prevalence data using a GLMM and logit-transformed proportions.

```
m.prop <- metaprop(event = event,
                   n = n,
                   studlab = author,
                   data = OpioidMisuse,
                   method = "GLMM",
                   sm = "PLOGIT",
                   comb.fixed = FALSE,
                   comb.random = TRUE,
```

```
                    hakn = TRUE,
                    title = "Opioid Misuse")

m.prop
```

```
## Review:     Opioid Misuse
##                   proportion            95%-CI
## Becker, 2008          0.1002 [0.0962; 0.1042]
## Boyd, 2009            0.0998 [0.0811; 0.1211]
## Boyd, 2007            0.1162 [0.0978; 0.1368]
## Cerda, 2014           0.0710 [0.0654; 0.0770]
## Fiellin, 2013         0.1176 [0.1150; 0.1204]
## [...]
##
## Number of studies combined: k = 15
##
##                         proportion            95%-CI
## Random effects model        0.0944 [0.0845; 0.1055]
##
## Quantifying heterogeneity:
##  tau^2 = 0.0558; tau = 0.2362; I^2 = 99.2%; H = 11.07
##
## Test of heterogeneity:
##       Q d.f.  p-value             Test
##  838.21   14 < 0.0001        Wald-type
##  826.87   14 < 0.0001 Likelihood-Ratio
##
## Details on meta-analytical method:
## - Random intercept logistic regression model
## - Maximum-likelihood estimator for tau^2
## - Hartung-Knapp adjustment for random effects model
## - Logit transformation
## - Clopper-Pearson confidence interval for individual studies
```

In the output, we see that the pooled 12-month prevalence of prescription opioid misuse in the selected studies is 9.4%, with the confidence interval ranging from 8.45 to 10.55%. As described before, the output does not display the individual weight of each effect. In the same vein, we get an estimate of the between-study heterogeneity ($\tau^2 = 0.056$), but no confidence interval around it.

□

4.3 Questions & Answers

Test your knowledge!

1. What is the difference between a fixed-effect model and a random-effects model?
2. Can you think of a case in which the results of the fixed- and random-effects model are identical?
3. What is τ^2? How can it be estimated?
4. On which distribution is the Knapp-Hartung adjustment based? What effect does it have?
5. What does "inverse-variance" pooling mean? When is this method *not* the best solution?
6. You want to meta-analyze binary outcome data. The number of observations in the study arms is roughly similar, the observed event is very rare, and you do no expect the treatment effect to be large. Which pooling method would you use?
7. For which outcome measures can GLMMs be used?

Answers to these questions are listed in Appendix A at the end of this book.

4.4 Summary

- In statistics, a model can be seen as a simplified "theory", describing the process through which observed data were generated. There are two alternative models in meta-analysis: the fixed-effect model and the random-effects model.
- While the fixed-effect model assumes that there is one true effect size, the random-effects model states that the true effect sizes also vary within meta-analyses. The goal of the random-effects model is therefore to find the mean of the true effect size distribution underlying our data.

- The variance of true effect sizes τ^2, also known as between-study heterogeneity variance, has to be estimated in random-effects meta-analyses. There are several methods for this, and which one works best depends on the context.
- The most common way to calculate a pooled effect size is through the inverse-variance method. However, for binary outcome data, other approaches such as the Mantel-Haenszel method may be preferable.
- In the *{meta}* package, there is a function to perform meta-analyses of pre-calculated effect size data, as well as a suite of functions that can be used for different types of "raw" outcome data.

5

Between-Study Heterogeneity

By now, we have already learned how to pool effect sizes in a meta-analysis. As we have seen, the aim of both the fixed- and random-effects model is to synthesize the effects of many different studies into one single number. This, however, only makes sense if we are not comparing apples and oranges. For example, it could be that while the overall effect we calculate in the meta-analysis is small, there are still a few outliers with very high effect sizes. Such information is lost in the aggregate effect, and we do not know if all studies yielded small effect sizes, or if there were exceptions.

The extent to which true effect sizes vary within a meta-analysis is called *between-study heterogeneity*. We already mentioned this concept briefly in the last chapter in connection with the random-effects model. The random-effects model assumes that between-study heterogeneity causes the true effect sizes of studies to differ. It therefore includes an estimate of τ^2, which quantifies this variance in true effects. This allows to calculate the pooled effect, defined as the mean of the true effect size distribution.

The random-effects model always allows us to calculate a pooled effect size, even if the studies are very heterogeneous. Yet, it does not tell us if this pooled effect can be *interpreted* in a meaningful way. There are many scenarios in which the pooled effect alone is not a good representation of the data in our meta-analysis.

Imagine a case where the heterogeneity is very high, meaning that the true effect sizes (e.g. of some treatment) range from highly positive to negative. If the pooled effect of such a meta-analysis is positive, this does not tell us that there were some studies with a true *negative* effect. The fact that the treatment had an adverse effect in some studies is lost. High heterogeneity can also be caused by the fact that there are two or more *subgroups* of studies in our data that have a different true effect. Such information can be very valuable for researchers, because it might allow us to find certain contexts in which effects are lower or higher. Yet, if we look at the pooled effect in isolation, this detail will likely be missed. In extreme cases, very high heterogeneity can mean that the studies have *nothing in common*, and that it makes no sense to interpret the pooled effect at all.

Therefore, meta-analysts must always take into account the variation in the analyzed studies. Every good meta-analysis should not only report an overall effect but also state how trustworthy this estimate is. An essential part of this is to quantify and analyze the between-study heterogeneity. In this chapter, we will have a closer

DOI: 10.1201/9781003107347-5

look at different ways to measure heterogeneity, and how they can be interpreted. We will also cover a few tools which allow us to detect studies that contribute to the heterogeneity in our data. Lastly, we discuss ways to address large amounts of heterogeneity in "real-world" meta-analyses.

5.1 Measures of Heterogeneity

Before we start discussing heterogeneity measures, we should first clarify that heterogeneity can mean different things. Rücker and colleagues (2008), for example, differentiate between *baseline* or *design-related* heterogeneity, and *statistical* heterogeneity.

- *Baseline* or *design-related* heterogeneity arises when the population or research design of studies differs across studies. We have discussed this type of heterogeneity when we talked about the "Apples and Oranges" problem (Chapter 1.3), and ways to define the research questions (Chapter 1.4.1). Design-related heterogeneity can be reduced *a priori* by setting up a suitable PICO that determines which types of populations and designs are eligible for the meta-analysis.
- *Statistical* heterogeneity, on the other hand, is a quantifiable property, influenced by the spread and precision of the effect size estimates included in a meta-analysis. Baseline heterogeneity *can* lead to statistical heterogeneity (for example, if effects differ between included populations) but does not have to. It is also possible for a meta-analysis to display high statistical heterogeneity, even if the included studies themselves are virtually identical. In this guide (and most other meta-analysis texts), the term "between-study heterogeneity" only refers to *statistical* heterogeneity.

5.1.1 Cochran's Q

Based on the random-effects model, we know that there are two sources of variation causing observed effects to differ from study to study. There is the sampling error ϵ_k, and the error caused by between-study heterogeneity, ζ_k (Chapter 4.1.2). When we want to quantify between-study heterogeneity, the difficulty is to identify how much of the variation can be attributed to the sampling error, and how much to true effect size differences.

Traditionally, meta-analysts have used *Cochran's* Q (Cochran, 1954) to distinguish studies' sampling error from actual between-study heterogeneity. Cochran's Q is defined as a *weighted sum of squares* (*WSS*). It uses the deviation of each study's observed effect $\hat{\theta}_k$ from the summary effect $\hat{\theta}$, weighted by the inverse of the study's variance, w_k:

$$Q = \sum_{k=1}^{K} w_k(\hat\theta_k - \hat\theta)^2 \tag{5.1}$$

Let us take a closer look at the formula. First of all, we see that it uses the same type of inverse-variance weighting that is also applied to pool effect sizes. The mean $\hat\theta$ in the formula is the pooled effect according to the fixed-effect model. The amount to which individual effects deviate from the summary effect, the *residuals*, is squared (so that the value is always positive), weighted, and then summed. The resulting value is Cochran's Q.

Because of the weighting by w_k, the value of Q does not only depend on how much $\hat\theta_k$'s deviate from $\hat\theta$ but also on the precision of studies. If the standard error of an effect size is very low (and thus the precision very high), even small deviations from the summary effect will be given a higher weight, leading to higher values of Q.

The value of Q can be used to check if there is *excess variation* in our data, meaning more variation than can be expected from sampling error alone. If this is the case, we can assume that the rest of the variation is due to between-study heterogeneity. We will illustrate this with a little simulation.

In our simulation, we want to inspect how Q behaves under two different scenarios: when there is no between-study heterogeneity, and when heterogeneity exists. Let us begin with the no-heterogeneity case. This implies that $\zeta_k = 0$, and that the residuals $\hat\theta_k - \hat\theta$ are only product of the sampling error ϵ_k. We can use the `rnorm` function to simulate deviates from some mean effect size $\hat\theta$ (assuming that they follow a normal distribution). Because they are centered around $\hat\theta$, we can expect the mean of these "residuals" to be zero ($\mu = 0$). For this example, let us assume that the population standard deviation is $\sigma = 1$, which leads to a *standard* normal distribution. Normal distributions are usually denoted with $\mathcal{N}$, and we can symbolize that the residuals are draws from a normal distribution with $\mu = 0$ and $\sigma = 1$ like this:

$$\hat\theta_k - \hat\theta \sim \mathcal{N}(0, 1) \tag{5.2}$$

Let us try this out in R, and draw K=40 effect size residuals $\hat\theta_k - \hat\theta$ using `rnorm`.

```
set.seed(123) # needed to reproduce results
rnorm(n = 40, mean = 0, sd = 1)
```

```
##  [1] -0.56048 -0.23018  1.55871  0.07051  0.12929
##  [6]  1.71506  0.46092 -1.26506 -0.68685 -0.44566
##  [...]
```

Because the standard normal distribution is the default for `rnorm`, we could have also used the simpler code `rnorm(40)`. Now, let us simulate that we repeat this process of drawing $n = 40$ samples many, many times. We can achieve this using the `replicate` function, which we tell to repeat the rnorm call ten thousand times. We save the resulting values in an object called error_fixed.

```
set.seed(123)
error_fixed <- replicate(n = 10000, rnorm(40))
```

We continue with a second scenario, in which we assume that *between-study heterogeneity* (ζ_k errors) exists in addition to the sampling error ϵ_k. We can simulate this by adding a second call to rnorm, representing the variance in true effect sizes. In this example, we also assume that the true effect sizes follow a standard normal distribution. We can simulate the residuals of ten thousand meta-analyses with K=40 studies and substantial between-study heterogeneity using this code:

```
set.seed(123)
error_random <- replicate(n = 10000, rnorm(40) + rnorm(40))
```

Now that we simulated $\hat{\theta}_k - \hat{\theta}$ residuals for meta-analyses *with* and *without* heterogeneity, let us do the same for values of Q. For this simulation, we can simplify the formula of Q a little by assuming that the variance, and thus the weight w_k of every study, is *one*, resulting in w_k to drop out of the equation. This means that we only have to use our calls to rnorm from before, square and sum the result, and replicate this process ten thousand times. Here is the code for that:

```
set.seed(123)
Q_fixed <- replicate(10000, sum(rnorm(40)^2))
Q_random <- replicate(10000, sum((rnorm(40) + rnorm(40))^2))
```

An important property of Q is that it is assumed to (approximately) follow a χ^2 distribution. A χ^2 distribution, like the weighted squared sum, can only take positive values. It is defined by its *degrees of freedom*, or d.f.; χ^2 distributions are right-skewed for small d.f., but get closer and closer to a normal distribution when the degrees of freedom become larger. At the same time, the degrees of freedom are also the *expected value*, or mean of the respective χ^2 distribution.

It is assumed that Q will approximately follow a χ^2 distribution with $K - 1$ degrees of freedom (with K being the number of studies in our meta-analysis)–*if* effect size differences are *only* caused by sampling error. This means that the mean of a χ^2 distribution with $K - 1$ degrees of freedom tells us the value of Q we can expect through sampling error alone.

This explanation was very abstract, so let us have a look at the distribution of our simulated values to make this more concrete. In the following code, we use the hist function to plot a histogram of the effect size "residuals" and Q values. We also add a line to each plot, showing the idealized distribution. Such distributions can be generated by the dnorm function for normal distributions, and using dchisq for χ^2 distributions, with df specifying the degrees of freedom.

```
# Histogram of the residuals (theta_k - theta)
# - We produce a histogram for both the simulated values in
#   error_fixed and error_random
# - `lines` is used to add a normal distribution in blue.

hist(error_fixed,
     xlab = expression(hat(theta[k])~-~hat(theta)), prob = TRUE,
     breaks = 100, ylim = c(0, .45), xlim = c(-4,4),
     main = "No Heterogeneity")
lines(seq(-4, 4, 0.01), dnorm(seq(-4, 4, 0.01)),
      col = "blue", lwd = 2)

hist(error_random,
     xlab = expression(hat(theta[k])~-~hat(theta)), prob = TRUE,
     breaks = 100,ylim = c(0, .45), xlim = c(-4,4),
     main = "Heterogeneity")
lines(seq(-4, 4, 0.01), dnorm(seq(-4, 4, 0.01)),
      col = "blue", lwd = 2)


# Histogram of simulated Q-values
# - We produce a histogram for both the simulated values in
#   Q_fixed and Q_random
# - `lines` is used to add a chi-squared distribution in blue.

# First, we calculate the degrees of freedom (k-1)
# remember: k=40 studies were used for each simulation
df <- 40-1

hist(Q_fixed, xlab = expression(italic("Q")), prob = TRUE,
     breaks = 100, ylim = c(0, .06),xlim = c(0,160),
     main = "No Heterogeneity")
lines(seq(0, 100, 0.01), dchisq(seq(0, 100, 0.01), df = df),
      col = "blue", lwd = 2)

hist(Q_random,  xlab = expression(italic("Q")), prob = TRUE,
     breaks = 100, ylim = c(0, .06), xlim = c(0,160),
     main = "Heterogeneity")
lines(seq(0, 100, 0.01), dchisq(seq(0, 100, 0.01), df = df),
      col = "blue", lwd = 2)
```

These are the plots that R draws for us:

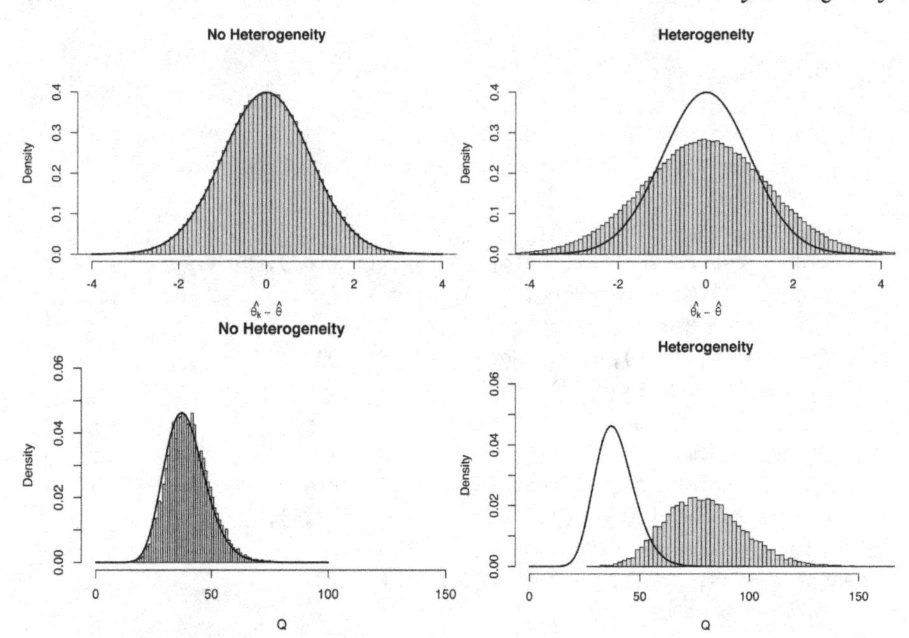

If you find the code we used to generate the plots difficult to understand, do not worry. We only used it for this simulation, and these are not plots one would produce as part of an actual meta-analysis.

Let us go through what we see in the four histograms. In the first row, we see the distribution of effect size "residuals", with and without heterogeneity. The no-heterogeneity data, as we can see, closely follows the line of the standard normal distribution we included in the plot. This is quite logical since the data was generated by `rnorm` assuming this exact distribution. The data in which we added extra heterogeneity does not follow the standard normal distribution. The dispersion of data is larger, resulting in a distribution with heavier tails.

Now, let us explore how this relates to the distribution of Q values in the second row. When there is no heterogeneity, the values of Q follow a characteristic, right-skewed χ^2 distribution. In the plot, the solid line shows the shape of a χ^2 distribution with 39 degrees of freedom (since d.f. = $K-1$, and K = 40 was used in each simulation). We see that the simulated data follows this curve pretty well. This is no great surprise. We have learned that Q follows a χ^2 distribution with $K-1$ degrees of freedom when there is no heterogeneity. Exactly this is the case in our simulated data: variation exists only due to the sampling error.

The distribution looks entirely different for our example *with* heterogeneity. The simulated data do not seem to follow the expected distribution at all. Values are shifted visibly to the right; the mean of the distribution is approximately twice as high. We can conclude that, when there is substantial between-study heterogeneity, the values of Q are considerably higher than the value of $K-1$ we expect under the assumption of no heterogeneity. This comes as no surprise, since we added extra variation to our data to simulate the presence of between-study heterogeneity.

This was a somewhat lengthy explanation, yet it may have helped us to better understand how we can exploit the statistical properties of Q. Cochran's Q can be used to *test* if the variation in a meta-analysis significantly exceeds the amount we would expect under the null hypothesis of no heterogeneity. This *test of heterogeneity* is commonly used in meta-analyses, and if you go back to Chapter 4, you will see that *{meta}* also provides us with it by default. It is often referred to as *Cochran's Q test*, but this is actually a misnomer. Cochran himself never intended Q to be used in this way (Hoaglin, 2016).

Cochran's Q is a very important statistic, mostly because other common ways to quantify heterogeneity, such as Higgins and Thompson's I^2 statistic and H^2, are based on it. We will get to these measures in the next sections. Cochran's Q is also used by some heterogeneity variance estimators to calculate τ^2, most famously by the DerSimonian-Laird estimator.

!

Problems With Q & the Q-Test

Although Q is commonly used and reported in meta-analyses, it has several flaws. Hoaglin (2016), for example, argues that the assumption of Q following a χ^2 distribution with $K-1$ degrees of freedom does not reflect Q's actual behavior in meta-analysis, and that related procedures such as the DerSimonian-Laird method may therefore be biased.

A more practical concern is that Q increases both when the number of studies K, and when the precision (i.e. the sample size of a study) increases. Therefore, Q and whether it is significant highly depends on the size of your meta-analysis, and thus its statistical power. From this follows that we should not only rely on the significance of a Q-test when assessing heterogeneity. Sometimes, meta-analysts decide whether to apply a fixed-effect or random-effects model based on the significance of the Q-test. For the reasons we stated here, this approach is highly discouraged.

5.1.2 Higgins & Thompson's I^2 Statistic

The I^2 statistic (Higgins and Thompson, 2002) is another way to quantify between-study heterogeneity, and directly based on Cochran's Q. It is defined as the percentage of variability in the effect sizes that is not caused by sampling error. I^2 draws on the assumption that Q follows a χ^2 distribution with $K-1$ degrees of freedom under the null hypothesis of no heterogeneity. It quantifies, in percent, how much the *observed* value of Q *exceeds* the *expected* Q value when there is no heterogeneity (i.e. $K-1$). The formula of I^2 looks like this:

$$I^2 = \frac{Q-(K-1)}{Q} \tag{5.3}$$

where K is the total number of studies. The value of I^2 cannot be lower than 0%, so if Q happens to be smaller than $K-1$, we simply use 0 instead of a negative value.

We can use our simulated values of Q from before to illustrate how I^2 is calculated. First, let us randomly pick the tenth simulated value in `Q_fixed`, where we assumed no heterogeneity. Then, we use the formula above to calculate I^2.

```
# Display the value of the 10th simulation of Q
Q_fixed[10]
```

```
## [1] 35.86
```

```
# Define k
k <- 40

# Calculate I^2
(Q_fixed[10] - (k-1))/Q_fixed[10]
```

```
## [1] -0.08763
```

Since the result is negative, we round up to zero, resulting in I^2 = 0%. This value tells us that zero percent of the variation in effect sizes is due to between-study heterogeneity. This is in line with the settings used for our simulation.

Now, we do the same with the tenth simulated value in `Q_random`.

```
(Q_random[10] - (k-1))/Q_random[10]
```

```
## [1] 0.5692
```

We see that the I^2 value of this simulation is approximately 50%, meaning that about half of the variation is due to between-study heterogeneity. This is also in line with our expectations since the variation in this example is based, in equal parts, on the simulated sampling error and between-study heterogeneity.

It is common to use the I^2 statistic to report the between-study heterogeneity in meta-analyses, and I^2 is included by default in the output we get from *{meta}*. The popularity of this statistic may be associated with the fact that there is a "rule of thumb" on how we can interpret it (Higgins and Thompson, 2002):

- I^2 = 25%: low heterogeneity
- I^2 = 50%: moderate heterogeneity

- $I^2 = 75\%$: substantial heterogeneity.

5.1.3 The H^2 Statistic

The H^2 statistic (Higgins and Thompson, 2002) is also derived from Cochran's Q, and similar to I^2. It describes the ratio of the observed variation, measured by Q, and the expected variance due to sampling error:

$$H^2 = \frac{Q}{K-1} \tag{5.4}$$

The computation of H^2 is a little more elegant than the one of I^2 because we do not have to artificially correct its value when Q is smaller than $K-1$. When there is no between-study heterogeneity, H^2 equals one (or smaller). Values greater than one indicate the presence of between-study heterogeneity.

Compared to I^2, it is far less common to find this statistic reported in published meta-analyses. However, H^2 is also included by default in the output of *{meta}*'s meta-analysis functions.

5.1.4 Heterogeneity Variance τ^2 & Standard Deviation τ

We already discussed the heterogeneity variance τ^2 in detail in Chapter 4.1.2. As we mentioned there, τ^2 quantifies the *variance* of the true effect sizes underlying our data. When we take the square root of τ^2, we obtain τ, which is the *standard deviation* of the true effect sizes.

A great asset of τ is that it is expressed on the same scale as the effect size metric. This means that we can interpret it in the same as one would interpret, for example, the mean and standard deviation of the sample's age in a primary study. The value of τ tells us something about the *range* of the true effect sizes.

We can, for example, calculate the 95% confidence interval of the true effect sizes by multiplying τ with 1.96, and then adding and subtracting this value from the pooled effect size. We can try this out using the `m.gen` meta-analysis we calculated in Chapter 4.2.1. Let us have a look again what the pooled effect and τ estimate in this meta-analysis were:

```
# Pooled effect
m.gen$TE.random
```

```
## [1] 0.5771
```

```
# Estimate of tau
m.gen$tau
```

```
## [1] 0.2863
```

We see that $g = 0.58$ and $\tau = 0.29$. Based on this data, we can calculate the lower and upper bound of the 95% true effect size confidence interval: $0.58 - 1.96 \times 0.29 = 0.01$ and $0.58 + 1.96 \times 0.29 = 1.15$.

"What's the Uncertainty of Our Uncertainty?" Calculation of Confidence Intervals Around τ^2

Methods to quantify the uncertainty of our between-study heterogeneity variance estimate (i.e. the confidence intervals around τ^2) remain a field of ongoing investigation. Several approaches are possible, and their adequateness depends on the type of τ^2 estimator (Chapter 4.1.2.1).

The *{meta}* package follows the recommendations of Veronikki (2016) and uses the *Q-Profile* method (Viechtbauer, 2007) for most estimators.

The Q-Profile method is based on an altered Q version , the *generalized Q-statistic* Q_{gen}. While the standard version of Q uses the pooled effect based on the fixed-effect model, Q_{gen} is based on the random-effects model. It uses the overall effect according to the random-effects model, $\hat{\mu}$, to calculate the deviates, as well as weights based on the random-effects model:

$$Q_{\text{gen}} = \sum_{k=1}^{K} w^*_k (\hat{\theta}_k - \hat{\mu})^2 \tag{5.5}$$

where w^*_k is the random-effects weight (see Chapter 4.1.2.1):

$$w^*_k = \frac{1}{s^2_k + \tau^2} \tag{5.6}$$

Q_{gen} has also been shown to follow a χ^2 distribution with $K - 1$ degrees of freedom. We can think of the generalized Q statistic as a function $Q_{\text{gen}}(\tau^2)$ which returns different values of Q_{gen} for higher or lower values of τ^2. The results of this function have a χ^2 distribution.

Since the χ^2 distribution follows a clearly predictable pattern, it is easy to determine confidence intervals with, for example, 95% coverage. We only have to get the value of χ^2 for the 2.5th and 97.5th percentile, based on its $K - 1$ degrees of freedom. In R, this can be easily done using the *quantile function* `qchisq`, for example: `qchisq(0.975, df=5)`. The Q-Profile method exploits this relationship to calculate confidence intervals around τ^2 using an iterative process (so-called "profiling"). In this approach, $Q_{\text{gen}}(\tilde{\tau}^2)$ is calculated repeatedly while increasing the value of τ^2, until the expected value of the lower and upper bound of the confidence interval based on the χ^2 distribution is reached.

The Q-Profile method can be specified in *{meta}* functions through the argument `method.tau.ci = "QP"`. This is the default setting, meaning that we do not have to add this argument manually. The only exception is when we use the DerSimonian-Laird estimator (`method.tau = "DL"`). In this case, a different method, the one by Jackson (2013), is used automatically (we can do this manually by specifying `method.tau.ci = "J"`). Usually, there is no necessity to deviate from *{meta}*'s default behavior, but it may be helpful for others to report which method has been used to calculate the confidence intervals around τ^2 in your meta-analysis.

5.2 Which Measure Should I Use?

When we assess and report heterogeneity in a meta-analysis, we need a measure which is robust, and not too heavily influenced by statistical power. Cochran's Q increases both when the number of studies increases, and when the precision (i.e. the sample size of a study) increases. Therefore, Q and whether it is significant highly depends on the size of your meta-analysis, and thus its statistical power. We should therefore not only rely on Q, and particularly the Q-test, when assessing between-study heterogeneity.

I^2, on the other hand, is not sensitive to changes in the number of studies in the analysis. It is relatively easy to interpret, and many researchers understand what it means. Generally, it is not a bad idea to include I^2 as a heterogeneity measure in our meta-analysis report, especially if we also provide a confidence interval for this statistic so that others can assess how certain the estimate is.

However, despite its common use in the literature, I^2 is not a perfect measure for heterogeneity either. It still heavily depends on the precision of the included studies (Borenstein et al., 2017; Rücker et al., 2008). As said before, I^2 is simply the percentage of variability not caused by sampling error ϵ. If our studies become increasingly large, the sampling error tends to zero, while at the same time, I^2 tends to 100%–simply because the studies have a greater sample size. *Only* relying on I^2 is therefore not a good option either. Since H^2 behaves similarly to I^2, the same caveats also apply to this statistic.

The value of τ^2 and τ, on the other hand, is insensitive to the number of studies, *and* their precision. Yet, it is often hard to interpret how relevant τ^2 is from a practical standpoint. Imagine, for example, that we found that the variance of true effect sizes in our study was $\tau^2 = 0.08$. It is often difficult for ourselves, and others, to determine if this amount of variance is meaningful or not.

Prediction intervals (PIs) are a good way to overcome this limitation (IntHout et al., 2016). Prediction intervals give us a range into which we can expect the effects of future studies to fall based on present evidence. Imagine that our prediction interval lies completely on the "positive" side favoring the intervention. This means that, despite varying effects, the intervention is expected to be beneficial in the future across the contexts we studied. If the prediction interval includes zero, we can be less sure about this, although it should be noted that broad prediction intervals are quite common.

To calculate prediction intervals around the overall effect $\hat{\mu}$, we use both the estimated between-study heterogeneity variance $\hat{\tau}^2$, as well as the standard error of the pooled effect, $SE_{\hat{\mu}}$. We sum the squared standard error and $\hat{\tau}^2$ value, and then take the square root of the result. This leaves us with the standard deviation of the prediction interval, SD_{PI}. A t distribution with $K-1$ degrees of freedom is assumed for the prediction range, which is why we multiply SD_{PI} with the 97.5th percentile value of t_{K-1}, and then add and subtract the result from $\hat{\mu}$. This gives us the 95% prediction interval of our pooled effect. The formula for 95% prediction intervals looks like this:

$$
\begin{aligned}
&\hat{\mu} \pm t_{K-1,0.975}\sqrt{SE^2_{\hat{\mu}} + \hat{\tau}^2} \\
&\hat{\mu} \pm t_{K-1,0.975}SD_{\text{PI}}
\end{aligned}
\tag{5.7}
$$

All of *{meta}*'s functions can provide us with a prediction interval around the pooled effect, but they do not do so by default. When running a meta-analysis, we have to add the argument `prediction = TRUE` so that prediction intervals appear in the output.

In sum, it is advisable to not resort to one measure only when characterizing the heterogeneity of a meta-analysis. It is recommended to at least always report I^2 (with confidence intervals), as well as prediction intervals, and interpret the results accordingly.

5.3 Assessing Heterogeneity in R

Let us see how we can use the things we learned about heterogeneity measures in practice. As an illustration, let us examine the heterogeneity of our `m.gen` meta-analysis object a little closer (we generated this object in Chapter 4.2.1).

Because the default output of metagen objects does not include prediction intervals, we have to update it first. We simply use the update.meta function, and tell it that we want prediction intervals to be printed out additionally.

```
m.gen <- update.meta(m.gen, prediction = TRUE)
```

Now we can reinspect the results:

```
m.gen
```

```
## Review:     Third Wave Psychotherapies
##
## [...]
##
## Number of studies combined: k = 18
##
##                             SMD            95%-CI    t  p-value
## Random effects model 0.5771 [ 0.3782; 0.7760] 6.12 < 0.0001
## Prediction interval         [-0.0619; 1.2162]
##
## Quantifying heterogeneity:
##  tau^2 = 0.0820 [0.0295; 0.3533]; tau = 0.2863 [0.1717; 0.5944];
##  I^2 = 62.6% [37.9%; 77.5%]; H = 1.64 [1.27; 2.11]
##
## Test of heterogeneity:
##      Q d.f. p-value
##  45.50   17  0.0002
##
## Details on meta-analytical method:
## - Inverse variance method
## - Restricted maximum-likelihood estimator for tau^2
## - Q-profile method for confidence interval of tau^2 and tau
## - Hartung-Knapp adjustment for random effects model
```

In the output, we see results for all heterogeneity measures we defined before. Let us begin with the Quantifying heterogeneity section. Here, we see that $\tau^2 = 0.08$. The confidence interval around τ^2 (0.03 - 0.35) does not contain zero, indicating that some between-study heterogeneity exists in our data. The value of τ is 0.29, meaning that the true effect sizes have an estimated standard deviation of $SD = 0.29$, expressed on the scale of the effect size metric (here, Hedges' g).

A look at the second line reveals that $I^2 = 63\%$ and that H (the square root of H^2) is 1.64. This means that more than half of the variation in our data is estimated to stem from true effect size differences. Using Higgins and Thompson's "rule of thumb", we can characterize this amount of heterogeneity as moderate to large.

Directly under the pooled effect, we see the prediction interval. It ranges from $g = -0.06$ to 1.21. This means that it is possible that some future studies will find a negative treatment effect based on present evidence. However, the interval is quite broad, meaning that very high effects are possible as well.

Lastly, we are also presented with Q and the `Test of heterogeneity`. We see that Q=45.5. This is a lot more than what we would expect based on the $K-1=17$ degrees of freedom in this analysis. Consequentially, the heterogeneity test is significant ($p < 0.001$). However, as we mentioned before, we should not base our assessment on the Q test alone, given its known deficiencies.

Reporting the Amount of Heterogeneity In Your Meta-Analysis

Here is how we could report the amount of heterogeneity we found in our example:

"The between-study heterogeneity variance was estimated at $\hat{\tau}^2 = 0.08$ (95%CI: 0.03-0.35), with an I^2 value of 63% (95%CI: 38-78%). The prediction interval ranged from $g = -0.06$ to 1.21, indicating that negative intervention effects cannot be ruled out for future studies."

So, what do we make out of these results? Overall, our indicators tell us that moderate to substantial heterogeneity is present in our data. The effects in our meta-analysis are not completely heterogeneous, but there are clearly some differences in the true effect sizes between studies.

It may therefore be a good idea to explore what causes this heterogeneity. It is possible that there are one or two studies that do not really "fit in", because they have a much higher effect size. This could have inflated the heterogeneity in our analysis, and even worse: it may have lead to an *overestimation* of the true effect. On the other hand, it is also possible that our pooled effect is influenced heavily by one study with a very large sample size reporting an unexpectedly small effect size. This could mean that the pooled effect *underestimates* the true benefits of the treatment.

To address these concerns, we will now turn to procedures which allow us to assess the robustness of our pooled results: *outlier* and *influence analyses*.

The I^2 > 50% "Guideline"

There are no iron-clad rules determining when exactly further analyses of the between-study heterogeneity are warranted. An approach that is sometimes used in practice is to check for outliers and influential cases when I^2 is greater than 50%. When this threshold is reached, we can assume at least moderate heterogeneity, and that (more than) half of the variation is due to true effect size differences.

This "rule of thumb" is somewhat arbitrary, and, knowing the problems of I^2 we discussed, in no way perfect. However, it can still be helpful from a practical perspective, because we can specify *a priori*, and in a consistent way, when we will try to get a more robust version of the pooled effect in our meta-analysis.

What should be avoided at any cost is to remove outlying and/or influential cases without any stringent rationale, just because we like the results. Such outcomes will be heavily biased by our "researcher agenda" (see Chapter 1.3), even if we did not consciously try to bend the results into a "favorable" direction.

5.4 Outliers & Influential Cases

As mentioned before, between-study heterogeneity can be caused by one or more studies with extreme effect sizes that do not quite "fit in". This may distort our pooled effect estimate, and it is a good idea to reinspect the pooled effect after such *outliers* have been removed from the analysis.

On the other hand, we also want to know if the pooled effect estimate we found is robust, meaning that it does not depend heavily on one single study. Therefore, we also want to know whether there are studies which heavily push the effect of our analysis into one direction. Such studies are called *influential cases*, and we will devote some time to this topic later in this chapter.

5.4.1 Basic Outlier Removal

There are several ways to define the effect of a study as "outlying" (Viechtbauer and Cheung, 2010). An easy, and somewhat brute force approach, is to view a study as an outlier if its confidence interval does not overlap with the confidence interval of the pooled effect. The effect size of an outlier is so *extreme* that it differs significantly from the overall effect. To detect such outliers, we can search for all studies:

- for which the *upper bound* of the 95% confidence interval is *lower* than the *lower bound* of the pooled effect confidence interval (i.e. extremely *small* effects)
- for which the *lower bound* of the 95% confidence interval is *higher* than the *upper bound* of the pooled effect confidence interval (i.e. extremely *large* effects).

The idea behind this method is quite straightforward. Studies with a high sampling error are expected to deviate substantially from the pooled effect. However, because the confidence interval of such studies will also be large, this increases the likelihood that the confidence intervals will overlap with the one of the pooled effect. Yet, if a study has a *low* standard error and *still* (unexpectedly) deviates substantially from the pooled effect, there is a good chance that the confidence intervals will not overlap, and that the study is classified as an outlier.

The *{dmetar}* package contains a function called `find.outliers`, which implements this simple outlier removal algorithm. It searches for outlying studies in a *{meta}* object, removes them, and then recalculates the results.

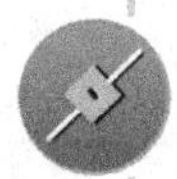

The "find.outliers" Function

The `find.outliers` function is included in the *{dmetar}* package. Once *{dmetar}* is installed and loaded on your computer, the function is ready to be used. If you did not install *{dmetar}*, follow these instructions:

1. Access the source code of the function online[a].
2. Let R "learn" the function by copying and pasting the source code in its entirety into the console (bottom left pane of R Studio), and then hit "Enter".
3. Make sure that the *{meta}* and *{metafor}* package is installed and loaded.

[a] https://raw.githubusercontent.com/MathiasHarrer/dmetar/master/R/find.outliers.R

The `find.outliers` function only needs an object created by a *{meta}* meta-analysis function as input. Let us see the what results we get for our `m.gen` object.

```
find.outliers(m.gen)
```

```
## Identified outliers (random-effects model)
## ----------------------------------------
```

```
## "DanitzOrsillo", "Shapiro et al."
##
## Results with outliers removed
## ----------------------------------
## Review:  Third Wave Psychotherapies
##                         SMD          95%-CI %W(random) exclude
## Call et al.             0.70 [ 0.19; 1.22]        4.4
## Cavanagh et al.         0.35 [-0.03; 0.73]        6.9
## DanitzOrsillo           1.79 [ 1.11; 2.46]        0.0       *
## de Vibe et al.          0.18 [-0.04; 0.41]       13.1
## Frazier et al.          0.42 [ 0.13; 0.70]       10.4
## Frogeli et al.          0.63 [ 0.24; 1.01]        6.9
## Gallego et al.          0.72 [ 0.28; 1.16]        5.6
## Hazlett-Stevens & Oren 0.52 [ 0.11; 0.94]        6.2
## Hintz et al.            0.28 [-0.04; 0.61]        8.6
## Kang et al.             1.27 [ 0.61; 1.93]        2.8
## Kuhlmann et al.         0.10 [-0.27; 0.48]        7.0
## Lever Taylor et al.     0.38 [-0.06; 0.84]        5.4
## Phang et al.            0.54 [ 0.06; 1.01]        4.9
## Rasanen et al.          0.42 [-0.07; 0.93]        4.5
## Ratanasiripong          0.51 [-0.17; 1.20]        2.6
## Shapiro et al.          1.47 [ 0.86; 2.09]        0.0       *
## Song & Lindquist        0.61 [ 0.16; 1.05]        5.6
## Warnecke et al.         0.60 [ 0.11; 1.08]        4.8
##
## Number of studies combined: k = 16
##
##                           SMD            95%-CI    t  p-value
## Random effects model 0.4528 [0.3257; 0.5800] 7.59 < 0.0001
## Prediction interval         [0.1693; 0.7363]
##
## Quantifying heterogeneity:
##  tau^2 = 0.0139 [0.0000; 0.1032]; tau = 0.1180 [0.0000; 0.3213];
##  I^2 = 24.8% [0.0%; 58.7%]; H = 1.15 [1.00; 1.56]
##
## Test of heterogeneity:
##      Q d.f. p-value
##  19.95   15  0.1739
##
## [...]
```

We see that the `find.outliers` function has detected two outliers, "DanitzOrsillo" and "Shapiro et al.". The function has also automatically rerun our analysis while excluding the identified studies. In the column displaying the random-effects weight of each study, `%W(random)`, we see that the weight of the outlying studies has been set to zero, thus removing them from the analysis.

Based on the output, we see that the I^2 heterogeneity shrinks considerably when the two studies are excluded, from $I^2 = 63\%$ to 25%. The confidence interval around τ^2 now also includes zero, and the Q-test of heterogeneity is not significant anymore. Consequentially, the prediction interval of our estimate has also narrowed. Now, it only contains positive values, providing much more certainty of the robustness of the pooled effect across future studies.

5.4.2 Influence Analysis

We have now learned a basic way to detect and remove outliers in meta-analyses. However, it is not only extreme effect sizes which can cause concerns regarding the robustness of the pooled effect. Some studies, even if their effect size is not particularly high or low, can still exert a very high *influence* on our overall results. For example, it could be that we find an overall effect in our meta-analysis, but that its significance depends on a single large study. This would mean that the pooled effect is not statistically significant anymore once the influential study is removed. Such information is very important if we want to communicate to the public how robust our results are.

Outlying and influential studies have an overlapping but slightly different meaning. Outliers are defined through the magnitude of their effect but do not necessarily need to have a substantial impact on the results of our meta-analysis. It is perfectly possible that removal of an outlier as defined before neither changes the average effect size, nor the heterogeneity in our data substantially. Influential cases, on the other hand, are those studies which–by definition–have a large impact on the pooled effect or heterogeneity, regardless of how high or low the effect is. This does not mean, of course, that a study with an extreme effect size cannot be an influential case. In fact, outliers are often also influential, as our example in the last chapter illustrated. But they do not *have* to be.

There are several techniques to identify influential studies, and they are a little more sophisticated than the basic outlier removal we discussed previously. They are based on the *leave-one-out* method. In this approach, we recalculate the results of our meta-analysis K times, each time *leaving out* one study. Based on this data, we can calculate different *influence diagnostics*. Influence diagnostics allow us to detect the studies which influence the overall estimate of our meta-analysis the most, and let us assess if this large influence distorts our pooled effect (Viechtbauer and Cheung, 2010).

The *{dmetar}* package contains a function called `InfluenceAnalysis`, which allows us to calculate these various influence diagnostics using one function. The function can be used for any type of meta-analysis object created by *{meta}* functions.

The "InfluenceAnalysis" function

The `InfluenceAnalysis` function is included in the *{dmetar}* package. Once *{dmetar}* is installed and loaded on your computer, the function is ready to be used. If you did not install *{dmetar}*, follow these instructions:

1. Access the source code of the function online[a].
2. Let R "learn" the function by copying and pasting the source code in its entirety into the console (bottom left pane of R Studio), and then hit "Enter".
3. Make sure that the *{meta}*, *{metafor}*, *{ggplot2}* and *{gridExtra}* package is installed and loaded.

[a] `https://raw.githubusercontent.com/MathiasHarrer/dmetar/master/R/influence.analysis.R`

Using the `InfluenceAnalysis` function is relatively straightforward. We only have to specify the name of the meta-analysis object for which we want to conduct the influence analysis. Here, we again use the `m.gen` object. Because `InfluenceAnalysis` uses the fixed-effect model by default, we also have to set `random = TRUE`, so that the random-effects model will be used. The function can also take other arguments, which primarily control the type of plots generated by the function. Those arguments are detailed in the function documentation.

We save the results of the function in an object called `m.gen.inf`.

```
m.gen.inf <- InfluenceAnalysis(m.gen, random = TRUE)
```

The `InfluenceAnalysis` function creates four influence diagnostic plots: a *Baujat* plot, *influence diagnostics* according to Viechtbauer and Cheung (2010), and the leave-one-out meta-analysis results, sorted by effect size and I^2 value. We can open each of these plots individually using the `plot` function. Let us go through them one after another.

5.4.2.1 Baujat Plot

A Baujat plot can be printed using the `plot` function and by specifying `"baujat"` in the second argument:

```
plot(m.gen.inf, "baujat")
```

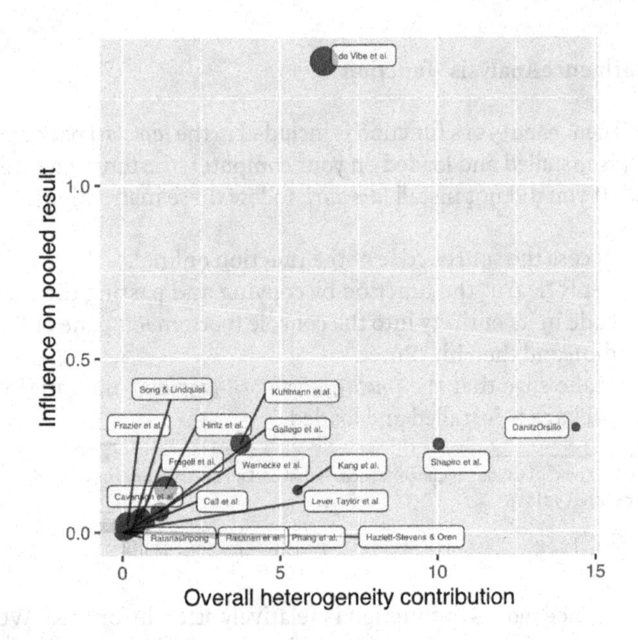

Baujat plots (Baujat et al., 2002) are diagnostic plots to detect studies which overly contribute to the heterogeneity in a meta-analysis. The plot shows the contribution of each study to the overall *heterogeneity* (as measured by Cochran's Q) on the *horizontal* axis, and its *influence* on the *pooled effect size* on the *vertical* axis. This "influence" value is determined through the leave-one-out method, and expresses the standardized difference of the overall effect when the study is included in the meta-analysis, versus when it is not included.

Studies on the right side of the plot can be regarded as potentially relevant cases since they contribute heavily to the overall heterogeneity in our meta-analysis. Studies in the upper right corner of the plot may be particularly influential since they have a large impact on both the estimated heterogeneity, and the pooled effect.

As you may have recognized, the two studies we find on the right side of the plot are the ones we already detected before ("DanitzOrsillo" and "Shapiro et al."). These studies do not have a large impact on the overall results (presumably because they have a small sample size), but they do add substantially to the heterogeneity we find in the meta-analysis.

5.4.2.2 Influence Diagnostics

The next plot contains several influence diagnostics for each of our studies. These can be plotted using this code:

```
plot(m.gen.inf, "influence")
```

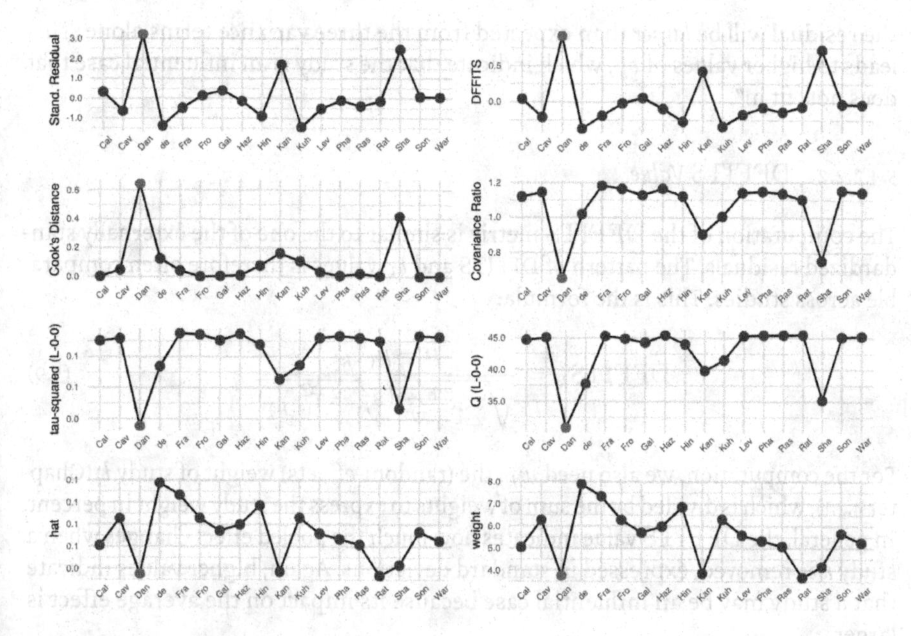

We see that the plot displays, for each study, the value of different influence measures. These measures are used to characterize which studies fit well into our meta-analysis model, and which do not. To understand what the diagnostics mean, let us briefly go through them from left to right, top to bottom.

5.4.2.2.1 *Externally Standardized Residuals*

The first plot displays the externally standardized residual of each study. As is says in the name, these residuals are the deviation of each observed effect size $\hat{\theta}_k$ from the pooled effect size. The residuals are standardized, and we use an "external" estimate of the pooled effect without the study to calculate the deviations. The "external" pooled effect $\hat{\mu}_{(-k)}$ is obtained by calculating the overall effect without study k, along with the principles of the leave-one-out method. The resulting residual is then standardized by (1) the *variance* of the external effect (i.e. the squared standard error of $\hat{\mu}_{(-k)}$), (2) the τ^2 estimate of the external pooled effect, and (3) the variance of k.

$$t_k = \frac{\hat{\theta}_k - \hat{\mu}_{(-k)}}{\sqrt{\mathrm{Var}(\hat{\mu}_{(-k)}) + \hat{\tau}^2_{(-k)} + s^2_k}} \tag{5.8}$$

Assuming that a study k fits well into the meta-analysis, the three terms in the denominator capture the sources of variability which determine how much an effect size differs from the average effect. These sources of variability are the sampling error of k, the variance of true effect sizes, and the imprecision in our pooled effect size estimate. If a study does *not* fit into the overall population, we can assume that the residual will be *larger* than expected from the three variance terms alone. This leads to higher values of t_k, which indicate that the study is an influential case that does not "fit in".

5.4.2.2.2 DFFITS *Value*

The computation of the DFFITS metric is similar to the one of the externally standardized residuals. The pattern of DFFITS and t_k values is therefore often comparable across studies. This is the formula:

$$\text{DFFITS}_k = \frac{\hat{\mu} - \hat{\mu}_{(-k)}}{\sqrt{\frac{w_k}{\sum_{k=1}^{K} w_k}(s_k^2 + \hat{\tau}^2_{(-k)})}} \tag{5.9}$$

For the computation, we also need w_k, the (random-effects) weight of study k (Chapter 4.1.1), which is divided by the sum of weights to express the study weight in percent. In general, the DFFITS value indicates how much the pooled effect changes when a study k is removed, expressed in standard deviations. Again, higher values indicate that a study may be an influential case because its impact on the average effect is larger.

5.4.2.2.3 *Cook's Distance*

The Cook's distance value D_k of a study can be calculated by a formula very similar to the one of the DFFITS value, with the largest difference being that for D_k, the difference of the pooled effect with and without k is *squared*. This results in D_k only taking positive values. The pattern across studies, however, is often similar to the DFFITS value. Here is the formula:

$$D_k = \frac{(\hat{\mu} - \hat{\mu}_{(-k)})^2}{\sqrt{s_k^2 + \hat{\tau}^2}} \tag{5.10}$$

5.4.2.2.4 *Covariance Ratio*

The covariance ratio of a study k can be calculated by dividing the variance of the pooled effect (i.e. its squared standard error) without k by the variance of the initial average effect.

$$\text{CovRatio}_k = \frac{\text{Var}(\hat{\mu}_{(-k)})}{\text{Var}(\hat{\mu})} \tag{5.11}$$

A CovRatio_k value below 1 indicates that removing study k results in a more precise estimate of the pooled effect size $\hat{\mu}$.

5.4.2.2.5 *Leave-One-Out τ^2 and Q Values*

The values in this row are quite easy to interpret: they simply display the estimated heterogeneity as measured by τ^2 and Cochran's Q, if study k is removed. Lower values of Q, but particularly of τ^2 are desirable, since this indicates lower heterogeneity.

5.4.2.2.6 *Hat Value and Study Weight*

In the last row, we see the study weight and hat value of each study. We already covered the calculation and meaning of study weights extensively in Chapter 4.1.1, so this measure does not need much more explanation. The hat value, on the other hand, is simply another metric that is equivalent to the study weight. The pattern of the hat values and weights will therefore be identical in our influence analyses.

All of these metrics provide us with a value which, if extreme, indicates that the study is an influential case, and may negatively affect the robustness of our pooled result. However, it is less clear when this point is reached. There is no strict rule which DFFITS, Cook's distance or standardized residual value is *too* high. It is always necessary to evaluate the results of the influence analysis in the context of the research question to determine if it is indicated to remove a study.

Yet, there are a few helpful "rules of thumb" which can guide our decision. The `InfluenceAnalysis` function regards a study as an influential case if one of these conditions is fulfilled:

$$\text{DFFITS}_k > 3\sqrt{\frac{1}{k-1}} \tag{5.12}$$

$$D_k > 0.45 \tag{5.13}$$

$$\text{hat}_k > 3\frac{1}{k}. \tag{5.14}$$

Studies determined to be influential are displayed in red in the plot generated by the `InfluenceAnalysis` function.

In our example, this is only the case for "Dan", the "DanitzOrsillo" study. Yet, while only this study was defined as influential, there are actually *two* spikes in most plots. We could also decide to define "Sha" (Shapiro et al.) as an influential case because the values of this study are very extreme too.

We found that the studies "DanitzOrsillo" and "Shapiro et al." might be influential. This is an interesting finding, as we selected the same studies based on the Baujat plot, and when only looking at statistical outliers. This further corroborates that the two studies could have distorted our pooled effect estimate and cause parts of the between-study heterogeneity we found in our initial meta-analysis.

5.4.2.3 Leave-One-Out Meta-Analysis Results

Lastly, we can also plot the overall effect and I^2 heterogeneity of all meta-analyses that were conducted using the leave-one-out method. We can print two *forest plots* (a type of plot we will get to know better in Chapter 6.2), one sorted by the pooled effect size, and the other by the I^2 value of the leave-one-out meta-analyses. The code to produce the plots looks like this:

```
plot(m.gen.inf, "es")
plot(m.gen.inf, "i2")
```

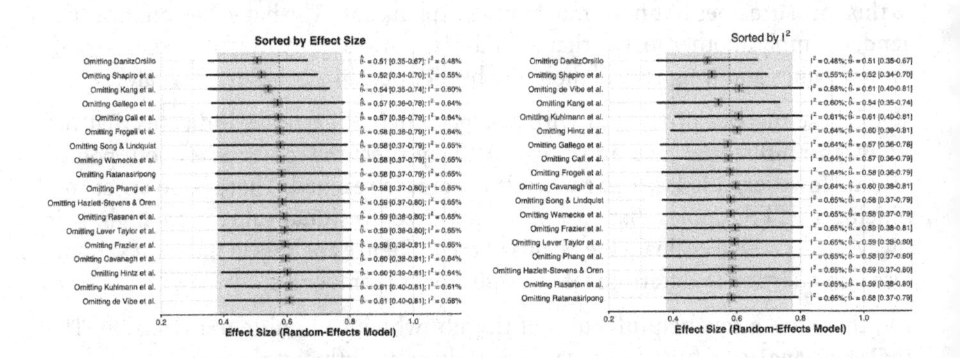

In these two forest plots, we see the recalculated pooled effects, with one study omitted each time. In both plots, there is a shaded area with a dashed line in its center. This represents the 95% confidence interval of the original pooled effect size, and the estimated pooled effect itself.

The first plot is ordered by effect size (low to high). Here, we see how the overall effect estimate changes when different studies are removed. Since the two outlying and influential studies "DanitzOrsillo" and "Shapiro et al." have very high effect sizes, we find that the overall effect is smallest when they are removed.

The second plot is ordered by heterogeneity (low to high), as measured by I^2. This plot illustrates that the lowest I^2 heterogeneity is reached by omitting the studies "DanitzOrsillo" and "Shapiro et al.". This corroborates our finding that these two studies were the main "culprits" for the between-study heterogeneity we found in the meta-analysis.

All in all, the results of our outlier and influence analysis in this example point into the same direction. There are two studies which are likely influential outliers. These two studies may distort the effect size estimate, as well as its precision. We should therefore also conduct and report the results of a sensitivity analysis in which both studies are excluded.

5.4.3 GOSH Plot Analysis

In the previous chapter, we explored the robustness of our meta-analysis using influence analyses based on the leave-one-out method. Another way to explore patterns of heterogeneity in our data are so-called *Graphic Display of Heterogeneity* (GOSH) plots (Olkin et al., 2012). For those plots, we fit the same meta-analysis model to *all possible subsets* of our included studies. In contrast to the leave-one-out method, we therefore not only fit K models, but a model for all 2^{k-1} possible study combinations. This means that creating GOSH plots can become quite computationally expensive when the total number of studies is large. The R implementation we cover here therefore only fits a maximum of 1 million randomly selected models.

Once the models are calculated, we can plot them, displaying the pooled effect size on the x-axis and the between-study heterogeneity on the y-axis. This allows us to look for specific patterns, for example, clusters with different effect sizes and amounts of heterogeneity. A GOSH plot with several distinct clusters indicates that there might be more than one effect size "population" in our data, warranting a subgroup analysis. If the effect sizes in our sample are homogeneous, on the other hand, the GOSH plot displays a roughly symmetric, homogeneous distribution.

To generate GOSH plots, we can use the `gosh` function in the *{metafor}* package. If you have not installed the package yet, do so now and then load it from the library.

```
library(metafor)
```

Let us generate a GOSH plot for our `m.gen` meta-analysis object. To do that, we have to "transform" this object created by the *{meta}* package into a *{metafor}* meta-analysis object first, because only those can be used by the `gosh` function.

The function used to perform a meta-analysis in *{metafor}* is called `rma`. It is not very complicated to translate a *{meta}* object to a `rma` meta-analysis. We only have to provide the function with the effect size (`TE`), Standard Error (`seTE`), and between-study heterogeneity estimator (`method.tau`) stored in `m.gen`. We can specify that the Knapp-Hartung adjustment should be used by specifying the argument `test = "knha"`. We save the newly generated *{metafor}*-based meta-analysis under the name `m.rma`.

```
m.rma <- rma(yi = m.gen$TE,
             sei = m.gen$seTE,
             method = m.gen$method.tau,
             test = "knha")
```

Please note that if you used the fixed-effect model in *{meta}*, it is not possible to simply copy `method.tau` to your `rma` call. Instead, this requires one to set the `method` argument to `"FE"` in `rma`.

We can then use the `m.rma` object to generate the GOSH plot. Depending on the number of studies in your analysis, this can take some time, even up to a few hours. We save the results as `res.gosh`.

```
res.gosh <- gosh(m.rma)
```

We can then display the plot by plugging the `res.gosh` object into the `plot` function. The additional `alpha` argument controls how transparent the dots in the plot are, with 1 indicating that they are completely opaque. Because there are many, many data points in the graph, it makes sense to use a small alpha value to make it clearer where the values "pile up".

```
plot(res.gosh, alpha = 0.01)
```

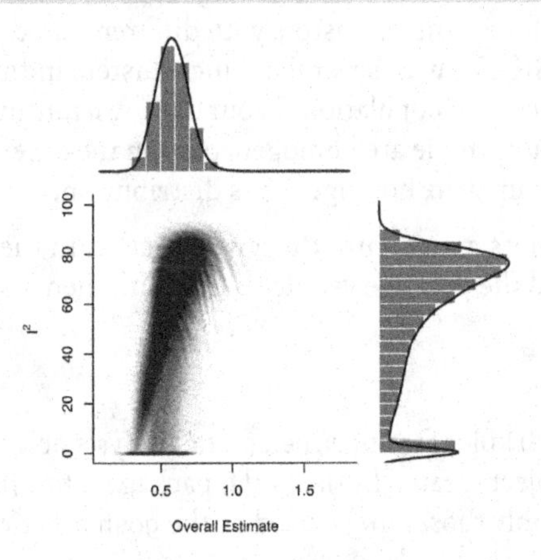

We see an interesting pattern in our data: while most values are concentrated in a cluster with relatively high effects and high heterogeneity, the distribution of I^2 values is heavily right-skewed and bi-modal. There seem to be some study combinations for which the estimated heterogeneity is much lower, but where the pooled effect size is also smaller, resulting in a shape with a "comet-like" tail.

Having seen the effect size–heterogeneity pattern in our data, the really important question is: which studies cause this shape? To answer this question, we can use the `gosh.diagnostics` function. This function uses three clustering or *unsupervised machine learning* algorithms to detect clusters in the GOSH plot data. Based on the identified clusters, the function automatically determines which studies contribute most to each cluster. If we find, for example, that one or several studies are over-represented in a cluster with high heterogeneity, this indicates that these studies, alone or in combination, may *cause* the high heterogeneity.

The "gosh.diagnostics" function

The `gosh.diagnostics` function is included in the *{dmetar}* package. Once *{dmetar}* is installed and loaded on your computer, the function is ready to be used. If you did not install *{dmetar}*, follow these instructions:

1. Access the source code of the function online[a].
2. Let R "learn" the function by copying and pasting the source code in its entirety into the console (bottom left pane of R Studio), and then hit "Enter".
3. Make sure that the *{gridExtra}*, *{ggplot2}*, *{fpc}* and *{mclust}* package are installed and loaded.

[a] https://raw.githubusercontent.com/MathiasHarrer/dmetar/master/R/gosh.diagnostics.R

The `gosh.diagnostics` function uses three cluster algorithms to detect patterns in our data: the k-means algorithm (Hartigan and Wong, 1979), *density reachability and connectivity clustering*, or DBSCAN (Schubert et al., 2017) and *gaussian mixture models* (Fraley and Raftery, 2002). It is possible to tune some of the parameters of these algorithms. In the arguments `km.params`, `db.params` and `gmm.params`, we can add a list element which contains specifications controlling the behavior of each algorithm. In our example, we will tweak a few details of the k-means and DBSCAN algorithm. We specify that the k-means algorithm should search for two clusters ("centers") in our data. In `db.params`, we change the eps, or ϵ value used by DBSCAN. We also specify the `MinPts` value, which determines the minimum number of points needed for each cluster.

You can learn more about the parameters of the algorithms in the `gosh.diagnostics` documentation. There is no clear rule when which parameter specification works best, so it can be helpful to tweak details about each algorithm several times and see how this affects the results.

The code for our `gosh.diagnostics` call looks like this:

```
res.gosh.diag <- gosh.diagnostics(res.gosh,
                                   km.params = list(centers = 2),
                                   db.params = list(eps = 0.08,
                                                    MinPts = 50))
res.gosh.diag
```

```
## GOSH Diagnostics
## ================================
##
```

```
##   - Number of K-means clusters detected: 2
##   - Number of DBSCAN clusters detected: 4
##   - Number of GMM clusters detected: 7
##
##  Identification of potential outliers
##  ---------------------------------
##
##   - K-means: Study 3, Study 16
##   - DBSCAN: Study 3, Study 4, Study 16
##   - Gaussian Mixture Model: Study 3, Study 4, Study 16
```

In the output, we see the number of clusters that each algorithm has detected. Because each approach uses a different mathematical strategy to segment the data, it is normal that the number of clusters is not identical. In the `Identification of potential outliers` section, we see that the procedure was able to identify three studies with a large impact on the cluster make-up: study 3, study 4 and study 16.

We can also plot the `gosh.diagnostics` object to inspect the results a little closer.

```
plot(res.gosh.diag)
```

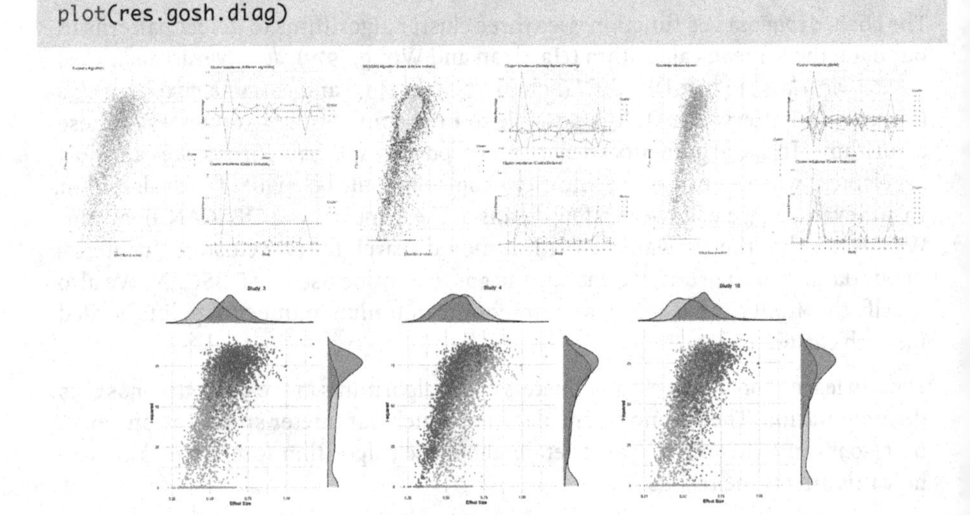

This produces several plots. The first three plots display the clustering solution found by each algorithm and the amount of cluster imbalance pertaining to each study in each cluster. Based on this information, a Cook's distance value is calculated for each study, which is used to determine if a study might have a large impact on the detected cluster (and may therefore be an influential case).

The other plots show a GOSH plot again, but there are now shaded points which represent the analyses in which a selected study was included. For example, we see that nearly all results in which study 3 was included are part of a cluster with high heterogeneity values and higher effect sizes. Results in which study 4 was included vary in their heterogeneity, but generally show a somewhat *smaller* average effect.

Results in which study 16 was included are similar to the ones found for study 3, but a little more dispersed.

Let us see what happens if we rerun the meta-analysis while removing the three studies that the `gosh.diagnostics` function has identified.

```
update.meta(m.gen, exclude = c(3, 4, 16))
```

```
## Review:     Third Wave Psychotherapies
##                            SMD            95%-CI %W(random) exclude
## Call et al.             0.7091 [ 0.1979; 1.2203]        4.6
## Cavanagh et al.         0.3549 [-0.0300; 0.7397]        8.1
## DanitzOrsillo           1.7912 [ 1.1139; 2.4685]        0.0       *
## de Vibe et al.          0.1825 [-0.0484; 0.4133]        0.0       *
## Frazier et al.          0.4219 [ 0.1380; 0.7057]       14.8
## Frogeli et al.          0.6300 [ 0.2458; 1.0142]        8.1
## Gallego et al.          0.7249 [ 0.2846; 1.1652]        6.2
## Hazlett-Stevens & Oren 0.5287 [ 0.1162; 0.9412]        7.0
## Hintz et al.            0.2840 [-0.0453; 0.6133]       11.0
## Kang et al.             1.2751 [ 0.6142; 1.9360]        2.7
## Kuhlmann et al.         0.1036 [-0.2781; 0.4853]        8.2
## Lever Taylor et al.     0.3884 [-0.0639; 0.8407]        5.8
## Phang et al.            0.5407 [ 0.0619; 1.0196]        5.2
## Rasanen et al.          0.4262 [-0.0794; 0.9317]        4.7
## Ratanasiripong          0.5154 [-0.1731; 1.2039]        2.5
## Shapiro et al.          1.4797 [ 0.8618; 2.0977]        0.0       *
## Song & Lindquist        0.6126 [ 0.1683; 1.0569]        6.1
## Warnecke et al.         0.6000 [ 0.1120; 1.0880]        5.0
##
## Number of studies combined: k = 15
##
##                           SMD            95%-CI    t  p-value
## Random effects model 0.4819 [0.3595; 0.6043] 8.44 < 0.0001
## Prediction interval          [0.3586; 0.6053]
##
## Quantifying heterogeneity:
##  tau^2 < 0.0001 [0.0000; 0.0955]; tau = 0.0012 [0.0000; 0.3091];
##  I^2 = 4.6% [0.0%; 55.7%]; H = 1.02 [1.00; 1.50]
##
## Test of heterogeneity:
##      Q d.f. p-value
##  14.67   14  0.4011
## [...]
```

We see that studies number 3 and 16 are "DanitzOrsillo" and "Shapiro et al.". These two studies were also found to be influential in previous analyses. Study 4 is the one by "de Vibe". This study does not have a particularly extreme effect size, but the narrow confidence intervals indicate that it has a *high weight*, despite its observed effect size being smaller than the average. This could explain why this study is also influential.

We see that removing the three studies has a large impact on the estimated heterogeneity. The value of τ^2 nearly drops to zero, and the I^2 value is also very low, indicating that only 4.6% of the variability in effect sizes is due to true effect size differences. The pooled effect of $g = 0.48$ is somewhat smaller than our initial estimate $g = 0.58$, but still within the same orders of magnitude. Overall, this indicates that the average effect we initially calculated is not *too* heavily biased by outliers and influential studies.

Reporting the Results of Influence Analyses

Let us assume we determined that "DanitzOrsillo", "de Vibe et al.", and "Shapiro et al." are influential studies in our meta-analysis. In this case, it makes sense to also report the results of a sensitivity analysis in which these studies are excluded.

To make it easy for readers to see the changes associated with removing the influential studies, we can create a table in which both the original results, as well as the results of the sensitivity analysis are displayed. This table should at least include the pooled effect, its confidence interval and p-value, as well as a few measures of heterogeneity, such as prediction intervals and the I^2 statistic (as well as the confidence interval thereof).

It is also important to specify which studies were removed as influential cases, so that others understand on which data the new results are based. Below is an example of how such a table looks like for our `m.gen` meta-analysis from before:

	g	95%CI	p	95%PI	I^2	95%CI
Main Analysis	0.58	0.38-0.78	<0.001	-0.06-1.22	0.63	0.38-0.78
Influential Cases Removed[1]	0.48	0.36-0.60	<0.001	0.36-0.61	0.05	0.00-0.56

[1] Removed as outliers: DanitzOrsillo, de Vibe, Shapiro.

This type of table is very convenient because we can also add further rows with results of other sensitivity analyses. For example, if we conduct an analysis in which only studies with a low risk of bias (Chapter 1.4.5) were considered, we could report the results in a third row.

5.5 Questions & Answers

Test your knowledge!

1. Why is it important to examine the between-study heterogeneity of a meta-analysis?
2. Can you name the two types of heterogeneity? Which one is relevant in the context of calculating a meta-analysis?
3. Why is the *significance* of Cochran's Q not a sufficient measure of between-study heterogeneity?
4. What are the advantages of using prediction intervals to express the amount of heterogeneity in a meta-analysis?
5. What is the difference between statistical outliers and influential studies?
6. For what can GOSH plots be used?

Answers to these questions are listed in Appendix A at the end of this book.

5.6 Summary

- In meta-analyses, we do not only have to pay attention to the pooled effect size but also to the *heterogeneity* of the data on which this average effect is based. The overall effect does not capture that the true effects in some studies may differ substantially from our point estimate.
- Cochran's Q is commonly used to quantify the variability in our data. Because we know that Q follows a χ^2 distribution, this measure allows us to detect if more variation is present than what can be expected based on sampling error alone. This *excess variability* represents true differences in the effect sizes of studies.
- A statistical test of Q, however, heavily depends on the type of data at hand. We should not only rely on Q to assess the amount of heterogeneity. There are other measures, such as I^2, τ or prediction intervals, which may be used additionally.

- The average effect in a meta-analysis can be biased when there are *outliers* in our data. Outliers do not always have a large impact on the results of a meta-analysis. But when they do, we speak of *influential cases*.
- There are various methods to identify outlying and influential cases. If such studies are detected, it is advisable to recalculate our meta-analysis without them to see if this changes the interpretation of our results.

6

Forest Plots

In the last chapters, we learned how we can pool effect sizes in R, and how to assess the heterogeneity in a meta-analysis. We now come to a somewhat more pleasant part of meta-analyses, in which we visualize the results we obtained in previous steps.

The most common way to visualize meta-analyses is through *forest plots*. Such plots provide a graphical display of the observed effect, confidence interval, and usually also the weight of each study. They also display the pooled effect we have calculated in a meta-analysis. Overall, this allows others to quickly examine the precision and spread of the included studies, and how the pooled effect relates to the observed effect sizes.

The *{meta}* package has an in-built function which makes it very easy to produce beautiful forest plots directly in R. The function has a broad functionality and allows one to change the appearance of the plot as desired. This forest plot function, and how we can use it in practice, will be the main focus of this chapter. Furthermore, we will also briefly discuss an alternative approach to visualize the results of a meta-analysis.

6.1 What Is a Forest Plot?

Figure 6.1 shows the main components of a forest plot. On the left side, forest plots display the name of each study included in the meta-analysis. For each study, a graphical representation of the effect size is provided, usually in the center of the plot. This visualization shows the point estimate of a study on the x-axis. This point estimate is supplemented by a line, which represents the range of the confidence interval calculated for the observed effect size. Usually, the point estimate is surrounded by a square. The size of this square is determined by the weight (Chapter 4.1.1) of the effect size: studies with a larger weight are given a larger square, while studies with a lower weight have a smaller square. Conventionally, a forest plot should also contain the effect size data that was used to perform the meta-analysis. This provides others with the data needed to replicate our results.

At the bottom of the plot, a diamond shape represents the average effect. The length of the diamond symbolizes the confidence interval of the pooled result on the x-axis. Typically, forest plots also include a vertical *reference line*, which indicates the point

DOI: 10.1201/9781003107347-6

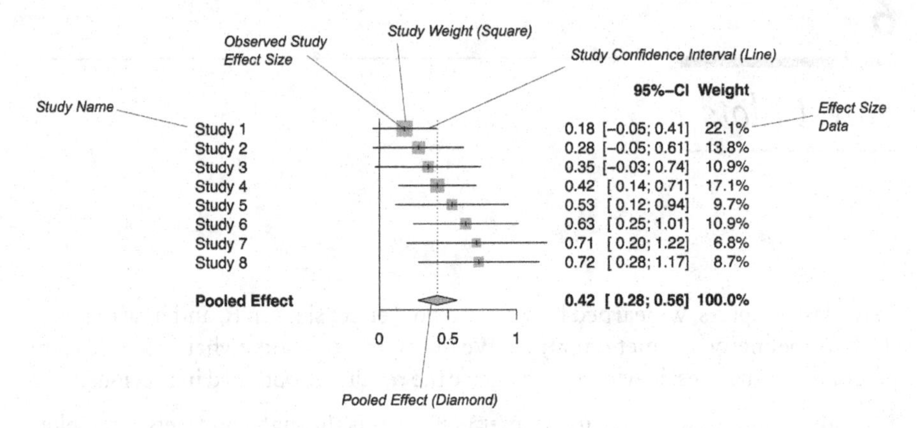

FIGURE 6.1: Key elements of a forest plot.

on the x-axis equal to no effect. As we will see in the coming examples, forest plots can be enhanced by also displaying, for example, a heterogeneity measure such as I^2 or τ^2.

Effect size and confidence intervals in forest plots are usually displayed on a linear scale. Yet, when the summary measure is a *ratio* (such as odds ratios or risk ratios), it is common to use a *logarithmic* scale on the x-axis instead. This means that values around 1 are more closely together than values which are much lower or higher than 1. This makes sense for ratios since these effect size metrics cannot be interpreted in a "linear" fashion (i.e. the "opposite" of RR = 0.50 is 2, not 1.5; see Chapter 3.3.2). The reference line for such effect sizes is usually 1, which indicates no effect.

6.2 Forest Plots in R

We can produce a forest plot for any type of *{meta}* meta-analysis object (e.g. results of `metagen`, `metacont`, or `metabin`) using the `forest.meta` function. We simply have to provide `forest.meta` with our *{meta}* object, and a plot will be created. Usually, these forest plots already look very good by default, but the function also has countless additional arguments to further adapt the appearance. All of these arguments are described in the function documentation (which can be accessed by running `?forest.meta`). Here is a list of the more important ones:

- `sortvar`. The variable in the meta-analysis data set by which studies are sorted in the forest plot. If we want to order the results by effect size, for example, we can use the code `sortvar = TE`.

- `comb.fixed`. Logical, indicating if the fixed-effect model estimate should be included in the plot.
- `comb.random`. Logical, indicating if the random-effects model estimate should be included in the plot.
- `text.fixed`. The *label* for the pooled effect according to the fixed-effect model. By default, `"Fixed effect model"` is printed.
- `text.random`. The label for the pooled effect according to the random-effects model. By default, `"Random effects model"` is printed.
- `prediction`. Logical, indicating if the prediction interval should be added to the plot.
- `label.left` and `label.right`. Label added to the left and right side of the forest plot. This can be used to specify that, for example, effects on this side favor the treatment (e.g. `label.left = "Favors treatment"`).
- `smlab`. A label displayed on top of the plot. This can be used to show which effect size metric was used.
- `xlim`. The limits of the x-axis, or the character "s" to produce symmetric forest plots. This argument is particularly relevant when your results deviate substantially from zero, or if you also want to have outliers depicted. If we want that the x-axis ranges from 0 to 2, for example, the code is `xlim = c(0,2)`.
- `ref`. The reference line in the plot. Depending on the summary measure we used, this is either 0 or 1 by default.
- `leftcols` and `rightcols`. Here, you can specify which variables should be displayed on the left and right side of the forest plot. There are a few in-built elements that the function uses by default. For example, `"studlab"` stands for the study label, `"effect"` for the observed effect size, and `effect.ci` for both the effect size and its confidence interval. It is also possible to add user-defined columns, as long as these were included in the `data.frame` we initially provided to the *{meta}* function. In this case, we only have to add the name of the column as a character string.
- `leftlabs` and `rightlabs`. The labels that should be used for the columns displayed to the left and right of the forest plot.
- `print.I2` and `print.I2.ci`. Logical, indicating if the I^2 value and its confidence interval should be printed. This is `TRUE` by default.
- `print.tau2` and `print.tau`. Logical, indicating if the τ^2 and τ value should be printed. The value of τ^2 is printed by default.
- `col.square`, `col.diamond` and `col.predict`. The color (e.g. `"blue"`) of the square, diamond and prediction interval, respectively.

Time to generate our first forest plot. In this example, we plot the `m.gen` object that we also used in the previous examples. We sort the studies in the forest plot by effect size, add a prediction interval, and user-defined labels to the left. The `forest.meta` function prints the τ^2 value by default, which we do not want here, so we set `print.tau2` to `FALSE`.

This is how our code looks in the end:

```
forest.meta(m.gen,
            sortvar = TE,
            predict = TRUE,
            print.tau2 = FALSE,
            leftlabs = c("Author", "g", "SE"))
```

Author	g	SE	Standardised Mean Difference	SMD	95%-CI	Weight
Kuhlmann et al.	0.10	0.1947		0.10	[-0.28; 0.49]	6.3%
de Vibe et al.	0.18	0.1178		0.18	[-0.05; 0.41]	7.9%
Hintz et al.	0.28	0.1680		0.28	[-0.05; 0.61]	6.9%
Cavanagh et al.	0.35	0.1964		0.35	[-0.03; 0.74]	6.3%
Lever Taylor et al.	0.39	0.2308		0.39	[-0.06; 0.84]	5.6%
Frazier et al.	0.42	0.1448		0.42	[0.14; 0.71]	7.3%
Rasanen et al.	0.43	0.2579		0.43	[-0.08; 0.93]	5.1%
Ratanasiripong	0.52	0.3513		0.52	[-0.17; 1.20]	3.7%
Hazlett-Stevens & Oren	0.53	0.2105		0.53	[0.12; 0.94]	6.0%
Phang et al.	0.54	0.2443		0.54	[0.06; 1.02]	5.3%
Warnecke et al.	0.60	0.2490		0.60	[0.11; 1.09]	5.2%
Song & Lindquist	0.61	0.2267		0.61	[0.17; 1.06]	5.7%
Frogeli et al.	0.63	0.1960		0.63	[0.25; 1.01]	6.3%
Call et al.	0.71	0.2608		0.71	[0.20; 1.22]	5.0%
Gallego et al.	0.72	0.2247		0.72	[0.28; 1.17]	5.7%
Kang et al.	1.28	0.3372		1.28	[0.61; 1.94]	3.9%
Shapiro et al.	1.48	0.3153		1.48	[0.86; 2.10]	4.2%
DanitzOrsillo	1.79	0.3456		1.79	[1.11; 2.47]	3.8%
Random effects model				**0.58**	**[0.38; 0.78]**	**100.0%**
Prediction interval					**[-0.06; 1.22]**	
Heterogeneity: $I^2 = 63\%$, $p < 0.01$			-2 -1 0 1 2			

The plot that `forest.meta` provides us with already looks quite decent. We also see that a think black line has been added to the plot, representing the prediction interval around our pooled effect.

We could enhance the plot by adding a column displaying the risk of bias of each study. The `ThirdWave` data set, which we used to generate `m.gen`, contains a column called `RiskOfBias`, in which the risk of bias assessment of each study is stored. When we used `metagen` to calculate the meta-analysis (Chapter 4.2.1), the function automatically saved this data within `m.gen`. Therefore, we can use the `leftcols` argument to add the column to the plot. This results in the following code:

```
forest.meta(m.gen,
            sortvar = TE,
            predict = TRUE,
```

```
print.tau2 = FALSE,
leftcols = c("studlab", "TE", "seTE", "RiskOfBias"),
leftlabs = c("Author", "g", "SE", "Risk of Bias"))
```

Author	g	SE	Risk of Bias	Standardised Mean Difference	SMD	95%-CI	Weight
Kuhlmann et al.	0.10	0.1947	high		0.10	[-0.28; 0.49]	6.3%
de Vibe et al.	0.18	0.1178	low		0.18	[-0.05; 0.41]	7.9%
Hintz et al.	0.28	0.1680	low		0.28	[-0.05; 0.61]	6.9%
Cavanagh et al.	0.35	0.1964	low		0.35	[-0.03; 0.74]	6.3%
Lever Taylor et al.	0.39	0.2308	low		0.39	[-0.06; 0.84]	5.6%
Frazier et al.	0.42	0.1448	low		0.42	[0.14; 0.71]	7.3%
Rasanen et al.	0.43	0.2579	low		0.43	[-0.08; 0.93]	5.1%
Ratanasiripong	0.52	0.3513	high		0.52	[-0.17; 1.20]	3.7%
Hazlett-Stevens & Oren	0.53	0.2105	low		0.53	[0.12; 0.94]	6.0%
Phang et al.	0.54	0.2443	low		0.54	[0.06; 1.02]	5.3%
Warnecke et al.	0.60	0.2490	low		0.60	[0.11; 1.09]	5.2%
Song & Lindquist	0.61	0.2267	high		0.61	[0.17; 1.06]	5.7%
Frogeli et al.	0.63	0.1960	low		0.63	[0.25; 1.01]	6.3%
Call et al.	0.71	0.2608	high		0.71	[0.20; 1.22]	5.0%
Gallego et al.	0.72	0.2247	high		0.72	[0.28; 1.17]	5.7%
Kang et al.	1.28	0.3372	low		1.28	[0.61; 1.94]	3.9%
Shapiro et al.	1.48	0.3153	high		1.48	[0.86; 2.10]	4.2%
DanitzOrsillo	1.79	0.3456	high		1.79	[1.11; 2.47]	3.8%
Random effects model					**0.58**	**[0.38; 0.78]**	**100.0%**
Prediction interval						**[-0.06; 1.22]**	

Heterogeneity: $I^2 = 63\%$, $p < 0.01$

-2 -1 0 1 2

We see that now, the risk of bias information of each study has been added to the forest plot.

6.2.1 Layout Types

The `forest.meta` function has two "pre-packaged" layouts, which we can use to bring our forest plot into a specific format without having to specify numerous arguments. One of them is the `"JAMA"` layout, which gives us a forest plot according to the guidelines of the *Journal of the American Medical Association*. This layout may be used if you want to publish your meta-analysis in a medical journal.

```
forest.meta(m.gen, layout = "JAMA")
```

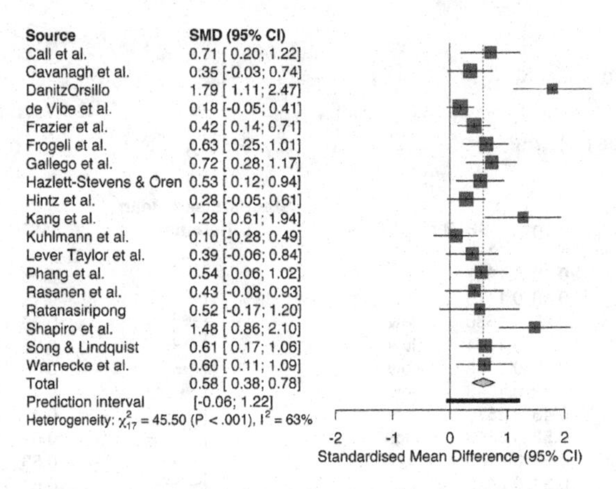

The other layout is "RevMan5", which produces a forest plot similar to the ones generated by Cochrane's *Review Manager 5*.

```
forest.meta(m.gen, layout = "RevMan5")
```

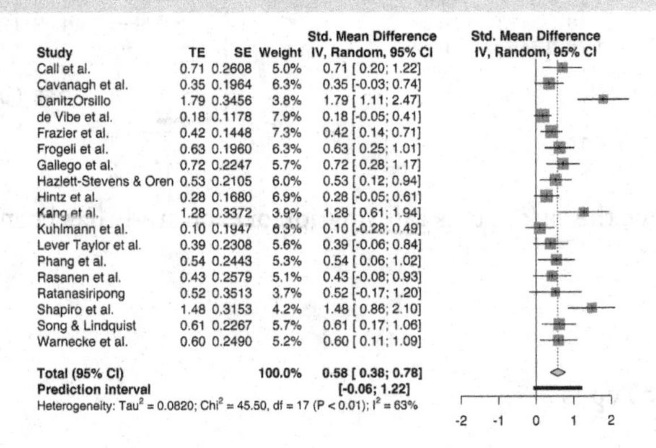

6.2.2 Saving the Forest Plots

Forest plots generated by `forest.meta` can be saved as a PDF, PNG, or *scalable vector graphic* (SVG) file. In contrast to other plots generated through base R or the *ggplot2* package, the output of `forest.meta` is not automatically re-scaled when we save it as a file. This means that forest plots are sometimes cut off on two or four sides, and we have to adjust the width and height manually so that everything is visible.

The `pdf`, `png` and `svg` function can be used to save plots via R code. We have to start with a call to one of these functions, which tells R that the output of the following code should be saved in the document. Then, we add our call to the `forest.meta` function. In the last line, we have to include `dev.off()`, which will save the generated output to the file we specified above.

All three functions require us to specify the `file` argument, which should contain the name of the file. The file is then automatically saved in the working directory under that name. Additionally, we can use the `width` and `height` argument to control the size of the plot, which can be helpful when the output is cut off.

Assuming we want to save our initial forest plot under the name "forestplot", we can use the following code to generate a PDF, PNG and SVG file.

PDF

```
pdf(file = "forestplot.pdf", width = 8, height = 7)

forest.meta(m.gen,
            sortvar = TE,
            predict = TRUE,
            print.tau2 = FALSE,
            leftlabs = c("Author", "g", "SE"))

dev.off()
```

PNG

```
png(file = "forestplot.png", width = 2800, height = 2400, res = 300)

forest.meta(m.gen,
            sortvar = TE,
            predict = TRUE,
            print.tau2 = FALSE,
            leftlabs = c("Author", "g", "SE"))

dev.off()
```

SVG

```
svg(file = "forestplot.svg", width = 8, height = 7)

forest.meta(m.gen,
            sortvar = TE,
            predict = TRUE,
            print.tau2 = FALSE,
            leftlabs = c("Author", "g", "SE"))

dev.off()
```

6.3 Drapery Plots

Forest plots are, by far, the most common way to visualize meta-analyses. Most published meta-analyses contain a forest plot, and many researchers understand how they are interpreted. It is advisable that you also include one in your meta-analysis report since forest plots provide a comprehensive and easily understandable summary of your findings.

However, forest plots are not the only way to illustrate our results. Meta-analyses can also be visualized, for example, through *drapery plots* (Rücker and Schwarzer, 2021). A drawback of forest plots is that they can only display confidence intervals assuming a fixed significance threshold, conventionally $p < 0.05$. It is based on these confidence intervals that researchers decide if an effect is significant or not. There has been a controversy around the use of p-values in recent years (Wellek, 2017), and some have argued that hypothesis testing based on p-values has contributed to the "replication crisis" in many research areas (Nuzzo, 2014).

Drapery plots are based on *p-value functions*. Such p-value functions have been proposed to prevent us from solely relying on the p<0.05 significance threshold when interpreting the results of an analysis (Infanger and Schmidt-Trucksäss, 2019). Therefore, instead of only calculating the 95% confidence interval, p-value functions provide a continuous curve which shows the confidence interval for varying values of p. In a drapery plot, a confidence curve is plotted for each study, as well as for the average effect. The x-axis shows the effect size metric, and the y-axis the assumed p-value.

Drapery plots can be generated through the `drapery` function in *{meta}*. Like `forest.meta`, this function automatically generates the plot once we provide it with a *{meta}* meta-analysis object. There are a few additional arguments, with the most important ones being:

- `type`: Defines the type of value to be plotted on the y-axis. This can be `"zvalue"` (default) for the test statistic, or the p-value (`"pvalue"`).
- `study.results`: Logical, indicating if the results of each study should be included in the plot. If `FALSE`, only the summary effect is printed.
- `labels`: When we set this argument to `"studlab"`, the study labels will be included in the plot.
- `legend`: Logical, indicating if a legend should be printed.
- `pos.legend`. The position of the legend. Either `"bottomright"`, `"bottom"`, `"bottomleft"`, `"left"`, `"topleft"`, `"top"`, `"topright"`, `"right"`, or `"center"`.

Let us try out the `drapery` function in an example using our `m.gen` meta-analysis object.

```
drapery(m.gen,
        labels = "studlab",
        type = "pval",
        legend = FALSE)
```

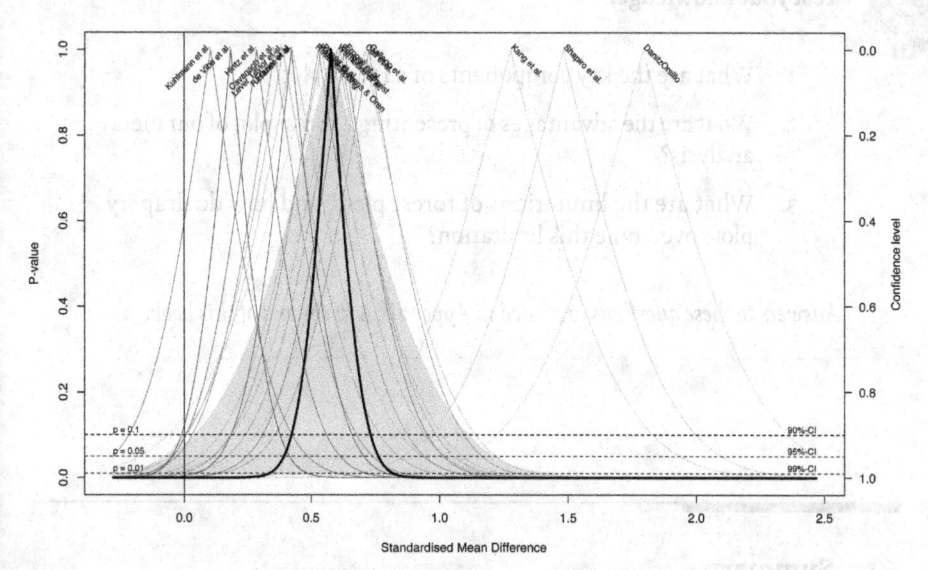

The resulting plot contains a p-value curve for each effect size, all in the shape of an upside down V. The thick line represents the average effect according to the random-effects model. The shaded area we see in the plot represents the prediction interval, which is considerably wider than the confidence interval of the pooled effect.

The "peak" of the p-value functions represents the exact value of the effect size in our meta-analysis. As we go down the y-axis, the p-value becomes smaller, and the confidence intervals wider and wider, until we reach conventional significance thresholds, indicated by the dashed horizontal lines. Based on the plot, we see that we can be quite confident in the pooled effect size being greater than zero, given that the thick line reaches zero on the x-axis when p is already very, very small (<0.01).

Rücker et al. (2021) recommend that drapery plots should mainly be used *in addition* to forest plots. Simply replacing the forest with a drapery plot may be not a good idea, because the latter does not contain much of the effect size information that might be needed by others to reproduce our results.

□

6.4 Questions & Answers

Test your knowledge!

1. What are the key components of a forest plot?
2. What are the advantages of presenting a forest plot of our meta-analysis?
3. What are the limitations of forest plots, and how do drapery plots overcome this limitation?

Answers to these questions are listed in Appendix A at the end of this book.

6.5 Summary

- It is conventional to visualize the results of meta-analyses through forest plots.
- Forest plots contain a graphical representation of each study's effect size and confidence interval, and also show the calculated overall effect. Furthermore, they contain the effect size data that was used for pooling.
- It is also possible to add other kinds of information to a forest plot, for example, the quality rating that each study received.
- Forest plots can only display results assuming a fixed significance threshold, usually $p < 0.05$. To visualize how results change for varying significance thresholds, drapery plots can be generated in addition.

7

Subgroup Analyses

In Chapter 5, we discussed the concept of between-study heterogeneity, and why it is so important in meta-analyses. We also learned methods that allow us to identify which studies contribute to the observed heterogeneity as part of outlier and influence analyses. In these analyses, we approach our meta-analysis from a purely statistical standpoint. We "measure" considerable heterogeneity in our data, and therefore exclude studies with unfitting statistical properties (i.e. outlying and influential studies) to improve the robustness of our model.

This approach can be seen as a *post hoc* procedure. Outlier and influence analyses are performed *after* seeing the data, and often *because* of the results we found. Also, they do not pay attention to anything else than the data itself. An influence analysis method may tell us that some study does not properly follow the expectations of our model, but not *why* this is the case. It might be because this study used just a slightly different research method or treatment. Yet, we are not able to know this based on the study's influence alone.

Imagine that you perform a meta-analysis investigating the effectiveness of a medical treatment. You find out that, overall, the treatment has no effect. However, there are three studies in which a considerable treatment effect was found. It may be possible to detect these studies in influence analyses, but this will not tell you why they are influential. It could be that all three studies used a treatment which varied slightly from the one used in all the other studies, and that this little detail had a profound impact on the treatment's effectiveness. This would be a groundbreaking discovery. However, it is one which cannot be made using outlier and influence analyses alone.

This makes it clear that we need a different approach, one that allows us to identify *why* a specific heterogeneity pattern can be found in our data. *Subgroup analyses*, also known as *moderator analyses*, are one way to do this. They allow us to test specific hypotheses, describing why some type of study produces lower or higher effects than another.

As we learned in Chapter 1.4.2, subgroup tests should be defined *a priori*. Before we begin with our meta-analysis, we should define different study characteristics which may influence the observed effect size, and code each study accordingly. There are countless reasons why effect sizes may differ, but we should restrict ourselves to the ones that matter in the context of our analysis. We can, for example, examine if some type of medication yields higher effects than another one. Or we might compare studies in which the follow-up period was rather short to studies in which it was long.

DOI: 10.1201/9781003107347-7

We can also examine if observed effects vary depending on the cultural region in which a study was conducted. As a meta-analyst, it helps to have some subject-specific expertise, because this allows to find questions that are actually relevant to other scientists or practitioners in the field.

The idea behind subgroup analyses is that meta-analysis is not only about calculating an average effect size but that it can also be a tool to investigate variation in our evidence. In subgroup analyses, we see heterogeneity not merely as a nuisance but as interesting variation which may or may not be explainable by a scientific hypothesis. In the best case, this can further our understanding of the world around us, or at least produce practical insights that guide future decision-making.

In this chapter, we will describe the statistical model behind subgroup analyses, and how we can conduct one directly in R.

7.1 The Fixed-Effects (Plural) Model

In subgroup analyses, we hypothesize that studies in our meta-analysis do not stem from one overall population. Instead, we assume that they fall into different *subgroups* and that each subgroup has its own true overall effect. Our aim is to reject the null hypothesis that there is no difference in effect sizes between subgroups.

The calculation of a subgroup analysis consists of two parts: first, we pool the effect in each subgroup. Subsequently, the effects of the subgroups are compared using a statistical test (Borenstein and Higgins, 2013).

7.1.1 Pooling the Effect in Subgroups

The first part is rather straightforward, as the same criteria as the ones for a meta-analysis without subgroups (see Chapter 4.1) apply. If we assume that all studies in a subgroup stem from the same population, and have one shared true effect, we can use the fixed-effect model. As we mentioned previously, it is often unrealistic that this assumption holds in practice, even when we partition our studies into smaller groups.

The alternative, therefore, is to use a random-effects model. This assumes that studies within a subgroup are drawn from a universe of populations, the mean of which we want to estimate. The difference to a normal meta-analysis is that we conduct *several* separate random-effects meta-analyses, one for each subgroup. Logically, this results in a pooled effect $\hat{\mu}_g$ for each subgroup g.

Since each subgroup gets its own separate meta-analysis, estimates of the τ^2 heterogeneity will also differ from subgroup to subgroup. In practice, however, the individual heterogeneity values $\hat{\tau}^2_g$ are often replaced with a version of τ^2 that was pooled across subgroups. This means that all subgroups are assumed to share a *common* estimate of the between-study heterogeneity. This is mostly done for practical reasons. When the number of studies in a subgroup is small, e.g. $k_g \leq 5$ (Borenstein et al., 2011, chapter 19), it is likely that the estimate of τ^2 will be imprecise. In this case, it is better to calculate a pooled version of τ^2 that is used across all subgroups, than to rely on a very imprecise estimate of the between-study heterogeneity in one subgroup.

7.1.2 Comparing the Subgroup Effects

In the next step, we assess if there is a *true* difference between the G subgroups. The assumption is that the subgroups are different, meaning that at least one subgroup is part of a different population of studies.

An elegant way to test this is to pretend that the pooled effect of a subgroup is actually nothing more than the *observed effect size* of *one large study* (see Borenstein et al., 2011, chapter 19). If we conduct a subgroup analysis with $G = 3$ subgroups, for example, we pretend that we have calculated the observed effect sizes (and standard errors) of three big studies. Once we look at the subgroups this way, it becomes clear that the question we ask ourselves is quite similar to the one we face when assessing the heterogeneity of a normal meta-analysis. We want to know if differences in effect sizes exist only due to sampling error, or because of *true* differences in the effect sizes.

Therefore, we use the value of Q to determine if the subgroup differences are large enough to not be explainable by sampling error alone. Pretending that the subgroup effects are *observed* effect sizes, we calculate the value of Q. This observed Q value is compared to its expected value assuming a χ^2 distribution with, in our case, $G - 1$ degrees of freedom (Chapter 5.1.1). If the observed value of Q is substantially larger than the expected one, the p-value of the Q test will become significant. This indicates that there is a difference in the true effect sizes between subgroups. This Q test is an *omnibus test*. It tests the null hypothesis that all subgroup effect sizes are equal, and is significant when at least two subgroups, or combinations thereof, differ.

While we normally assume that studies within the subgroups behave according to the random-effects model, the situation looks different on the pooled subgroup level. Borenstein and Higgins (2013) argue that in many fields, the subgroups we choose to analyze cannot be seen as random draws from a "universe" of possible subgroups, but represent *fixed* levels of a characteristic we want to examine. Take employment status as an example. This feature has two fixed subgroups, "employed" and "unemployed". Same also applies, for example, to studies in patients with and without a specific co-morbidity.

Borenstein and Higgins call the model for subgroup analyses the *fixed-effects (plural) model*. The word "plural" is added because we have to differentiate it from the standard fixed-effect model. The fixed-effects (plural) model can be seen like a hybrid creature, including features of both the fixed-effect model and the random-effects model. Like in the random-effects model, we assume that there is more than one true effect size, because there are subgroups in our data. However, we do not see the subgroups as random draws from a whole universe of subgroups. Our subgroup levels are fixed, and *exhaustive*, meaning that no generalization is needed. This makes it clear why we call the process generating our subgroup data a fixed-effects "plural" model: because there are *several* true effect sizes, but the true effect sizes represent subgroup levels that are assumed to be *fixed*.

Borenstein and colleagues (2011, chapter 19) argue that all of this may seem a little confusing to us because the word "fixed" can mean different things in statistics. In conventional meta-analyses, the term "fixed effect" is used synonymously with "common effect". In the context of subgroup analyses, however, we speak of "fixed effects" to underline that they are "not random". They are not simply random manifestations of an over-arching distribution, to which we aim to generalize, but the *real* and *only* categories into which a variable can fall. Figure 7.1 visualizes the fixed-effects (plural) model, assuming that studies within subgroups follow the random-effects model.

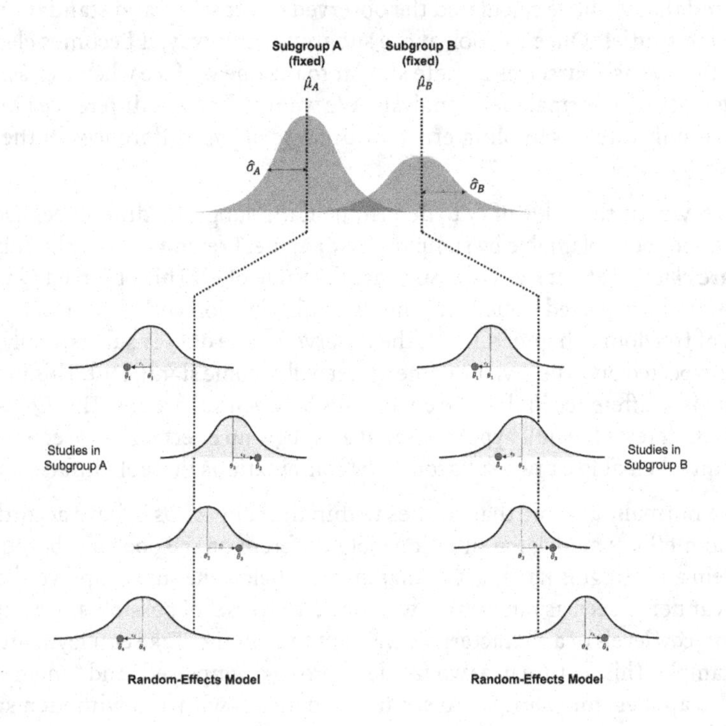

FIGURE 7.1: Visualization of the fixed-effects (plural) model, assuming a random-effects model within subgroups.

A Few Examples of Subgroup Variables With Fixed Levels

- *Age group*: children, young adults, adults, elderly persons.
- *Cultural background*: western, non-western.
- *Control group*: alternative treatment, minimal treatment, no treatment.
- *Tool used to measure the outcome*: self-report, expert-rated.
- *Study quality*: high, low, unclear.
- *Species*: plants, animals.
- *Setting*: schools, hospitals, private households.

Note that the concrete selection and definition of subgroups can and should be adapted based on the aim and scope of your meta-analysis.

Because the fixed-effects (plural) model contains both random effects (within subgroups) and fixed effects (since subgroups are assumed to be fixed), it is also known in the literature as a *mixed-effects model*. We already came across this term previously in Chapter 4.2.6, where we discussed a different type of (generalized) mixed-effects model that can be used to pool, for example, proportions.

The model we use for subgroup analyses is heavily related to other methods that are also often used in meta-analyses. In Chapter 8, we will show that subgroup analyses are just a special case of *meta-regression*, for which we also use a mixed-effects model. Furthermore, it is also possible that subgroup levels *can not* be assumed to be fixed. Imagine that we want to assess if effect sizes differ depending on the location in which the effect was observed. Some studies assessed the effect in Israel, some in Italy, others in Mexico, and some in mainland China. One can argue that "country of origin" is not a factor with fixed levels: there are many, many countries in the world, and our study simply includes a "random" selection. In this case, it makes sense to not model the subgroups as fixed, but also let our model estimate the variability between countries as a random effect. This leads to a *multi-level model*, which we will cover in Chapter 10.

7.2 Limitations & Pitfalls of Subgroup Analyses

Intuitively, one might think that subgroup analysis is an exceptionally good tool to detect effect moderators. The aim of meta-analyses, after all, is to study all available evidence. This means that the total number of individuals analyzed in meta-analyses will usually surpass the one of a primary study by orders of magnitude. Unfortunately, however, this does not necessarily provide us with more *statistical power* to detect subgroup differences. There are several reasons for this (Hedges and Pigott, 2004):

- First, remember that in subgroup analyses, the results within subgroups are usually pooled using the random-effects model. If there is substantial between-study heterogeneity within the subgroup, this will decrease the precision (i.e. increase the standard error) of the pooled effect. Yet, when the subgroup effect estimates are very imprecise, this means that their confidence intervals will have a large overlap. Consequentially, this makes it harder to find a significant difference between subgroups–even if this difference does exist.

- In the same vein, statistical power is also often low because the effects we want to detect in subgroup analyses are much lower than in normal meta-analyses. Imagine that we want to examine if effects differed between studies assessing an outcome of interest through *self-reports* versus *expert ratings*. Even if there is a difference, it is very likely to be small. It is often possible to find a significant difference between treatment and control groups. Yet, detecting effect size differences *between studies* is usually much harder, because the differences are smaller, and more statistical power is needed.

- From the points above follows an important caveat: *absence of evidence is not evidence of absence*. If we do *not* find a difference in effect sizes between subgroups, this does not automatically mean that the subgroups produce *equivalent* outcomes. As we argued above, there are various reasons why our subgroup analysis may not have the statistical power needed to ascertain a true difference in effects. If this is the case, it would be a gross misinterpretation to say that the subgroups have the same effect–we simply do not know if differences exist or not. This becomes particularly explosive when we want to assess if one treatment is better than the other. Some stakeholders, including corporations, often have a vested interest in showing the equivalence of a treatment. But subgroup analyses are usually not an adequate way to prove this.

- We can check if statistical power is a problem in our subgroup analysis by performing a *subgroup power analysis* beforehand. In such an analysis, we can check the minimum effect size difference we are able to detect in our subgroup analysis. In Chapter 14.3 in the "Helpful Tools" section, we cover how subgroup power analyses can be performed in R. But note that power analyses can at best be seen as a helpful diagnostic, not as proof that the power of our analysis is high enough to show that the subgroups are equivalent. Schwarzer and colleagues (Schwarzer et al., 2015, chapter 4.3) mention, as a general rule of thumb, that subgroup analyses only make sense when your meta-analysis contains at least $K = 10$ studies.

Another important limitation of subgroup analyses is that they are purely observational (Borenstein and Higgins, 2013). Meta-analyses often only include randomized controlled trials (RCTs), in which participants were randomly allocated to either a treatment or control group. When properly conducted, such RCTs can provide evidence that the treatment *caused* the difference observed in the study. This is because all relevant variables that may influence the assessed outcomes are equal in the two groups. The only difference is that one group received the treatment, while the other did not. Subgroup analyses, even when consisting solely of randomized studies, cannot show causality. Imagine that our subgroup analysis finds that one type of treatment is more effective than the other. There are countless reasons why this finding may be spurious; for example, it could be that studies investigating treatment A used other control groups than the ones examining treatment B. This means that both treatments could be equally effective–we just see a difference because the treatment type is *confounded* with methodological factors. This example should underline that one should always appraise the results of subgroup analyses critically.

A last important pitfall involves the way the subgroups are defined. Often, it may be tempting to sort studies into subgroups based on *aggregate information*. Schwarzer and colleagues (Schwarzer et al., 2015, chapter 4.3) name the mean age of a study as a common example. Imagine you want to assess if effects differ between elderly individuals (65+ years of age) and general adult populations. Therefore, you sort studies into these two categories, depending on whether the reported mean age is above or below 65.

If we find that effects are higher in the subgroup with higher mean age, we may intuitively think that this shows that the effects are higher in older individuals. But this reasoning is deeply flawed. When the *mean* age of a primary study is above 65, it is still possible that it included a substantial proportion of individuals who were *younger* than that. *Vice versa*, it is also perfectly possible that a study included a large share of individuals *older* than 65, even when the *mean* age is *lower*. This means that the higher effects found in the "elderly" subgroup could *solely* be driven by individuals who are actually younger than 65. Conversely, it is possible that in the "younger" subgroup, the lower effects were caused by the individuals in the studies that were older than 65. This leads to a paradoxical situation: on the aggregate level, we find that studies with a higher mean age have higher effects. But on the individual level, the opposite is true: a person will experience *lower* effects with rising age.

The scenario we just described is caused by so-called *ecological bias* (Thompson and Higgins, 2002; Piantadosi et al., 1988). It arises whenever we want to use relationships on an aggregate (*macro*) level to predict associations on the individual (*micro*) level. The best way to avoid it is to *never, ever* use aggregate information in subgroup analyses and meta-regression. The situation is different, however, if we know that *all* individuals in a study fall into one category. If, for example, we have a few studies in which *only* adolescents under the age of 18 were included, and others in which *only* adults (18+ years) could participate, the risk of ecological bias is largely eliminated. However, it is still possible that effect differences were caused by confounding variables and not by the age of the participants.

Subgroup Analysis: Summary of the Dos & Don'ts

1. Subgroup analyses depend on the statistical power, so it usually makes no sense to conduct one when the number of studies is small (i.e. $K < 10$).

2. If you do not find a difference in effect sizes between subgroups, this does *not* automatically mean that the subgroups produce *equivalent* results.

3. Subgroup analyses are purely *observational*, so we should always keep in mind that effect differences may also be caused by confounding variables.

4. It is a bad idea to use aggregate study information in subgroup analyses, because this may introduce ecological bias.

7.3 Subgroup Analysis in R

Time to implement what we learned in R. Conducting a subgroup analysis using the *{meta}* package is relatively straightforward. In every meta-analysis function in *{meta}*, the `byvar` argument can be specified. This tells the function which effect size falls into which subgroup and runs a subgroup analysis. The `byvar` argument accepts `character`, `factor`, `logical` or `numeric` variables. The only thing we have to take care of is that studies in the same subgroup have absolutely identical labels.

In this example, we use our `m.gen` meta-analysis object again. The `ThirdWave` data set, which we used to calculate the meta-analysis, contains a few columns with subgroup information. Here, we want to examine if there are effect size differences between studies with a high versus low risk of bias. The risk of bias information is stored in the `RiskOfBias` column. Let us have a look at this column first. In our code, we use the `head` function so that only the first few rows of the data set are shown.

```
# Show first entries of study name and 'RiskOfBias' column
head(ThirdWave[,c("Author", "RiskOfBias")])
```

```
##            Author RiskOfBias
## 1     Call et al.       high
```

```
## 2 Cavanagh et al.          low
## 3   DanitzOrsillo         high
## 4  de Vibe et al.          low
## 5  Frazier et al.          low
## 6  Frogeli et al.          low
```

We see that every study in our data set has a label specifying its risk of bias assessment. When we calculated the meta-analysis using `metagen`, this information was saved internally in the `m.gen` object. To conduct a subgroup analysis, we can therefore use the `update.meta` function, provide it with the `m.gen` object, and use the `byvar` argument to specify which column in our data set contains the subgroup labels.

Previously, we also covered that subgroup analyses can be conducted with or without a common estimate of τ^2 across subgroups. This can be controlled in *{meta}* by setting `tau.common` to `TRUE` or `FALSE`. For now, let us use separate estimates of the between-study heterogeneity variance in each subgroup.

In our example, we want to apply the fixed-effects (plural) model and assume that studies within subgroups are pooled using the random-effects model. Given that `m.gen` contains results for the random-effects model (because we set `comb.fixed` to `FALSE` and `comb.random` to `TRUE`), there is nothing we have to change. Because the original meta-analysis was performed using the random-effects model, `update.meta` automatically assumes that studies within subgroups should also be pooled using the random-effects model. Therefore, the resulting code looks like this:

```
update.meta(m.gen,
            byvar = RiskOfBias,
            tau.common = FALSE)
```

```
## Review:     Third Wave Psychotherapies
##                         SMD       95%-CI %W(random) RiskOfBias
## Call et al.            0.70 [ 0.19; 1.22]       5.0       high
## Cavanagh et al.        0.35 [-0.03; 0.73]       6.3        low
## DanitzOrsillo          1.79 [ 1.11; 2.46]       3.8       high
## [...]
##
## Number of studies combined: k = 18
##
##                         SMD            95%-CI    t  p-value
## Random effects model 0.5771 [0.3782; 0.7760] 6.12 < 0.0001
##
## Quantifying heterogeneity:
##  tau^2 = 0.0820 [0.0295; 0.3533]; tau = 0.2863 [0.1717; 0.5944];
##  I^2 = 62.6% [37.9%; 77.5%]; H = 1.64 [1.27; 2.11]
##
## Quantifying residual heterogeneity:
##  I^2 = 59.3% [30.6%; 76.1%]; H = 1.57 [1.20; 2.05]
```

```
##
## Test of heterogeneity:
##      Q d.f. p-value
##  45.50   17  0.0002
##
## Results for subgroups (random effects model):
##                      k    SMD         95%-CI  tau^2    tau     Q   I^2
## RiskOfBias = high    7 0.8126 [0.28; 1.34] 0.2423 0.4922 25.89 76.8%
## RiskOfBias = low    11 0.4300 [0.28; 0.58] 0.0099 0.0997 13.42 25.5%
##
## Test for subgroup differences (random effects model):
##                     Q d.f. p-value
## Between groups   2.84    1  0.0917
##
## Details on meta-analytical method:
## - Inverse variance method
## - Restricted maximum-likelihood estimator for tau^2
## [...]
```

In the output, we see a new section called `Results for subgroups`. This part of the output shows the pooled effect size separately for each subgroup. We see that there are $k = 7$ studies with a high risk of bias, and 11 with a low risk of bias. The estimated between-study heterogeneity differs considerably, with $I^2 = 77\%$ in high risk of bias studies, but only 26% in studies with a low risk.

The effect sizes of the subgroups also differ. With $g = 0.43$, the effect estimate in low risk of bias studies is smaller than in studies with a high risk of bias. This is a common finding because biased studies are more likely to overestimate the effects of a treatment. But is the difference statistically significant? We can check this by looking at the results of the `Test for subgroup differences`. This shows us the Q-test, which, in our example with 2 subgroups, is based on one degree of freedom. The p-value of the test is 0.09, which is larger than the conventional significance threshold, but still indicates a difference on a trend level.

We can also check the results if were to assume a *common* τ^2 estimate in both subgroups. We only have to set `tau.common` to `TRUE`.

```
update.meta(m.gen, byvar = RiskOfBias, tau.common = TRUE)
```

```
## [...]
##                     k  SMD        95%-CI  tau^2    tau     Q   I^2
## RiskOfBias = high 7 0.76 [0.25; 1.28] 0.0691 0.2630 25.89 76.8%
## RiskOfBias = low 11 0.46 [0.30; 0.63] 0.0691 0.2630 13.42 25.5%
##
## Test for subgroup differences (random effects model):
##                    Q d.f. p-value
## Between groups  1.79    1  0.1814
```

```
## Within groups  39.31   16  0.0010
##
## Details on meta-analytical method:
## - Inverse variance method
## - Restricted maximum-likelihood estimator for tau^2
##   (assuming common tau^2 in subgroups)
## [...]
```

In the output, we see that the estimated between-study heterogeneity variance is $\tau^2 = 0.069$, and identical in both subgroups. We are presented with two Q-tests: one *between* groups (the actual subgroup test) and another for the *within-subgroup* heterogeneity. Like in a normal meta-analysis, the latter simply indicates that there is excess variability in the subgroups ($p = 0.001$). The test of subgroup differences again indicates that there is not a significant difference between studies with a low versus high risk of bias ($p = 0.181$).

We now explored the results assuming either an independent or common estimate of τ^2. Since we are not aware of good reasons to assume that the heterogeneity in both subgroups is equal, and given that we have a minimum $k = 7$ studies in each subgroup, separate estimates of τ^2 may be appropriate. However, we saw that the interpretation of our results is similar for both approaches anyway, at least in our example.

Reporting the Results of a Subgroup Analysis

The results of subgroup analyses are usually reported in a table displaying the estimated effect and heterogeneity in each subgroup, as well as the p-value of the test for subgroup differences.

	g	95%CI	p	I^2	95%CI	p_{subgroup}
Risk of Bias						0.092
- High	0.81	0.28-1.34	0.009	0.77	0.51-0.89	
- Low	0.43	0.28-0.58	<0.001	0.25	0.00-0.63	

Further rows can be added to the table if more than one subgroup analysis was conducted.

□

7.4 Questions & Answers

Test your knowledge!

1. In the best case, what can a subgroup analysis tell us that influence and outlier analyses cannot?
2. Why is the model behind subgroup analyses called the fixed-effects (plural) model?
3. As part of your meta-analysis, you want to examine if the effect of an educational training program differs depending on the school district in which it was delivered. Is a subgroup analysis using the fixed-effects (plural) model appropriate to answer this question?
4. A friend of yours conducted a meta-analysis containing a total of nine studies. Five of these studies fall into one subgroup, four into the other. She asks you if it makes sense to perform a subgroup analysis. What would you recommend?
5. You found a meta-analysis in which the authors claim that the analyzed treatment is more effective in women than men. This finding is based on a subgroup analysis in which studies were divided into groups based on the *share* of females included in the study population. Is this finding credible, and why (not)?

Answers to these questions are listed in Appendix A at the end of this book.

7.5 Summary

- Although there are various ways to assess the heterogeneity of a meta-analysis, these approaches do not tell us *why* we find excess variability in our data. Subgroup analysis allows us to test hypotheses on why some studies have higher or lower true effect sizes than others.

- For subgroup analyses, we usually assume a fixed-effects (plural) model. Studies within subgroups are pooled, in most cases, using the random-effects model. Subsequently, a Q-test based on the overall subgroup results is used to determine if the groups differ significantly.
- The subgroup analysis model is called a "fixed-effects" model because the different categories themselves are assumed to be fixed. The subgroup levels are not seen as random draws from a universe of possible categories. They represent the only values that the subgroup variable can take.
- When calculating a subgroup analysis, we have to decide whether separate or common estimates of the between-study heterogeneity should be used to pool the results within subgroups.
- Subgroup analyses are not a panacea. They often lack the statistical power needed to detect subgroup differences. Therefore, a non-significant test for subgroup differences does not automatically mean that the subgroups produce equivalent results.

8

Meta-Regression

In the last chapter, we added subgroup analyses as a new method to our meta-analytic "toolbox". As we learned, subgroup analyses shift the focus of our analyses away from finding one overall effect. Instead, they allow us to investigate patterns of heterogeneity in our data, and what causes them.

We also mentioned that subgroup analyses are a special form of *meta-regression*. It is very likely that you have heard the term "regression" before. Regression analysis is one of the most common statistical methods and used in various disciplines. In its simplest form, a regression model tries to use the value of some variable x to predict the value of another variable y. Usually, regression models are based on data comprising individual persons or specimens, for which both the value of x and y is measured.

In meta-regression, this logic is applied to *entire studies*. The variable x represents characteristics of studies, for example, the year in which it was conducted. Based on this information, a meta-regression model tries to predict y, the study's effect size. The fact that effect sizes are used as predicted variables, however, adds some complexity. In Chapter 3.1, we already learned that observed effect sizes $\hat{\theta}$ can be more or less *precise* estimators of the study's true effect, depending on their standard error. In "normal" meta-analyses, we take this into account by giving studies a smaller or higher weight. In meta-regression, we also have to make sure that the model pays more attention to studies with a lower sampling error, since we can assume that their estimates are closer to the "truth".

Meta-regression achieves this by assuming a *mixed-effects model*. This model accounts for the fact that observed studies deviate from the true overall effect due to sampling error and between-study heterogeneity. More importantly, however, it also uses one or more variables x to predict differences in the true effect sizes. We already mentioned in the last chapter that subgroup analysis is also based on a mixed-effects model. In this chapter, we will delve a little deeper, and discuss why subgroup analysis and meta-regression are inherently related.

Meta-regression, although it has its own limitations, can be a very powerful tool in meta-analyses. It is also very versatile: *multiple meta-regression*, for example, allows us to include not only one, but several predictor variables, along with their interaction. In the second part of this chapter, we will therefore also have a look at multiple meta-regression, and how we can conduct one using R.

DOI: 10.1201/9781003107347-8

8.1 The Meta-Regression Model

In the past, you may have already performed a regression using primary study data, where participants are the unit of analysis. In meta-analyses, the individual data of each participant is usually not available, and we can only resort to aggregated results. This is why we have to perform meta-regression with predictors on a *study level*. This also means that, while we conduct analyses on samples much larger than usual for primary studies, it is still possible that we do not have enough data points for a meta-regression to be useful. In Chapter 7.2, we already covered that subgroup analyses often make no sense when $K < 10$. Borenstein and colleagues (2011, Chapter 20) mention that this guideline may also be applied to meta-regression models, but that it should not be seen as an iron-clad rule.

In a conventional regression, we want to estimate the value y_i of person i using a *predictor* (or *covariate*) x_i with a regression coefficient β. A standard regression equation, therefore, looks like this:

$$y_i = \beta_0 + \beta_1 x_i \tag{8.1}$$

In meta-regression, the variable y we want to predict is the observed effect size $\hat{\theta}_k$ of study k. The formula for a *meta-regression* looks similar to the one of a normal regression model:

$$\hat{\theta}_k = \theta + \beta x_k + \epsilon_k + \zeta_k \tag{8.2}$$

Note that this formula contains two extra terms, ϵ_k and ζ_k. The same terms can also be found in the equation for the random-effects-model (Chapter 4.1.2) and signify two types of independent errors. The first one, ϵ_k, is the sampling error through which the effect size of a study deviates from its true effect. The second error, ζ_k, denotes that even the true effect size of the study is only sampled from an overarching distribution of effect sizes. This means that between-study heterogeneity exists in our data, which is captured by the heterogeneity variance τ^2.

Since the equation above includes a *fixed* effect (the β coefficient) as well as a *random* effect (ζ_k), the model used in meta-regression is often called a *mixed-effects model*. Conceptually, this model is identical to the mixed-effects model we described in Chapter 7.1.2, where we explained how subgroup analyses work.

8.1.1 Meta-Regression with a Categorical Predictor

Indeed, as mentioned before, subgroup analysis is nothing else than a meta-regression with a categorical predictor. Such categorical variables can be included through *dummy-coding*, e.g.:

$$D_g = \begin{cases} 0: & \text{Subgroup A} \\ 1: & \text{Subgroup B.} \end{cases} \tag{8.3}$$

To specify a subgroup analysis in the form of a meta-regression, we simply have to replace the covariate x_k with D_g:

$$\hat{\theta}_k = \theta + \beta D_g + \epsilon_k + \zeta_k. \tag{8.4}$$

To understand this formula, we have to read it from the left to the right. The goal of the meta-regression model, like every statistical model, is to explain how the observed data was generated. In our case, this is the observed effect size $\hat{\theta}_k$ of some study k in our meta-analysis. The formula above works like a recipe, telling us which ingredients are needed to produce the observed effect.

First, we take θ, which serves as the *intercept* in our regression model. The value of θ is identical with the true overall effect size of subgroup A. To see why this is the case, we need to look at the next "ingredient", the term βD_g. The value of β in this term represents the effect size difference θ_Δ between subgroup A and subgroup B. The value of β is multiplied with D_g, which can be either 0 or 1, depending on whether the study is part of subgroup A ($D_g = 0$) or subgroup B ($D_g = 1$).

Because multiplying with zero gives zero, the βD_g term completely falls out of the equation when we are dealing with a study in subgroup A. When $D_g = 1$, on the other hand, we multiply by 1, meaning that β remains in the equation and is added to θ, which provides us with the overall effect size in subgroup B. Essentially, the dummy predictor is a way to integrate *two* formulas into *one*. We can easily see this when we write down the formula individually for each subgroup:

$$D_g = \begin{cases} 0: & \hat{\theta}_k = \theta_A + \epsilon_k + \zeta_k \\ 1: & \hat{\theta}_k = \theta_A + \theta_\Delta + \epsilon_k + \zeta_k \end{cases} \tag{8.5}$$

Written this way, it becomes clearer that our formula actually contains two models, one for subgroup A, and one for subgroup B. The main difference between the models is that the effect of the second subgroup is "shifted" up or down, depending on the value of β (which we denote as θ_Δ in the formula above).

This should make it clear that subgroup analyses work just like a normal regression: they use some variable x to predict the value of y, which, in our case, is the effect size of a study. The special thing is that βx_k is not continuous–it is a fixed value we add to the prediction, depending on whether a study belongs to a certain subgroup or not. This fixed value of β is the estimated difference in effect sizes between two subgroups.

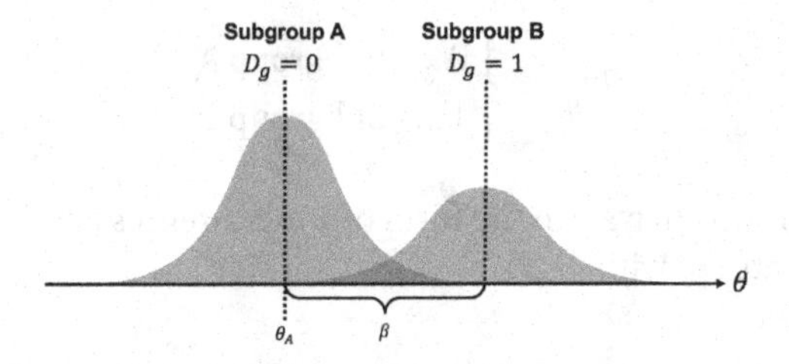

FIGURE 8.1: Meta-regression with a categorical predictor (subgroup analysis).

8.1.2 Meta-Regression with a Continuous Predictor

When people speak of a "meta-regression", however, they usually think of models in which a *continuous* variable was used as the predictor. This brings us back the generic meta-regression formula shown in equation 8.2. Here, the regression terms we discussed before are also used, but they serve a slightly different purpose. The term θ again stands for the intercept, but now represents the predicted effect size when $x = 0$. To the intercept, the term βx_k is added. This part produces a *regression slope*: the continuous variable x is multiplied with the *regression weight* β, thus lowering or elevating the predicted effect for different values of the covariate. The aim of the meta-regression model is to find values of θ and β which minimize the difference between the *predicted* effect size and the *true* effect size of studies (see Figure 8.2).

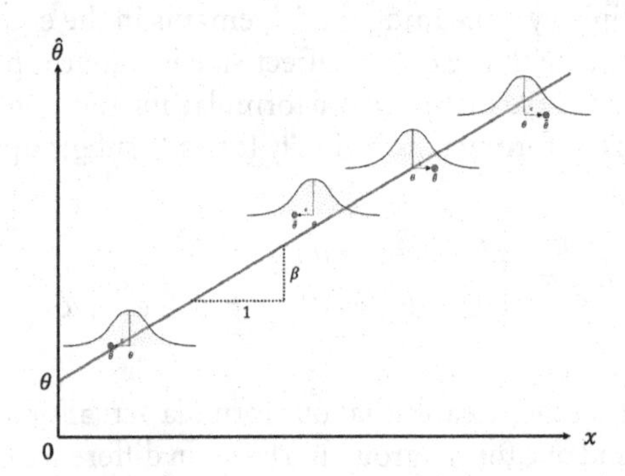

FIGURE 8.2: Meta-regression with a continuous predictor and four studies.

Looking closely at the meta-regression formula, we see that it contains two types of terms. Some terms include a subscript k, while others do not. A subscript k indicates that a value *varies* from study to study. When a term does not include a subscript k, this means that it stays the same for all studies.

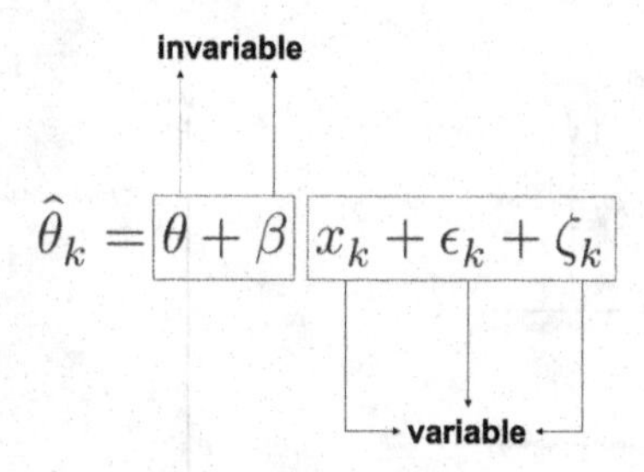

In a meta-regression, both θ and β are invariable, or fixed. This tells us something important about what a meta-regression does: based on the variation in a predictor variable and the observed effects, it tries to "distill" a *fixed pattern* underlying our data, in the form of a *regression line*. If the meta-regression model fits the data well, the estimated parameters θ and β can be used to predict the effect size of a study the model has *never seen before* (provided we know x). Taking into account both the sampling error ϵ_k and between-study heterogeneity ζ_k, meta-regression thus tries to find a model that *generalizes* well; not only to the observed effect sizes but to the "universe" of all possible studies of interest.

8.1.3 Assessing the Model Fit

An important detail about meta-regression models is that they can be seen as an extension of the "normal" random-effects model we use to pool effect sizes. The random-effects model is nothing but a meta-regression model *without a slope term*. Since it contains no slope, the random-effects model simply predicts the *same value* for each study: the estimate of the pooled effect size μ, which is equivalent to the intercept.

In the first step, the calculation of a meta-regression therefore closely resembles the one of a random-effects meta-analysis, in that the between-study heterogeneity τ^2 is estimated using one of the methods we described in Chapter 4.1.2.1 (e.g. the DerSimonian-Laird or REML method). In the next step, the fixed weights θ and β are estimated. Normal linear regression models use the *ordinary least squares* (OLS) method to find the regression line that fits the data best. In meta-regression, a modified method called *weighted least squares* (WLS) is used, which makes sure that studies with a smaller standard error are given a higher weight.

Once the optimal solution is found, we can check if the newly added regression term explains parts of the effect size heterogeneity. If the meta-regression model fits the data well, the true effect sizes should deviate less from the regression line compared to the pooled effect $\hat{\mu}$. If this is the case, the predictor x *explains* some of the heterogeneity variance in our meta-analysis.

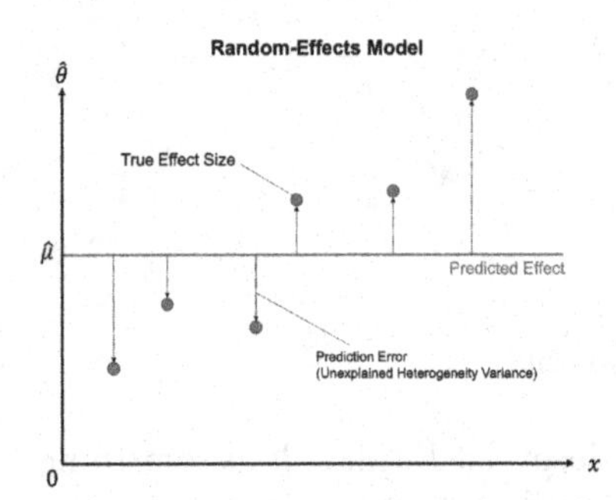

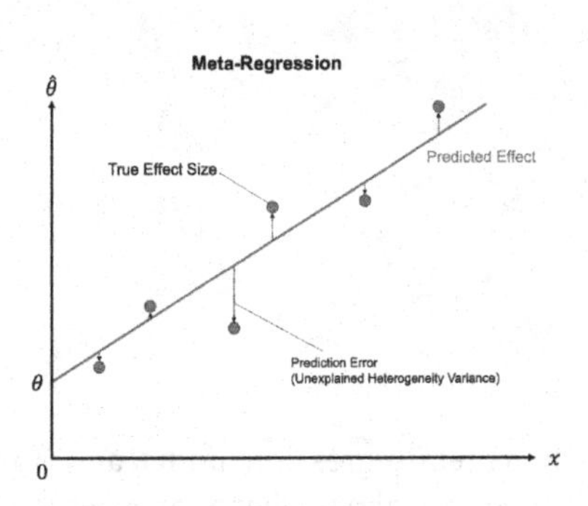

The fit of the meta-regression model can therefore be assessed by checking how much of the heterogeneity variance it explains. The predictors included in the mixed-effects model should minimize the amount of the *residual,* or unexplained, heterogeneity variance, which we denote with $\tau^2_{\text{unexplained}}$.

In regression analyses, the R^2 index is commonly used to quantify the percentage of variation explained by the model. An analogous index, R^2_*, can also be calculated for meta-regression. We add an asterisk here to indicate that the R^2 in meta-regression is slightly different to the one used in conventional regressions, because we deal with *true effect sizes* instead of observed data points. The formula for R^2_* looks like this:

$$R^2_* = 1 - \frac{\hat{\tau}^2_{\text{unexplained}}}{\hat{\tau}^2_{\text{(total)}}} \tag{8.6}$$

R^2_* uses the amount of residual heterogeneity variance that even the meta-regression slope cannot explain, and puts it in relation to the total heterogeneity that we initially found in our meta-analysis. Subtracting this fraction from 1 leaves us with the percentage of between-study heterogeneity explained by the predictor.

There is also another way to formulate R^2_*. We can say that it expresses how much the mixed-effects model has *reduced* the heterogeneity variance compared to the initial random-effects pooling model, in percent. This results in the following formula:

$$R^2_* = \frac{\hat{\tau}^2_{\text{REM}} - \hat{\tau}^2_{\text{MEM}}}{\hat{\tau}^2_{\text{REM}}} \tag{8.7}$$

In this formula, $\hat{\tau}^2_{\text{REM}}$ represents the amount of between-study heterogeneity found in the random-effects pooling model, and $\hat{\tau}^2_{\text{MEM}}$ represents the (residual) variance in the mixed-effects meta-regression model (i.e. the "prediction error" with respect to the true effect sizes).

Usually, we are not only interested in the amount of heterogeneity explained by the regression model but also if the regression weight of our predictor x is significant. If this is the case, we can be quite confident that x has an influence on the effect sizes of studies. Both in conventional and meta-regression, the significance of a regression weight is commonly assessed through a *Wald-type* test. This involves calculating the test statistic z, by dividing the estimate of β through its standard error:

$$z = \frac{\hat{\beta}}{SE_{\hat{\beta}}} \tag{8.8}$$

Under the null hypothesis that $\beta = 0$, this z-statistic follows a standard normal distribution. This allows us to calculate a corresponding p-value, which determines if the predictor is significant or not.

However, a test based on the z-statistic is not the only way to assess the significance of predictors. Like in normal meta-analysis models, we can also use the Knapp-Hartung adjustment, which results in a test statistic based on the t-distribution (see Chapter 4.1.2.2). As we learned previously, it is often advisable to use the Knapp-Hartung method, because it reduces the risk of false positives.

8.2 Meta-Regression in R

The *{meta}* package contains a function called `metareg`, which allows us to conduct a meta-regression. The `metareg` function only requires a *{meta}* meta-analysis object and the name of a covariate as input.

In this example, we will use our `m.gen` meta-analysis object again, which is based on the `ThirdWave` data set (see Chapter 4.2.1). Using meta-regression, we want to examine if the *publication year* of a study can be used to predict its effect size. By default, the `ThirdWave` data set does not contain a variable in which the publication year is stored, so we have to create a new `numeric` variable which contains this information. We simply concatenate the publication years of all studies, in the same order in which they appear in the `ThirdWave` data set. We save this variable under the name `year`[1].

```
year <- c(2014, 1998, 2010, 1999, 2005, 2014,
          2019, 2010, 1982, 2020, 1978, 2001,
          2018, 2002, 2009, 2011, 2011, 2013)
```

Now, we have all the information we need to run a meta-regression. In the `metareg` function, we specify the name of our meta-analysis object `m.gen` as the first argument, and the name of our predictor, `year`, as the second argument. We give the results the name `m.gen.reg`.

[1]The publication years we use in this example are made up, and only used for illustration purposes.

```
m.gen.reg <- metareg(m.gen, year)
```

Now, let us have a look at the results:

```
m.gen.reg
```

```
## Mixed-Effects Model (k = 18; tau^2 estimator: REML)
##
## tau^2 (estimated amount of residual heterogeneity):
##      0.019 (SE = 0.023)
## tau (square root of estimated tau^2 value):             0.1371
## I^2 (residual heterogeneity / unaccounted variability): 29.26%
## H^2 (unaccounted variability / sampling variability):   1.41
## R^2 (amount of heterogeneity accounted for):            77.08%
##
## Test for Residual Heterogeneity:
## QE(df = 16) = 27.8273, p-val = 0.0332
##
## Test of Moderators (coefficient 2):
## F(df1 = 1, df2 = 16) = 9.3755, p-val = 0.0075
##
## Model Results:
##
##          estimate     se   tval   pval    ci.lb    ci.ub
## intrcpt    -36.15  11.98  -3.01  0.008  -61.551  -10.758  **
## year         0.01   0.00   3.06  0.007    0.005    0.031  **
##
## ---
## Signif. codes:  0 '***' 0.001 '**' 0.01 '*' 0.05 '.' 0.1 ' ' 1
```

Let us go through what we can see here. In the first line, the output tells us that a mixed-effects model has been fitted to the data, just as intended. The next few lines provide details on the amount of heterogeneity explained by the model. We see that the estimate of the residual heterogeneity variance, the variance that is not explained by the predictor, is $\hat{\tau}^2_{\text{unexplained}} = 0.019$. The output also provides us with an I^2 equivalent, which tells us that after inclusion of the predictor, 29.26% of the variability in our data can be attributed to the remaining between-study heterogeneity. In the normal random-effects meta-analysis model, we found that the I^2 heterogeneity was 63%, which means that the predictor was able to "explain away" a substantial amount of the differences in true effect sizes. In the last line, we see the value of R^2_*, which in our example is 77%. This means that 77% of the difference in true effect sizes can be explained by the publication year, a value that is quite substantial.

The next section contains a `Test for Residual Heterogeneity`, which is essentially the Q-test we already got to know previously (see Chapter 5.1.1). Now, however, we test if the heterogeneity not explained by the predictor is significant. We see that this is the case, with $p = 0.03$. However, we know the limitations of the Q-test (Chapter 5.1.1) and should therefore not rely too heavily on this result.

The next part shows the `Test of Moderators`. We see that this test is also significant ($p = 0.0075$). This means that our predictor, the publication year, does indeed influence the studies' effect size.

The last section provides more details on the estimated regression coefficients. The first line shows the results for the intercept (`intrcpt`). This is the expected effect size (in our case: Hedges' g) when our predictor publication year is zero. In our example, this represents a scenario which is, arguably, a little contrived: it shows the predicted effect of a study conducted in the year 0, which is $\hat{g} = -36.15$. This serves as yet another reminder that good statistical models do not have to be a perfect representation of reality; they just have to be *useful*.

The coefficient we are primarily interested in is the one in the second row. We see that the model's estimate of the regression weight for `year` is 0.01. This means that for every additional year, the effect size g of a study is expected to rise by 0.01. Therefore, we can say that the effect sizes of studies have increased over time. The 95% confidence interval ranges from 0.005 to 0.3, showing that the effect is significant. Importantly, we are also presented with the corresponding t-statistic for each regression coefficient (`tval`). This tells us that the Knapp-Hartung method was used to calculate the confidence interval and p-value. Since we also used this adjustment in our initial meta-analysis model, `metareg` automatically used it again here. Otherwise, z values and Wald-type confidence intervals would have been provided.

The *{meta}* package allows us to visualize a meta-regression using the `bubble` function. This creates a *bubble plot*, which shows the estimated regression slope, as well as the effect size of each study. To indicate the weight of a study, the bubbles have different sizes, with a greater size representing a higher weight. To produce a bubble plot, we only have to plug our meta-regression object into the `bubble` function. Because we also want study labels to be displayed, we set `studlab` to `TRUE`.

```
bubble(m.gen.reg, studlab = TRUE)
```

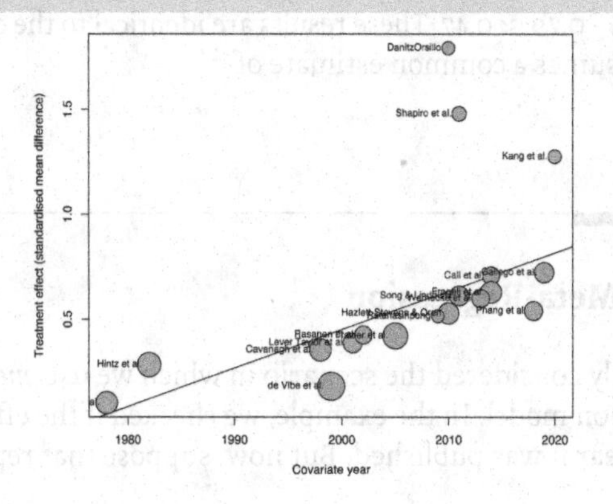

For the sake of completeness, we can also try to repeat our subgroup analysis from the previous chapter (Chapter 7.3), but this time within a meta-regression framework. This means that we use the risk of bias assessment as a categorical predictor. Since the variable `RiskOfBias` is already included in the `ThirdWave` data set, we do not have to save this information in an additional object. It suffices to simply run the `metareg` function again, but this time, we use `RiskOfBias` as the second function argument.

```
metareg(m.gen, RiskOfBias)
```

```
## [...]
## R^2 (amount of heterogeneity accounted for):             15.66%
##
## Test for Residual Heterogeneity:
## QE(df = 16) = 39.3084, p-val = 0.0010
##
## Test of Moderators (coefficient 2):
## F(df1 = 1, df2 = 16) = 2.5066, p-val = 0.1329
##
## Model Results:
##
##                  estimate    se   tval    pval  ci.lb ci.ub
## intrcpt              0.76  0.15   5.00  0.0001   0.44  1.09  ***
## RiskOfBiaslow       -0.29  0.18  -1.58  0.1329  -0.69  0.10
## [...]
```

In the output, we see that the value of R^2_*, with 15.66%, is considerably smaller than the one of `year`. Consistent with our previous results, we see that the risk of bias variable is not a significant effect size predictor (p = 0.13).

Under `Model Results`, we see that `metareg` has automatically transformed `RiskOfBias` into a dummy variable. The estimate of the intercept, which represents the pooled effect of the "high risk" subgroup, is g=0.76. The estimate of the regression coefficient representing studies with a *low* risk of bias is -0.29. To get the effect size for this subgroup, we have to add the regression weight to the intercept, which results in $g = 0.76 - 0.29 \approx 0.47$. These results are identical to the ones of a subgroup analysis which assumes a common estimate of τ^2.

8.3 Multiple Meta-Regression

Previously, we only considered the scenario in which we use *one* predictor βx_k in our meta-regression model. In the example, we checked if the effect size of a study depends on the year it was published. But now, suppose that reported effect sizes also depend on the *prestige* of the scientific journal in which the study was published. We think that it might be possible that studies in journals with a high reputation report higher effects. This could be because prestigious journals are more selective and mostly publish studies with "ground-breaking" findings.

On the other hand, it is also plausible that journals with a good reputation generally publish studies of a *higher quality*. Maybe it is just the better study quality that is associated with higher effect sizes. So, to check if journal reputation is indeed associated with higher effects, we have to make sure that this relationship is not *confounded* by the fact that prestigious journals are more likely to publish high-quality evidence. This means we have to *control* for study quality when examining the relationship between journal prestige and effect size.

This, and many other research questions, can be dealt with using *multiple meta-regression*. In multiple meta-regression, we use several predictors instead of just one to explain variation in effects. To allow for multiple predictors, we need to modify our previous meta-regression formula (see Equation 8.2), so that it looks like this:

$$\hat{\theta}_k = \theta + \beta_1 x_{1k} + ... + \beta_n x_{nk} + \epsilon_k + \zeta_k \tag{8.9}$$

This formula tells us that we can add $n-1$ more predictors x to our meta-regression model, thus turning it into a multiple meta-regression. The three dots in the formula symbolize that, in theory, we can add as many predictors as desired. In reality, however, things are usually more tricky. In the following, we will discuss a few important pitfalls in multiple meta-regression, and how we can build models that are robust and trustworthy. But first, let us cover another important feature of multiple meta-regression, *interactions*.

8.3.1 Interactions

So far, we only considered the case where we have multiple predictor variables $x_1, x_2, ... x_n$ in our model, which are added together along with their regression weights β. Multiple meta-regression models, however, are not only restricted to such *additive* relationships. They can also model predictor *interactions*. An interaction means that the *relationship* between one predictor (e.g. x_1) and the estimated effect size *changes* for different values of another covariate (e.g. x_2).

Imagine that we want to model two predictors and how they are associated with effect sizes: the publication year (x_1) and the quality (x_2) of a study. The study quality is coded like this:

$$x_2 = \begin{cases} 0: & \text{low} \\ 1: & \text{moderate} \\ 2: & \text{high.} \end{cases} \tag{8.10}$$

When we assume that there is no interaction between publication year and study quality, we can build a meta-regression model by giving both x_1 and x_2 a regression weight β, and *adding* the terms together in our formula:

$$\hat{\theta}_k = \theta + \beta_1 x_{1k} + \beta_2 x_{2k} + \epsilon_k + \zeta_k \tag{8.11}$$

But what if the relationship between x_1 and x_2 is more complex? It is possible, like in our previous example, that a more recent publication year is positively associated with higher effects. But not all studies must follow this trend. Maybe the increase is most pronounced among high-quality studies, while the results of low-quality studies stayed largely the same over time. We can visualize this assumed relationship between effect size ($\hat{\theta}_k$), publication year (x_1), and study quality (x_2) in the following way:

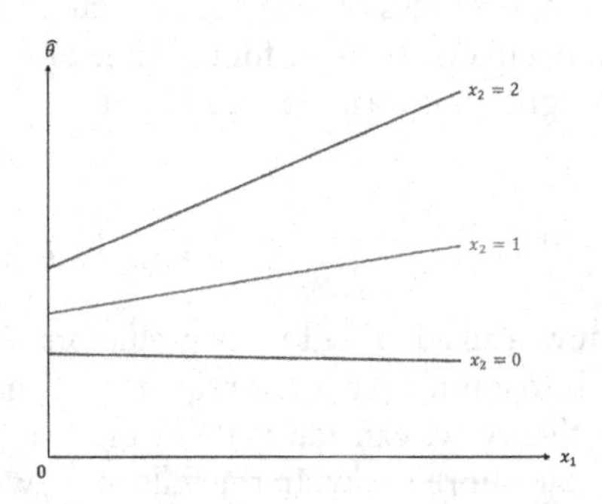

The graph shows a classic example of an interaction. We see that the steepness of the regression slope depends on the value of another predictor. While the slope for high-quality studies is very steep, indicating a strong relationship between year and effect, the situation is different for low-quality studies. The regression line in this subgroup is virtually horizontal, indicating that the publication year has no, or even a slightly negative effect on the results. This example shows one of the strengths of interactions: they allow us to examine if the influence of a predictor is the same across all studies, or if it is moderated by another characteristic.

To assess interactions via meta-regression, we need to add an *interaction term* to the model. In our example, this can be achieved by adding a third regression weight β_3, which captures the interaction $x_{1k}x_{2k}$ we want to test in our model. This gives the following formula:

$$\hat{\theta}_k = \theta + \beta_1 x_{1k} + \beta_2 x_{2k} + \beta_3 x_{1k} x_{2k} + \epsilon_k + \zeta_k \tag{8.12}$$

Although linear multiple meta-regression models only consist of these simple building blocks, they lend themselves to various applications. Before we start fitting multiple meta-regressions using R, however, we should first consider their limitations and pitfalls.

8.3.2 Common Pitfalls in Multiple Meta-Regression

Multiple meta-regression, while very useful when applied properly, comes with certain caveats. Some argue that (multiple) meta-regression is often improperly used and interpreted in practice, leading to a low validity of the results (Higgins and Thompson, 2004). There are some points we have to keep in mind when fitting multiple meta-regression models, which we describe in the following.

8.3.2.1 Overfitting: Seeing A Signal Where There Is None

To better understand the risks of (multiple) meta-regression models, we have to understand the concept of *overfitting*. Overfitting occurs when we build a statistical model that fits the data *too* closely. In essence, this means that we build a statistical model which can predict the data *at hand* very well, but performs badly at predicting *future* data. This happens when our model assumes that some variation in our data stems from a true "signal", when in fact we only capture random noise (Iniesta et al., 2016). As a result, the model produces *false positive* results: it sees relationships where there are none.

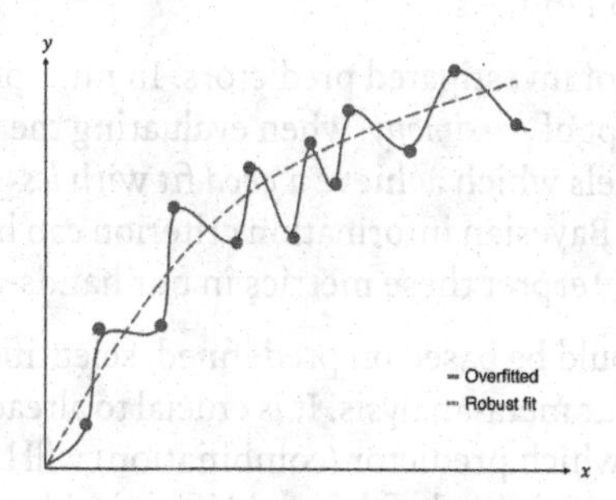

FIGURE 8.3: Predictions of an overfitted model versus model with a robust fit.

For model fitting, regression utilizes *optimization* techniques such as ordinary least squares or maximum likelihood estimation. As we learned, meta-regression uses a weighted version of ordinary least squares (see Chapter 8.1.3), and is, therefore, no exception. This "greedy" optimization, however, means that regression approaches can be prone to overfitting (Gigerenzer, 2004). Unfortunately, the risk of building a non-robust model is even higher once we go from conventional to meta-regression. There are several reasons for this (Higgins and Thompson, 2004):

1. In meta-regression, the number of data points is usually small, since we can only use the aggregated information of the included studies.
2. Because meta-analysis aims to be a comprehensive overview of all available evidence, we have no additional data on which we can "test" how well our regression model can predict unseen data.

3. In meta-regression, we have to deal with the potential presence of effect size heterogeneity. Imagine a case in which we have two studies with different effect sizes and non-overlapping confidence intervals. Every variable which has different values for the two studies might then be a potential explanation for the effect size difference. Yet, it seems clear that most of these explanations will be spurious.
4. Meta-regression in general, and multiple meta-regression in particular, makes it very easy to "play around" with predictors. We can test numerous meta-regression models, include more predictors or remove them, in an attempt to explain the heterogeneity in our data. Such an approach is tempting and often found in practice, because meta-analysts want to find an explanation why effect sizes differ (Higgins et al., 2002). However, such behavior has been shown to massively increase the risk of spurious findings, because we can change parts of our model indefinitely until we find a significant model, which is then very likely to be overfitted (i.e. it mostly models statistical noise).

Some guidelines have been proposed to avoid an excessive false positive rate when building meta-regression models:

- Minimize the number of investigated predictors. In multiple meta-regression, this translates to the concept of *parsimony*: when evaluating the fit of a meta-regression model, we prefer models which achieve a *good* fit with *less* predictors. Estimators such as the Akaike and Bayesian information criterion can help with such decisions. We will show how to interpret these metrics in our hands-on example.
- Predictor selection should be based on predefined, scientifically relevant questions we want to answer in our meta-analysis. It is crucial to already define in the analysis report (Chapter 1.4.2) which predictor (combination) will be included in the meta-regression model. It is not the end of the world if we decide to run a meta-regression that is not mentioned in our analysis plan. In this case, however, we should be honest and mention in our meta-analysis report that we decided to fit the model *after* seeing the data.
- When the number of studies is low (which is very likely to be the case), and we want to compute the significance of a predictor, we should use the Knapp-Hartung adjustment to obtain more robust estimates.
- We can use *permutation* to assess the robustness of our model in resampled data. We will describe the details of this method later.

8.3.2.2 Multi-Collinearity

Multi-collinearity means that one or more predictors in our regression model can be predicted by another model predictor with high accuracy (Mansfield and Helms, 1982). This typically means that we have two or more independent variables in our model which are highly correlated. Most of the dangers of multi-collinearity are associated with the problem of overfitting. High collinearity can cause our predictor coefficient estimates $\hat{\beta}$ to behave erratically and change considerably with minor changes in our data. It also limits the size of the explained variance by the model, in our case R^2_*.

Multi-collinearity in meta-regression is common (Berlin and Antman, 1994). Although multiple regression can handle low degrees of collinearity, we should check and, if necessary, control for very highly correlated predictors. There is no consolidated yes-no-rule to determine the presence of multi-collinearity. A crude, but often effective way is to check for very high predictor correlations (i.e. $r \geq 0.8$) before fitting the model. Multi-collinearity can then be reduced by either (1) removing one of the close-to-redundant predictors or (2) trying to combine the predictors into one single variable.

8.3.2.3 Model Fitting Approaches

When building a multiple meta-regression model, there are different approaches to select and include predictors. Here, we discuss the most important ones, along with their strengths and weaknesses:

- **Forced entry**. In forced entry methods, all relevant predictors are forced into the regression model simultaneously. For most functions in R, this is the default setting. Although this is a generally recommended procedure, keep in mind that all predictors to use via forced entry should still be based on a predefined, theory-led decision.

- **Hierarchical**. Hierarchical multiple regression means including predictors into our regression model step-wise, based on a clearly defined scientific rationale. First, only predictors which have been associated with effect size differences in previous research are included in the order of their importance. After this step, novel predictors can be added to explore if these variables explain heterogeneity which has not yet been captured by the known predictors.

- **Step-wise**. Step-wise entry means that variables/predictors are added to the model one after another. At first glance, this sounds a lot like hierarchical regression, but there is a crucial difference: step-wise regression methods select predictors based on a *statistical criterion*. In a procedure called *forward selection*, the variable explaining the largest amount of variability in the data is used as the first predictor. This process is then repeated for the remaining variables, each time selecting the variable which explains most of the residual unexplained variability in the data. There is also a procedure called *backward selection*, in which all variables are used as predictors in the model first, and then removed successively based on a predefined statistical criterion. There is an extensive literature discouraging the usage of step-wise methods (Chatfield, 1995; Whittingham et al., 2006). If we recall the common pitfalls of multiple regression models we presented above, it becomes clear that these methods have high risk of producing overfitted models with spurious findings. Nevertheless, step-wise methods are still frequently used in practice, which makes it important to know that these procedures exist. If we use stepwise methods ourselves, however, it is advised to primarily do so in an exploratory fashion and to keep the limitations of this procedure in mind.

- **Multi-model inference**. Multi-model methods differ from step-wise approaches because they do not try to successively build the one "best" model explaining most of the variance. Instead, in this technique, *all* possible predictor combinations are modeled. This means that several different meta-regressions are created, and subsequently evaluated. This enables a full examination of all possible predictor combinations, and how they perform. A common finding is that there are many different specifications which lead to a good model fit. The estimated coefficients of predictors can then be pooled across all fitted models to infer how important certain variables are overall.

8.3.3 Multiple Meta-Regression in R

After all this input, it is high time that we start fitting our first multiple meta-regression using R. The following examples will be the first ones in which we will not use the *{meta}* package. Instead, we will have a look at *{metafor}* (Viechtbauer, 2010). This package provides a vast array of advanced functionality for meta-analysis, along with a great documentation. So, to begin, make sure you have *{metafor}* installed, and loaded from the library.

```
library(metafor)
```

In our hands-on illustration, we will use the `MVRegressionData` data set. This is a "toy" data set, which we simulated for illustrative purposes.

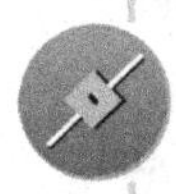

The "MVRegressionData" Data Set

The `MVRegressionData` data set is included directly in the *{dmetar}* package. If you have installed *{dmetar}*, and loaded it from your library, running `data(MVRegressionData)` automatically saves the data set in your R environment. The data set is then ready to be used. If you do not have *{dmetar}* installed, you can download the data set as an *.rda* file from the Internet[a], save it in your working directory, and then click on it in your R Studio window to import it.

[a] `https://www.protectlab.org/meta-analysis-in-r/data/MVRegressionData.rda`

First, let us have a look at the structure of the data frame:

```
library(tidyverse)
library(dmetar)
data(MVRegressionData)

glimpse(MVRegressionData)
```

```
## Rows: 36
## Columns: 6
## $ yi         <dbl> 0.09438, 0.09982, 0.16932, 0.17...
## $ sei        <dbl> 0.1959, 0.1919, 0.1193, 0.1162,...
## $ reputation <dbl> -11, 0, -11, 4, -10, -9, -8, -8...
## $ quality    <dbl> 6, 9, 5, 9, 2, 10, 6, 3, 10, 3,...
## $ pubyear    <dbl> -0.85475, -0.75277, -0.66048, -...
## $ continent  <fct> 1, 0, 1, 0, 1, 0, 0, 0, 0, 0, 0...
```

We see that there are six variables in our data set. The `yi` and `sei` columns store the effect size and standard error of a particular study. These columns correspond with the `TE` and `seTE` columns we used before. We have named the variables this way because this is the standard notation that *{metafor}* uses: `yi` represents the observed effect size y_i we want to predict in our (meta-)regression, while `sei` represents SE_i, the standard error of study i.

The other four variables are predictors to be used in the meta-regression. First, we have `reputation`, which is the (mean-centered) *impact factor* of the journal the study was published in. Impact factors quantify how often articles in a journal are cited, which we use as a proxy for the journal's prestige. The other variables are `quality`, the quality of the study rated from 0 to 10, `pubyear`, the (centered and scaled) publication year, and `continent`, the continent on which the study was performed. All of these variables are continuous, except for `continent`. The latter is a categorical variable with two levels: Europe and North America.

8.3.3.1 Checking for Multi-Collinearity

As we mentioned before, we have to check for multi-collinearity of our predictors to make sure that the meta-regression coefficient estimates are robust. A quick way to check for high correlations is to calculate a *intercorrelation matrix* for all continuous variables. This can be done using the cor function:

```
MVRegressionData[,c("reputation", "quality", "pubyear")] %>% cor()
```

```
##            reputation quality pubyear
## reputation     1.0000  0.3016  0.3347
## quality        0.3016  1.0000 -0.1551
## pubyear        0.3347 -0.1551  1.0000
```

The *{PerformanceAnalytics}* package (Peterson and Carl, 2020) contains a function called `chart.Correlation`, which we can use to visualize the correlation matrix. We have to install the `PerformanceAnalytics` package first, and then use this code:

```
library(PerformanceAnalytics)

MVRegressionData[,c("reputation", "quality", "pubyear")] %>%
  chart.Correlation()
```

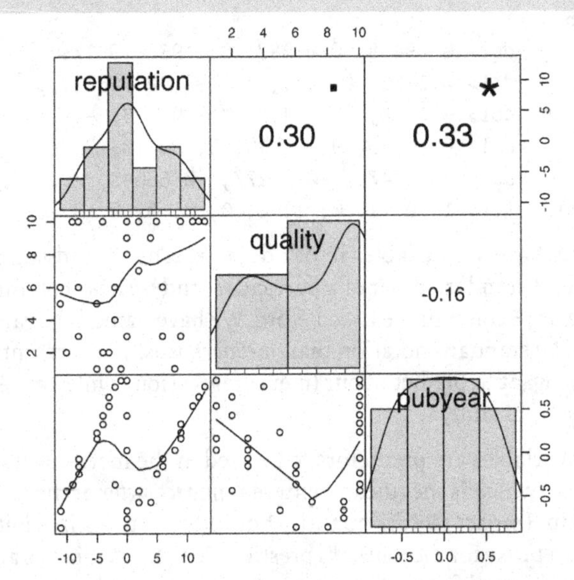

We see that our variables are indeed correlated, but probably not to a degree that would warrant excluding one of them.

8.3.3.2 Fitting a Multiple Meta-Regression Model

Now, we can fit our first meta-regression model using *{metafor}*. Previously, we wanted to explore if a high journal reputation predicts higher effect sizes, or if this is just an artifact caused by the fact that studies in prestigious journals have a higher quality. Let us assume we already know very well, for example, from previous research, that the quality of a study is predictive of its effect size. If this is the case, it makes sense to perform a hierarchical regression: we first include our known predictor `quality`, and then check if `reputation` explains heterogeneity beyond that. When this is true, we can say that journal reputation is indeed associated with higher effects, even if we *control* for the fact that studies in prestigious journals also tend to have a higher quality.

To do this, we use the `rma` function in *{metafor}*. This function runs a random-effects meta-analysis, which is extended to mixed-effects meta-regression models when moderators are added. The `rma` function can take countless arguments, which we can examine by running `?rma` in the R console. Usually, however, we only need to specify a few of them:

- `yi`. The column in our data frame in which the effect size of each study is stored.
- `sei`. The column in our data frame in which the standard error of the effect size of each study is stored.
- `data`. The name of the data frame containing all our meta-analysis data.
- `method`. The τ^2 estimator we want to use. The codes we can use for this argument are identical to the ones in *{meta}* (e.g. `"REML"` for restricted maximum likelihood). It is advisable to use `"ML"`, because this allows one to compare different meta-regression models later on.
- `mods`. This parameter defines our meta-regression model. First, we specify our model with ~ (a tilde). Then, we add the predictors we want to include, separating them with + (e.g. `variable1 + variable2`). Interactions between two variables are denoted by an asterisk (e.g. `variable1 * variable2`).
- `test`. The test we want to apply for our regression coefficients. We can choose from `"z"` (default) and `"knha"` (Knapp-Hartung method).

First, let us perform a meta-regression using only `quality` as a predictor. We save the results into an object called `m.qual`, and then inspect the output.

```
m.qual <- rma(yi = yi,
              sei = sei,
              data = MVRegressionData,
              method = "ML",
              mods = ~ quality,
              test = "knha")

m.qual
```

```
## Mixed-Effects Model (k = 36; tau^2 estimator: ML)
##
## tau^2 (estimated amount of residual heterogeneity):
     0.066 (SE = 0.023)
## tau (square root of estimated tau^2 value):             0.2583
## I^2 (residual heterogeneity / unaccounted variability): 60.04%
## H^2 (unaccounted variability / sampling variability):   2.50
## R^2 (amount of heterogeneity accounted for):            7.37%
##
## Test for Residual Heterogeneity:
## QE(df = 34) = 88.6130, p-val < .0001
##
## Test of Moderators (coefficient 2):
## F(df1 = 1, df2 = 34) = 3.5330, p-val = 0.0688
```

```
##
## Model Results:
##
##          estimate      se    tval    pval    ci.lb   ci.ub
## intrcpt    0.3429  0.1354  2.5318  0.0161   0.0677  0.6181  *
## quality    0.0356  0.0189  1.8796  0.0688  -0.0029  0.0740  .
##
## ---
## Signif. codes:  0 '***' 0.001 '**' 0.01 '*' 0.05 '.' 0.1 ' ' 1
```

In the output, we can inspect the results for our predictor `quality` under `Model Results`. We see that the regression weight is not significant ($p = 0.063$), although it is significant on a trend level ($p < 0.1$). In total, our model explains $R^2_* = 7.37\%$ of the heterogeneity.

Now, let us see what happens when we include `reputation` as a predictor. We add `+ reputation` to our input to `mods` and this time, we save the output as `m.qual.rep`.

```
m.qual.rep <- rma(yi = yi,
                  sei = sei,
                  data = MVRegressionData,
                  method = "ML",
                  mods = ~ quality + reputation,
                  test = "knha")

m.qual.rep
```

```
## Mixed-Effects Model (k = 36; tau^2 estimator: ML)
##
## tau^2 (estimated amount of residual heterogeneity):
     0.0238 (SE = 0.01)
## tau (square root of estimated tau^2 value):             0.1543
## I^2 (residual heterogeneity / unaccounted variability): 34.62%
## H^2 (unaccounted variability / sampling variability):   1.53
## R^2 (amount of heterogeneity accounted for):            66.95%
##
## Test for Residual Heterogeneity:
## QE(df = 33) = 58.3042, p-val = 0.0042
##
## Test of Moderators (coefficients 2:3):
## F(df1 = 2, df2 = 33) = 12.2476, p-val = 0.0001
##
## Model Results:
##
##             estimate      se    tval    pval   ci.lb   ci.ub
## intrcpt       0.5005  0.1090  4.5927  <.0001  0.2788  0.7222  ***
```

```
## quality       0.0110  0.0151 0.7312  0.4698 -0.0197 0.0417
## reputation    0.0343  0.0075 4.5435  <.0001  0.0189 0.0496  ***
##
## ---
## Signif. codes:  0 '***' 0.001 '**' 0.01 '*' 0.05 '.' 0.1 ' ' 1
```

We see that now, a new line appears in the `Model Results` section, displaying the results for our `reputation` predictor. The model estimated the regression weight to be 0.034, which is highly significant ($p < 0.001$). We also see that the meta-regression model as a whole explains a significant amount of heterogeneity, $R^2_* = 66.95\%$ to be precise. This means that journal reputation is associated with higher effect sizes, even when controlling for study quality.

But does our second model indeed provide a better fit than our first one? To assess this, we can use the *anova* function, providing it with the two models we want to compare. Note that this is only feasible because we fitted both mixed-effects models using maximum likelihood (`"ML"`) instead of restricted maximum likelihood (`"REML"`).

```
anova(m.qual, m.qual.rep)
```

```
##          df   AIC    BIC   AICc logLik   LRT   pval    QE tau^2    R^2
## Full      4 19.86 26.19 21.15  -5.93               58.30  0.03
## Reduced   3 36.98 41.73 37.73 -15.49 19.11 <.0001 88.61  0.06 48.32%
```

This function performs a model test and provides us with several statistics to assess if `m.qual.rep` has a better fit than `m.qual`. We compare our full model, `m.qual.rep`, which includes both `quality` and `reputation`, with the reduced model, which only includes `quality`.

The *anova* function performs a *likelihood ratio test*, the results of which we can see in the `LRT` column. The test is highly significant ($\chi^2_1 = 19.11, p < 0.001$), which means that that our full model indeed provides a better fit.

Another important statistic is reported in the `AICc` column. This provides us with Akaike's information criterion (AIC), corrected for small samples. As we mentioned before, AICc penalizes complex models with more predictors to avoid overfitting. It is important to note that lower values of AIC mean that a model performs better. In our output, we see that the full model (AICc = 21.15) has a better AIC value than our reduced model (AICc = 37.73), despite having more parameters. All of this suggests that our multiple regression model does indeed provide a good fit to our data.

8.3.3.3 Modeling Interactions

Let us say we want to model an interaction with our additional predictors `pubyear` (publication year) and `continent`. We assume that the relationship between publication year and effect size differs for European and North American studies. To model this assumption using the *rma* function, we have to connect our predictors with `*` in the `mods` parameter. Because we do not want to compare the models directly using the anova function, we use the "REML" (restricted maximum likelihood) τ^2 estimator this time.

To facilitate the interpretation, we add factor labels to the continent variable in MVRegressionData before running the model.

```
# Add factor labels to 'continent'
# 0 = Europe
# 1 = North America
levels(MVRegressionData$continent) = c("Europe", "North America")

# Fit the meta-regression model
m.qual.rep.int <- rma(yi = yi,
                      sei = sei,
                      data = MVRegressionData,
                      method = "REML",
                      mods = ~ pubyear * continent,
                      test = "knha")

m.qual.rep.int
```

```
## Mixed-Effects Model (k = 36; tau^2 estimator: REML)
##
## tau^2 (estimated amount of residual heterogeneity):     0 (SE = 0.01)
## tau (square root of estimated tau^2 value):             0
## I^2 (residual heterogeneity / unaccounted variability): 0.00%
## H^2 (unaccounted variability / sampling variability):   1.00
## R^2 (amount of heterogeneity accounted for):            100.00%
##
## Test for Residual Heterogeneity:
## QE(df = 32) = 24.8408, p-val = 0.8124
##
## Test of Moderators (coefficients 2:4):
## F(df1 = 3, df2 = 32) = 28.7778, p-val < .0001
##
## Model Results:
##
##                          estimate    se   tval    pval  ci.lb ci.ub
## intrcpt                      0.38  0.04   9.24  <.0001   0.30  0.47  ***
## pubyear                      0.16  0.08   2.01  0.0520  -0.00  0.33    .
## continentNorth America       0.39  0.06   6.05  <.0001   0.26  0.53  ***
## pubyear:continent            0.63  0.12   4.97  <.0001   0.37  0.89  ***
##    North America
## [...]
```

The last line, pubyear:continentNorth America, contains the coefficient for our interaction term. Note that *{metafor}* automatically includes not only the interaction term but also both `pubyear` and `continent` as "normal" lower-order predictors (as one should do). Also note that, since `continent` is a factor, `rma` detected that this is a dummy-coded predictor, and used our category "Europe" as the $D_g = 0$ baseline against which the North America category is compared. We see that our interaction term has a positive coefficient 0.63), and is highly significant ($p < 0.001$).

This indicates that there is an increase in effect sizes in recent years, but that it is stronger in studies conducted in North America. We also see that the model we fitted explains $R^2_* = 100\%$ of our heterogeneity. This is because our data was simulated for illustrative purposes. In practice, you will hardly ever explain all of the heterogeneity in your data–in fact, one should rather be concerned if one finds such results in real-life data, as this might mean that we have overfitted our model.

8.3.3.4 Permutation Test

Permutation is a mathematical operation in which we take a set containing numbers or objects, and iteratively draw elements from this set to put them in a sequential order. When we already have an ordered set of numbers, this equals a process in which we rearrange, or *shuffle*, the order of our data. As an example, imagine we have a set S containing three numbers: $S = \{1, 2, 3\}$. One possible permutation of this set is $(2, 1, 3)$; another is $(3, 2, 1)$. We see that the permuted results both contain all three numbers from before, but in a different order.

Permutation can also be used to perform *permutation tests*, which is a specific type of resampling method. Broadly speaking, resampling methods are used to validate the robustness of a statistical model by providing it with (slightly) different data sampled from the same source or generative process (Good, 2013, chapter 3.1). This is a way to better assess if the coefficients in our model indeed capture a true pattern underlying our data; or if we overfitted our model, thereby falsely assuming patterns in our data when they are in fact statistical noise.

Permutation tests do not require that we have a spare "test" data set on which we can evaluate how our meta-regression performs in predicting unseen effect sizes. For this reason, among others, permutation tests have been recommended to assess the robustness of our meta-regression models (Higgins and Thompson, 2004).

We will not go too much into the details of how a permutation test is performed for meta-regression models. The most important part is that we re-calculate the p-values of our model based on the test statistics obtained across all possible, or many randomly selected, permutations of our original data set. The crucial indicator here is *how often* the test statistic we obtain from in our permuted data is *equal to or greater* than our original test statistic. For example, if our test statistic is greater or equal to the original one in 50 of 1000 permuted data sets, we get a p-value of $p = 0.05$ (see also Viechtbauer et al., 2015).

To perform a permutation test on our meta-regression model, we can use *{metafor}*'s in-built `permutest` function. As an example, we recalculate the results of the `m.qual.rep` model we fitted before. We only have to provide the `permutest` function with the `rma` object. Be aware that the permutation test is computationally expensive, especially for large data sets. This means that the function might need some time to run.

```
permutest(m.qual.rep)
```

```
## Test of Moderators (coefficients 2:3):
## F(df1 = 2, df2 = 33) = 12.7844, p-val* = 0.0010
##
## Model Results:
##
##             estimate      se    tval   pval*    ci.lb   ci.ub
## intrcpt       0.4964  0.1096  4.5316  0.2240   0.2736  0.7193
## quality       0.0130  0.0152  0.8531  0.3640  -0.0179  0.0438
## reputation    0.0350  0.0076  4.5964  0.0010   0.0195  0.0505  ***
##
## ---
## Signif. codes:  0 '***' 0.001 '**' 0.01 '*' 0.05 '.' 0.1 ' ' 1
```

We again see our familiar output including the results for all predictors. Looking at the `pval*` column, we see that our p-value for the reputation predictor has decreased from $p < 0.001$ to $p_* = 0.001$. This, however, is still highly significant, indicating that the effect of the predictor is robust. It has been recommended to always use this permutation test before reporting the results of a meta-regression model (Higgins and Thompson, 2004).

Permutation Tests in Small Data Sets

Please note that when the number of studies K included in our model is small, conventionally used thresholds for statistical significance (i.e. $p < 0.05$) cannot be reached. For meta-regression models, a permutation test using `permutest` will only be able to reach statistical significance if $K > 4$.

8.3.3.5 Multi-Model Inference

We already mentioned that one can also try to model all possible predictor combinations in a procedure called *multi-model inference*. This allows to examine which possible predictor combination provides the best fit, and which predictors are the most important ones overall. To perform multi-model inference, we can use the `multimodel.inference` function.

The "multimodel.inference" Function

The `multimodel.inference` function is included in the *{dmetar}* package. Once *{dmetar}* is installed and loaded on your computer, the function is ready to be used. If you did not install *{dmetar}*, follow these instructions:

1. Access the source code of the function online[a].
2. Let R "learn" the function by copying and pasting the source code in its entirety into the console (bottom left pane of R Studio), and then hit "Enter".
3. Make sure that the *{metafor}*, *{ggplot2}* and *{MuMIn}* package is installed and loaded.

[a] `https://raw.githubusercontent.com/MathiasHarrer/dmetar/master/R/mreg.multimodel.inference.R`

In the function, the following parameters need to be specified:

- `TE`. The effect size of each study. Must be supplied as the name of the effect size column in the data set, in quotation marks (e.g. `TE = "effectsize"`).
- `seTE`. The standard error of the effect sizes. Must be supplied as the name of the standard error column in the data set (also in quotation marks, e.g. `seTE = "se"`).
- `data`. A data frame containing the effect size, standard error and meta-regression predictor(s).
- `predictors`. A concatenated array of characters specifying the predictors to be used for multi-model inference. Names of the predictors must be identical to the column names of the data frame supplied to `data`.
- `method`. Meta-analysis model to use for pooling effect sizes. `"FE"` is used for the fixed-effect model. Different random-effect models are available, for example, `"DL"`, `"SJ"`, `"ML"`, or `"REML"`. If `"FE"` is used, the `test` argument is automatically set to `"z"`, as the Knapp-Hartung method is not meant to be used with fixed-effect models. Default is `"REML"`.
- `test`. Method to use for computing test statistics and confidence intervals. Default is `"knha"`, which uses the Knapp-Hartung adjustment. Conventional Wald-type tests are calculated by setting this argument to `"z"`.
- `eval.criterion`. Evaluation criterion to apply to the fitted models. Can be either `"AICc"` (default; small sample-corrected Akaike's information criterion), `"AIC"` (Akaike's information criterion) or `"BIC"` (Bayesian information criterion).
- `interaction`. When set to `FALSE` (default), no interactions between predictors are considered. Setting this parameter to `TRUE` means that all interactions are modeled.

Now, let us perform multi-model inference using all predictors in the MVRegressionData data set, but *without* interactions. Be aware that running the multimodel.inference function can take some time, especially if the number of predictors is large.

```
multimodel.inference(TE = "yi",
                     seTE = "sei",
                     data = MVRegressionData,
                     predictors = c("pubyear", "quality",
                                    "reputation", "continent"),
                     interaction = FALSE)
```

```
## Multimodel Inference: Final Results
## --------------------------
##  - Number of fitted models: 16
##  - Full formula: ~ pubyear + quality + reputation + continent
##  - Coefficient significance test: knha
##  - Interactions modeled: no
##  - Evaluation criterion: AICc
##
##
## Best 5 Models
## --------------------------
## [...]
##    (Intrc) cntnn  pubyr   qulty   rpttn df logLik AICc delta weight
## 12       +     + 0.3533         0.02160  5  2.981  6.0  0.00  0.536
## 16       +     + 0.4028 0.02210 0.01754  6  4.071  6.8  0.72  0.375
## 8        +     + 0.4948 0.03574          5  0.646 10.7  4.67  0.052
## 11       +       0.2957         0.02725  4 -1.750 12.8  6.75  0.018
## 15       +       0.3547 0.02666 0.02296  5 -0.395 12.8  6.75  0.018
## Models ranked by AICc(x)
##
##
## Multimodel Inference Coefficients
## --------------------------
##                          Estimate  Std. Error   z value  Pr(>|z|)
## intrcpt                0.38614661 0.106983583 3.6094006 0.0003069
## continentNorth America 0.24743836 0.083113174 2.9771256 0.0029096
## pubyear                0.37816796 0.083045572 4.5537402 0.0000053
## reputation             0.01899347 0.007420427 2.5596198 0.0104787
## quality                0.01060060 0.014321158 0.7402055 0.4591753
##
##
## Predictor Importance
## --------------------------
```

```
##         model importance
## 1    pubyear  0.9988339
## 2  continent  0.9621839
## 3 reputation  0.9428750
## 4    quality  0.4432826
```

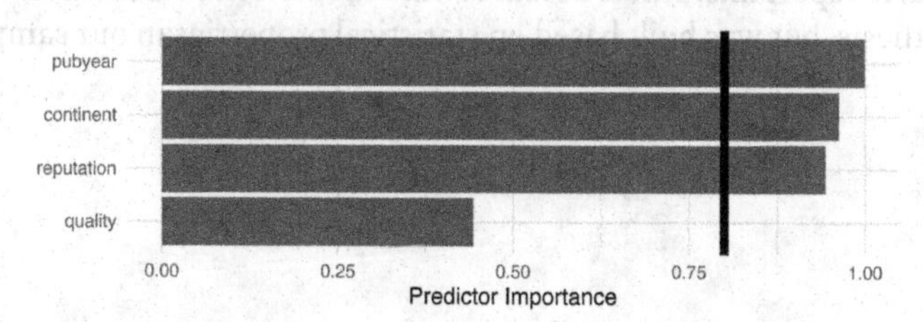

There is a lot to see here, so let us go through the output step by step.

- `Multimodel Inference: Final Results`. This part of the output provides us with details about the fitted models. We see that the total number of $2^4 = 16$ possible models have been fitted. We also see that the function used the corrected AIC (`aicc`) to compare the models.
- `Best 5 Models`. Displayed here are the five models with the lowest AICc, sorted from low to high. Predictors are shown in the columns of the table, and models in the rows. A number (weight) or + sign (for categorical predictors) indicates that a predictor/interaction term was used in the model, while empty cells indicate that the predictor was omitted. We see that `TE ~ 1 + continent + pubyear + reputation` shows the best fit (AICc = 6.0). But other predictor combinations come very close to this value. Thus, it is hard to say which model is really the "best" model. However, all top five models contain the predictor `pubyear`, suggesting that this variable might be particularly important.
- `Multimodel Inference Coefficients`. Here, we can see the coefficients of all predictors, aggregated over all models in which they appear. We see that the coefficient estimate is largest for `pubyear` (0.378), which corroborates our finding from before. Approximate confidence intervals can be obtained by subtracting and adding the value stored in `Std.Error`, multiplied by 1.96, from/to `Estimate`.
- **Model-averaged predictor importance plot**. In the plot, the averaged importance of each predictor across all models is displayed. We again see that `pubyear` is the most important predictor, followed by `reputation`, `continent`, and `quality`.

Limitations of Multi-Model Inference

This example should make clear that multi-model inference can be a useful way to obtain a comprehensive look at which predictors are important for predicting differences in effect sizes.

Despite avoiding some of the problems of step-wise regression methods, please note that this method should still be seen as exploratory, and may be used when we have no prior knowledge on how our predictors are related to effect sizes in the research field we analyze. If you decide to build a meta-regression model based on the results of multi-model inference, it is crucial to report this. This is because such a model is not based on an *a priori* hypothesis, but was built based on statistical properties in our sample.

□

8.4 Questions & Answers

Test your knowledge!

1. What is the difference between a conventional regression analysis as used in primary studies, and meta-regression?
2. Subgroup analyses and meta-regression are closely related. How can the meta-regression formula be adapted for subgroup data?
3. Which method is used in meta-regression to give individual studies a differing weight?
4. What characteristics mark a meta-regression model that fits our data well? Which index can be used to examine this?
5. When we calculate a subgroup analysis using meta-regression techniques, do we assume a separate or common value of τ^2 in the subgroups?
6. What are the limitations and pitfalls of (multiple) meta-regression?
7. Name two methods that can be used to improve the robustness of (multiple) meta-regression models, and why they are helpful.

Answers to these questions are listed in Appendix A at the end of this book.

8.5 Summary

- In meta-regression, we adapt conventional regression techniques to study-level data. Subgroup analyses can be seen as a special case of meta-regression with categorical predictors and a common estimate of τ^2.
- The aim of a meta-regression model is to explain (parts of) the true effect size differences in our data (i.e. the between-study heterogeneity variance τ^2). When a model fits the data well, the deviation of true effects from the regression line should be smaller than their initial deviation from the pooled effect. When this is the case, the unexplained, or residual, heterogeneity will be small. This is captured by the R^2_* index, which tells us the percentage of heterogeneity variation explained by our model.
- In multiple meta-regression, two or more predictors are used in the same meta-regression model. It is also possible to test if the predictions of one variable change for different values of another, by introducing interaction terms.
- Although (multiple) meta-regression is very versatile, it is not without limitations. Multiple meta-regression makes it very easy to overfit models, meaning that random noise instead of true relationships are modeled. Multi-collinearity of predictors may also pose a threat to the validity of our model.
- There are several approaches to ensure that our meta-regression model is robust. We can, for example, only fit models based on a predefined theoretical rationale, or use permutation tests. Multi-model inference can be used as an exploratory approach. This method can point us to potentially important predictors and can be used to derive hypotheses to be tested in future research.

9

Publication Bias

Looking back at the last chapters, we see that we already covered a vast range of meta-analytic techniques. Not only did we learn how to pool effect sizes, we also know now how to assess the robustness of our findings, inspect patterns of heterogeneity, and test hypotheses on why effects differ.

All of these approaches can help us to draw valid conclusions from our meta-analysis. This, however, rests on a tacit assumption concerning the nature of our data, which we have not challenged yet. When conducting a meta-analysis, we take it as a given that the data we collected is *comprehensive*, or at least *representative* of the research field under examination. Back in Chapter 1.4.3, we mentioned that meta-analyses usually try to include *all* available evidence, in order to derive a single effect size that adequately describes the research field. From a statistical perspective, we may be able to tolerate that a few studies are missing in our analysis–but only if these studies were "left out" by chance.

Unfortunately, meta-analyses are often unable to include all existing evidence. To make things worse, there are also good reasons to assume that some studies are not missing "at random" from our collected data. Our world is imperfect, and so are the incentives and "rules" that govern scientific practice. This means that there are systemic biases that can determine if a study ends up in our meta-analysis or not.

A good example of this problem can be found in a not-so-recent anecdote from pharmacotherapy research. Even back in the 1990s, it was considered secured knowledge that antidepressive medication (such as *selective serotonin re-uptake inhibitors*, or SSRIs) are effective in treating patients suffering from depression. Most of this evidence was provided by meta-analyses of published pharmocotherapy trials, in which an antidepressant is compared to a pill placebo. The question regarding the effects of antidepressive medication is an important one, considering that the antidepressant drug market is worth billions of dollars, and growing steadily. This may help to understand the turmoil caused by an article called *The Emperor's New Drugs*, written by Irving Kirsch and colleagues (2002), which argued that things may not look so bright after all.

Drawing on the "Freedom of Information Act", Kirsch and colleagues obtained previously unpublished antidepressant trial data which pharmaceutical companies had provided to the US Food and Drug Administration. They found that when this unpublished data was also considered, the benefits of antidepressants compared to placebos were at best minimal, and clinically negligible. Kirsch and colleagues argued that

DOI: 10.1201/9781003107347-9

this was because companies only published studies with favorable findings, while studies with "disappointing" evidence were withheld (Kirsch, 2010).

A contentious debate ensued, and Kirsch's claims have remained controversial until today. We have chosen this example not to pick sides, but to illustrate the potential threat that missing studies can pose to the validity of meta-analytic inferences. In the meta-analysis literature, such problems are usually summarized under the term *publication bias*.

The problem of publication bias underlines that every finding in meta-analyses can only be as good as the data it is based on. Meta-analytic techniques can only work with the data at hand. Therefore, if the collected data is distorted, even the best statistical model will only reproduce inherent biases. Maybe you recall that we already covered this fundamental caveat at the very beginning of this book, where we discussed the "File Drawer" problem (see Chapter 1.3). Indeed, the terms "file drawer problem" and "publication bias" are often used synonymously.

The consequences of publication bias and related issues on the results of meta-analyses can be enormous. It can cause us to overestimate the effects of treatments, overlook negative side effects, or reinforce the belief in theories that are actually invalid.

In this chapter, we will therefore discuss the various shapes and forms through which publication bias can distort our findings. We will also have a look at a few approaches that we as meta-analysts can use to examine the risk of publication bias in our data; and how publication bias can be mitigated in the first place.

9.1 What Is Publication Bias?

Publication bias exists when the probability of a study getting published is affected by its results (Rothstein et al., 2005, chapters 2 and 5). There is widespread evidence that a study is more likely to find its way into the public if its findings are statistically significant, or confirm the initial hypothesis (Schmucker et al., 2014; Scherer et al., 2018; Chan et al., 2014; Dechartres et al., 2018).

When searching for eligible studies, we are usually constrained to evidence that has been made public in some form or the other, for example, through peer-reviewed articles, preprints, books, or other kinds of accessible reports. In the presence of publication bias, this not only means that some studies are missing in our data set–it also means that the missing studies are likely the ones with unfavorable findings. Meta-analytic techniques allow us to find an unbiased estimate of the average effect size in the population. But if our sample itself is distorted, even an effect estimate that is "true" from a statistical standpoint will not be representative of the reality. It is like trying to estimate the size of an iceberg, but only measuring its tip: our finding will inevitably be wrong, even if we are able to measure the height above the water surface with perfect accuracy.

Publication bias is actually just one of many *reporting biases*. There are several other factors that can also distort the evidence that we obtain in our meta-analysis (Page et al., 2020), including:

- **Citation bias**: Even when published, studies with negative or inconclusive findings are less likely to be cited by related literature. This makes it harder to detect them through reference searches, for example.
- **Time-lag bias**: Studies with positive results are often published earlier than those with unfavorable findings. This means that findings of recently conducted studies with positive findings are often already available, while those with non-significant results are not.
- **Multiple publication bias**: Results of "successful" studies are more likely to be reported in several journal articles, which makes it easier to find at least one of them. The practice of reporting study findings across several articles is also known as "salami slicing".
- **Language bias**: In most disciplines, the primary language in which evidence is published is English. Publications in other languages are less likely to be detected, especially when the researchers themselves cannot understand the contents without translation. If studies in English systematically differ from the ones published in other languages, this may also introduce bias.
- **Outcome reporting bias**: Many studies, and clinical trials in particular, measure more than one outcome of interest. Some researchers exploit this, and only report those outcomes for which positive results were attained, while the ones that did not confirm the hypothesis are dropped. This can also lead to bias: technically speaking, the study has been published, but its (unfavorable) result will still be missing in our meta-analysis because it is not reported.

Non-reporting biases can be seen as systemic factors which make it harder for us to find existing evidence. However, even if we were able to include all relevant findings, our results may still be flawed. Bias may also exist due to *questionable research practices* (QRPs) that researchers have applied when analyzing and reporting their findings (Simonsohn et al., 2020).

We already mentioned the concept of "researcher degrees of freedom" previously (Chapter 1.3). QRPs can be defined as practices in which researchers abuse these degrees of freedom to "bend" results into the desired direction. Unfortunately, there is no clear consensus on what constitutes a QRP. There are, however, a few commonly suggested examples. One of the most prominent QRPs is *p-hacking*, in which analyses are tweaked until the conventional significance threshold of $p < 0.05$ is reached. This can include the way outliers are removed, analyses of subgroups, or the missing data handling.

Another QRP is *HARKing* (Kerr, 1998), which stands for *hypothesizing after the results are known*. One way of HARKing is to pretend that a finding in exploratory analyses has been an *a priori* hypothesis of the study all along. A researcher, for example, may run various tests on a data set, and then "invent" hypotheses around all the tests that were significant. This is a seriously flawed approach, which inflates the false discovery rate of a study, and thus increases the risk of spurious findings (to name just a few problems). Another type of HARKing is to drop all hypotheses that were not supported by the data, which can ultimately lead to outcome reporting bias.

9.2 Addressing Publication Bias in Meta-Analyses

It is quite clear that publication bias, other reporting biases and QRPs can have a strong and deleterious effect on the validity of our meta-analysis. They constitute major challenges since it is usually practically impossible to know the exact magnitude of the bias–or if it exists at all.

In meta-analyses, we can apply techniques which can, to some extent, reduce the risk of distortions due to publication and reporting bias, as well as QRPs. Some of these approaches pertain to the study search, while others are statistical methods.

- **Study search**. In Chapter 1.4.3, we discussed the process of searching for eligible studies. If publication bias exists, this step is of great import, because it means that that a search of the published literature may yield data that is not fully representative of all the evidence. We can counteract this by also searching for *grey literature*, which includes dissertations, preprints, government reports, or conference proceedings. Fortunately, pre-registration is also becoming more common in many disciplines. This makes it possible to search study registries such as the ICTRP or *OSF Registries* (see Table 1.1 in Chapter 1.4.3) for studies with unpublished data, and ask the authors if they can provide us with data that has not been made public (yet)[1]. Grey literature search can be tedious and frustrating, but it is worth the effort. One large study has found that the inclusion of grey and unpublished literature can help to avoid an overestimation of the true effects (McAuley et al., 2000).
- **Statistical methods**. It is also possible to examine the presence of publication through statistical procedures. None of these methods can identify publication bias directly, but they can examine certain properties of our data that may be indicative of it. Some methods can also be used to quantify the true overall effect when correcting for publication bias.

In this chapter, we will showcase common *statistical* methods to evaluate and control for publication bias. We begin with methods focusing on *small-study effects* (Sterne

[1] Mahmood and colleagues (2014) provide a detailed account of how a comprehensive grey literature search can be conducted, and what challenges this may entail. The article can be openly accessed online.

et al., 2000; Schwarzer et al., 2015, chapter 5). A common thread among these approaches is that they find indicators of publication bias by looking at the relationship between the precision and observed effect size of studies.

9.2.1 Small-Study Effect Methods

There are various small-study effect methods to assess and correct for publication bias in meta-analyses. Many of the techniques have been conventional for many years. As it says in the name, these approaches are particularly concerned with *small studies*. From a statistical standpoint, this translates to studies with a high standard error. Small-study effect methods assume that small studies are more likely to fall prey to publication bias. This assumption is based on three core ideas (see Borenstein et al., 2011, chapter 30):

1. Because they involve a large commitment of resources and time, large studies are likely to get published, no matter whether the results are significant or not.
2. Moderately sized studies are at greater risk of not being published. However, even when the statistical power is only moderate, this is still often sufficient to produce significant results. This means that only some studies will not get published because they delivered "undesirable" (i.e. non-significant) results.
3. Small studies are at the greatest risk of generating non-significant findings, and thus of remaining in the "file drawer". In small studies, only very large effects become significant. This means that only small studies with very high effect sizes will be published.

We see that the purported mechanism behind these assumptions is quite simple. Essentially, it says that publication bias exists because only significant effects are published. Since the probability of obtaining significant results rises with larger sample size, it follows that publication bias will disproportionately affect small studies.

9.2.1.1 The Funnel Plot

Earlier in this guide (Chapter 3.1), we learned that a study's sample size and standard error are closely related. Larger standard errors of an effect size result in wider confidence intervals and increase the chance that the effect is not statistically significant. Therefore, it is sensible to assume that small-study effects will largely affect studies with larger standard errors.

Suppose that our collected data is burdened by publication bias. If this is the case, we can assume that the studies with large standard errors have higher effect sizes than the ones with a low standard error. This is because the smaller studies with lower effects were not significant, and thus never considered for publication. Consequently, we never included them in our meta-analysis.

It is conventional to inspect small-study effects through *funnel plots*. A funnel plot is a scatter plot of the studies' observed effect sizes on the x-axis against a measure of their standard error on the y-axis. Usually, the y-axis in funnel plots is inverted (meaning that "higher" values on the y-axis represent *lower* standard errors).

When there is no publication bias, the data points in such a plot should form a roughly symmetrical, upside-down funnel. This is why they are called funnel plots. Studies in the top part of the plot (those with low standard errors) should lie closely together, and not far away from the pooled effect size. In the lower part of the plot, with increasing standard errors, the funnel "opens up", and effect sizes are expected to scatter more heavily to the left and right of the pooled effect.

It becomes easier to see why studies should form a funnel when we think back to what we learned about the behavior of effect sizes in Chapter 3.1, and when discussing the fixed-effect model in Chapter 4.1.1 (Figure 4.1). The standard error is indicative of a study's *precision*: with decreasing standard error, we expect the observed effect size to become an increasingly good estimator of the true effect size. When the standard error is high, the effect size has a low precision and is therefore much more likely to be far off from the actual effect in the population.

We will now make this more concrete by generating a funnel plot ourselves. In the *{meta}* package, the `funnel.meta` function can be used to print a funnel plot for a meta-analysis object. Here, we produce a funnel plot for our `m.gen` meta-analysis object. We specify two further arguments, `xlim` and `studlab`. The first controls the limits of the x-axis in the plot, while the latter tells the function to include study labels. A call to the `title` function after running `funnel` adds a title to the plot. Our code looks like this:

```
# Load 'meta' package
library(meta)

# Produce funnel plot
funnel.meta(m.gen,
            xlim = c(-0.5, 2),
            studlab = TRUE)

# Add title
title("Funnel Plot (Third Wave Psychotherapies)")
```

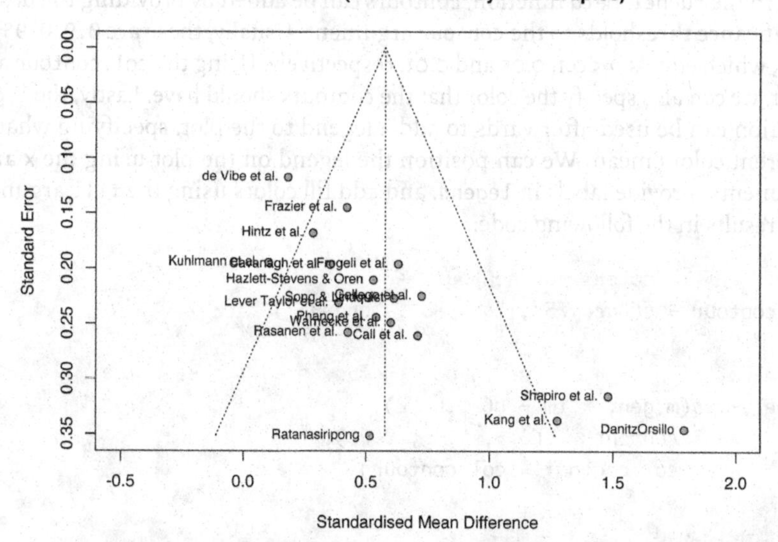

As discussed, the resulting funnel plot shows the effect size of each study (expressed as the standardized mean difference) on the x-axis, and the standard error (from large to small) on the y-axis. To facilitate the interpretation, the plot also includes the idealized funnel-shape that we expect our studies to follow. The vertical line in the middle of the funnel shows the average effect size. Because we used a random-effects model when generating `m.gen`, the funnel plot also uses the random-effects estimate.

In the absence of small-study effects, our studies should roughly follow the shape delineated by the funnel displayed in the plot. Is this the case in our example? Well, not really. We see that, while studies with lower standard errors lie more concentrated around the estimated true effect, the pattern overall looks asymmetrical. This is because there are three small studies with very high effect sizes in the bottom-right corner of the plot (the ones by Shapiro, Kang, and Danitz-Orsillo). These studies, however, have no equivalent in the bottom-left corner in the plot. There are no small studies with very low or negative effect sizes to "balance out" the ones with very high effects. Another worrisome detail is that the study with the greatest precision in our sample, the one by de Vibe, does not seem to follow the funnel pattern well either. Its effect size is considerably smaller than expected.

Overall, the data set shows an asymmetrical pattern in the funnel plot that *might* be indicative of publication bias. It could be that the three small studies are the ones that were lucky to find effects high enough to become significant, while there is an underbelly of unpublished studies with similar standard errors, but smaller and thus non-significant effects which did not make the cut.

A good way to inspect how asymmetry patterns relate to statistical significance is to generate *contour-enhanced funnel plots* (Peters et al., 2008). Such plots can help to distinguish publication bias from other forms of asymmetry. Contour-enhanced funnel plots include colors which signify the significance level of each study in the plot. In the `funnel.meta` function, contours can be added by providing the desired significance thresholds to the `contour` argument. Usually, these are `0.9`, `0.95` and `0.99`, which equals $p < 0.1$, 0.05 and 0.01, respectively. Using the `col.contour` argument, we can also specify the color that the contours should have. Lastly, the `legend` function can be used afterwards to add a legend to the plot, specifying what the different colors mean. We can position the legend on the plot using the `x` and `y` arguments, provide labels in `legend`, and add fill colors using the `fill` argument. This results in the following code:

```
# Define fill colors for contour
col.contour = c("gray75", "gray85", "gray95")

# Generate funnel plot (we do not include study labels here)
funnel.meta(m.gen, xlim = c(-0.5, 2),
            contour = c(0.9, 0.95, 0.99),
            col.contour = col.contour)

# Add a legend
legend(x = 1.6, y = 0.01,
       legend = c("p < 0.1", "p < 0.05", "p < 0.01"),
       fill = col.contour)

# Add a title
title("Contour-Enhanced Funnel Plot (Third Wave Psychotherapies)")
```

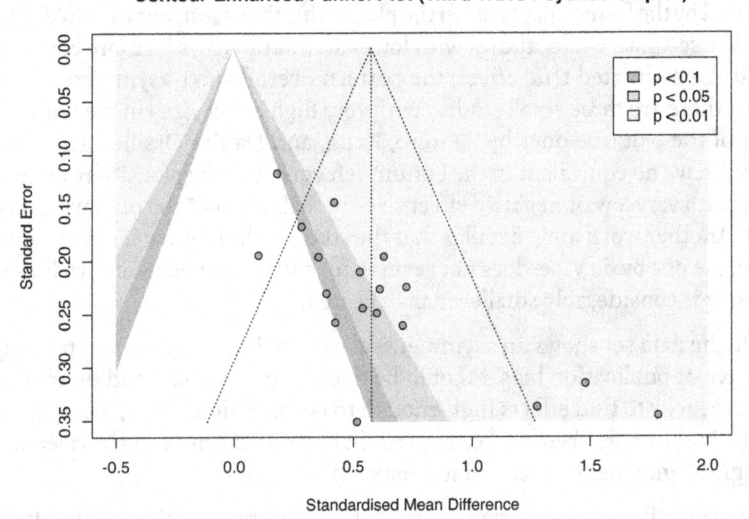

We see that the funnel plot now contains three shaded regions. We are particularly interested in the $p < 0.05$ and $p < 0.01$ regions, because effect sizes falling into this area are traditionally considered significant. Adding the contour regions is illuminating: it shows that the three small studies all have significant effects, despite having a large standard error. There is only one study with a similar standard error that is not significant. If we would "impute" the missing studies in the lower left corner of the plot to increase the symmetry, these studies would lie in the non-significance region of the plot; or they would actually have a significant negative effect.

The pattern looks a little different for the larger studies. We see that there are several studies for which $p > 0.05$, and the distribution of effects is less lopsided. What could be problematic though is that, while not strictly significant, all but one study are very close to the significance threshold (i.e. they lie in the $0.1 > p > 0.05$ region). It is possible that these studies simply calculated the effect size differently in the original paper, which led to a significant result. Or maybe, finding effects that are significant on a trend level was already convincing enough to get the study published.

In sum, inspection of the contour-enhanced funnel plot corroborates our initial hunch that there is asymmetry in the funnel plot and that this may be caused by publication bias. It is crucial, however, not to jump to conclusions, and interpret the funnel plot cautiously. We have to keep in mind that publication bias is just one of many possible reasons for funnel plot asymmetry.

!

Alternative Explanations for Funnel Plot Asymmetry

Although publication bias can lead to asymmetrical funnel plots, there are also other, rather "benign", causes that may produce similar patterns (Page et al., 2020):

- Asymmetry can also be caused by between-study heterogeneity. Funnel plots assume that the dispersion of effect sizes is caused by the studies' sampling error, but do not control for the fact the studies may be estimators of different true effects.
- It is possible that study procedures were different in small studies, and that this resulted in higher effects. In clinical studies, for example, it is easier to make sure that every participant receives the treatment as intended when the sample size is small. This may not be the case in large studies, resulting in a lower *treatment fidelity*, and thus lower effects. It can make sense to inspect the characteristics of the included studies in order to evaluate if such an alternative explanation is plausible.
- It is a common finding that low-quality studies tend to show larger effect sizes, because there is a higher risk of bias. Large studies require more investment, so it is likely that their methodology will also be more rigorous. This can also lead to funnel plot asymmetry, even when there is no publication bias.

- Lastly, it is perfectly possible that funnel plot asymmetry simply occurs by chance.

We see that visual inspection of the (contour-enhanced) funnel plot can already provide us with a few "red flags" that indicate that our results may be affected by publication bias. However, interpreting the funnel plot just by looking at it clearly also has its limitations. There is no explicit rule when our results are "too asymmetric", meaning that inferences from funnel plots are always somewhat subjective. Therefore, it is helpful to assess the presence of funnel plot asymmetry in a quantitative way. This is usually achieved through *Egger's regression test*, which we will discuss next.

9.2.1.2 Egger's Regression Test

Egger's regression test (Egger et al., 1997) is a commonly used quantitative method that tests for asymmetry in the funnel plot. Like visual inspection of the funnel plot, it can only identify small-study effects and not tell us directly if publication bias exists. The test is based on a simple linear regression model, the formula of which looks like this:

$$\frac{\hat\theta_k}{SE_{\hat\theta_k}} = \beta_0 + \beta_1 \frac{1}{SE_{\hat\theta_k}} \tag{9.1}$$

The responses y in this formula are the observed effect sizes $\hat\theta_k$ in our meta-analysis, divided through their standard error. The resulting values are equivalent to z-scores. These scores tell us directly if an effect size is significant; when $z \geq 1.96$ or ≤ -1.96, we know that the effect is significant ($p < 0.05$). This response is regressed on the inverse of the studies' standard error, which is equivalent to their precision.

When using Egger's test, however, we are not interested in the size and significance of the regression weight β_1, but in the *intercept* β_0. To evaluate the funnel asymmetry, we inspect the size of $\hat\beta_0$, and if it differs significantly from zero. When this is the case, Egger's test indicates funnel plot asymmetry.

Let us take a moment to understand why the size of the regression intercept tells us something about asymmetry in the funnel plot. In every linear regression model, the intercept represents the value of y when all other predictors are zero. The predictor in our model is the precision of a study, so the intercept shows the expected z-score when the precision is zero (i.e. when the standard error of a study is infinitely large).

When there is no publication bias, the expected z-score should be scattered around zero. This is because studies with extremely large standard errors have extremely large confidence intervals, making it nearly impossible to reach a value of $|z| \geq 1.96$. However, when the funnel plot is asymmetric, for example, due to publication bias, we expect that small studies with very high effect sizes will be considerably over-represented in our data, leading to a surprisingly high number of low-precision studies with z values greater or equal to 1.96. Due to this distortion, the predicted value of y for zero precision will be much larger than zero, resulting in a significant intercept.

The plots below illustrate the effects of funnel plot asymmetry on the regression slope and intercept underlying Egger's test.

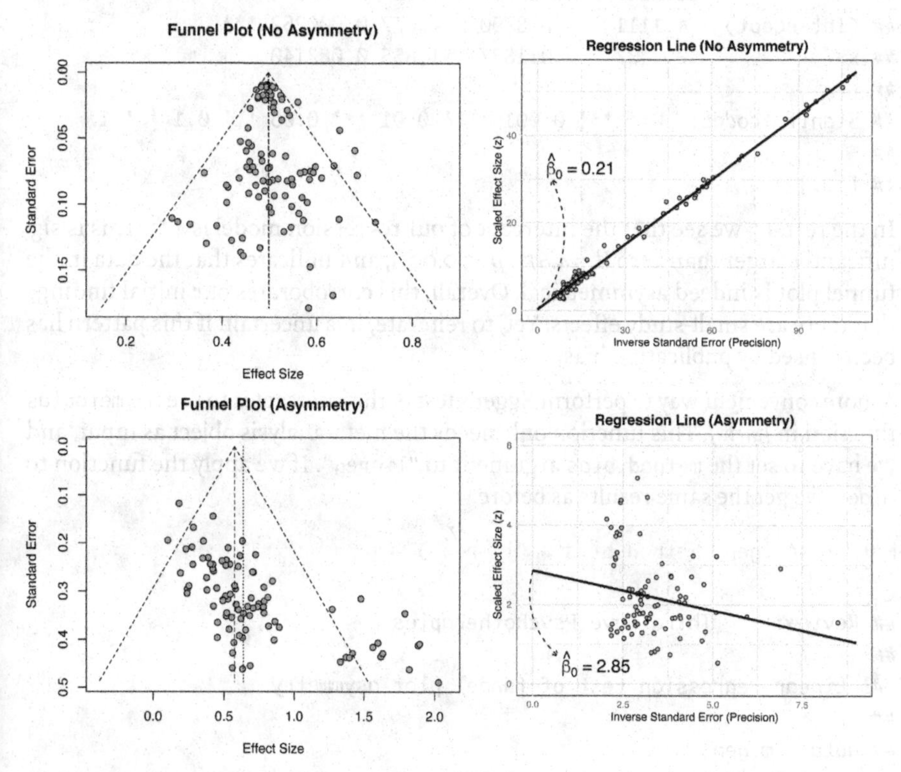

Let us see what results we get when we fit such a regression model to the data in `m.gen`. Using R, we can extract the original data in `m.gen` to calculate the response `y` and our predictor `x`. In the code below, we do this using a pipe (Chapter 2.5.3) and the `mutate` function, which is part of the *{tidyverse}*. After that, we use the linear model function `lm` to regress the z scores `y` on the precision `x`. In the last part of the pipe, we request a `summary` of the results.

```
# Load required package
library(tidyverse)

m.gen$data %>%
  mutate(y = TE/seTE, x = 1/seTE) %>%
```

```
lm(y ~ x, data = .) %>%
summary()
```

```
## [...]
##
## Coefficients:
##             Estimate Std. Error t value Pr(>|t|)
## (Intercept)   4.1111     0.8790   4.677 0.000252 ***
## x            -0.3407     0.1837  -1.855 0.082140 .
## ---
## Signif. codes:  0 '***' 0.001 '**' 0.01 '*' 0.05 '.' 0.1 ' ' 1
##
## [...]
```

In the results, we see that the intercept of our regression model is 4.11. This is significantly larger than zero ($t = 4.677$, $p < 0.001$), and indicates that the data in the funnel plot is indeed asymmetrical. Overall, this corroborates our initial findings that there are small-study effects. Yet, to reiterate, it is uncertain if this pattern has been caused by publication bias.

A more convenient way to perform Egger's test of the intercept is to use the `metabias` function in *{meta}*. This function only needs the meta-analysis object as input, and we have to set the `method.bias` argument to `"linreg"`. If we apply the function to `m.gen`, we get the same results as before.

```
metabias(m.gen, method.bias = "linreg")
```

```
## Review:     Third Wave Psychotherapies
##
##  Linear regression test of funnel plot asymmetry
##
## data:  m.gen
## t = 4.7, df = 16, p-value = 0.0003
## alternative hypothesis: asymmetry in funnel plot
## sample estimates:
##      bias   se.bias intercept
##    4.1111    0.8790   -0.3407
```

Reporting the Results of Egger's Test

For Egger's tests, it is usually sufficient to report the value of the intercept, its 95% confidence interval, as well as the t and p value. In the *{dmetar}* package, we included a convenience function called `eggers.test`. This function is a wrapper for `metabias`, and provides the results of Egger's test in a format suitable for reporting. In case you do not have *{dmetar}* installed, you can find the function's source code online[a]. Here is an example:

```
eggers.test(m.gen)
```

```
## Eggers' test of the intercept
## =============================
##
##  intercept       95% CI      t          p
##      4.111 2.39 - 5.83 4.677 0.0002525
##
## Eggers' test indicates the presence of funnel plot asymmetry.
```

[a] https://raw.githubusercontent.com/MathiasHarrer/dmetar/master/R/eggers.test.R

The effect size metric used in `m.gen` is the small sample bias-corrected SMD (Hedges' *g*). It has been argued that running Egger's test on SMDs can lead to an inflation of false positive results (Pustejovsky and Rodgers, 2019). This is because a study's standardized mean difference and standard error are not independent. We can easily see this by looking at the formula used to calculate the standard error of between-group SMDs (equation 3.18, Chapter 3.3.1.2). This formula includes the SMD itself, which means that a study's standard error changes for smaller or larger values of the observed effect (i.e. there is an artifactual correlation between the SMD and its standard error).

Pustejovsky and Rodgers (2019) propose to use a modified version of the standard error when testing for the funnel plot asymmetry of standardized mean differences. Only the first part of the standard error formula is used, which means that the observed effect size drops out of the equation. Thus, the formula looks like this:

$$SE^*_{\text{SMD}_{\text{between}}} = \sqrt{\frac{n_1 + n_2}{n_1 n_2}} \tag{9.2}$$

Where $SE^*_{\text{SMD}_{\text{between}}}$ is the modified version of the standard error. It might be a good idea to check if Egger's test gives the same results when using this improvement. In the following code, we add the sample size per group of each study to our initial data set, calculate the adapted standard error, and then use it to re-run the analyses.

```
# Add experimental (n1) and control group (n2) sample size
n1 <- c(62, 72, 44, 135, 103, 71, 69, 68, 95,
        43, 79, 61, 62, 60, 43, 42, 64, 63)
```

```
n2 <- c(51, 78, 41, 115, 100, 79, 62, 72, 80,
        44, 72, 67, 59, 54, 41, 51, 66, 55)

# Calculate modified SE
ThirdWave$seTE_c <- sqrt((n1+n2)/(n1*n2))

# Re-run 'metagen' with modified SE to get meta-analysis object
m.gen.c <- metagen(TE = TE, seTE = seTE_c,
                   studlab = Author, data = ThirdWave, sm = "SMD",
                   comb.fixed = FALSE, comb.random = TRUE,
                   method.tau = "REML", hakn = TRUE,
                   title = "Third Wave Psychotherapies")

# Egger's test
metabias(m.gen.c, method = "linreg")
```

```
## Review:     Third Wave Psychotherapies
##
##  Linear regression test of funnel plot asymmetry
##
## data:  m.gen.c
## t = 4.4, df = 16, p-value = 0.0005
## alternative hypothesis: asymmetry in funnel plot
## sample estimates:
##      bias   se.bias intercept
##    11.190     2.567    -1.353
```

We see that, although the exact values differ, the interpretation of the results remains the same. This points to the robustness of our previous finding.

9.2.1.3 Peters' Regression Test

The dependence of effect size and standard error not only applies to standardized mean differences. This mathematical association also exists in effect sizes based on binary outcome data, such as (log) odds ratios (Chapter 3.3.2.2), risk ratios (Chapter 3.3.2.1) or proportions (Chapter 3.2.2).

To avoid an inflated risk of false positives when using binary effect size data, we can use another type of regression test, proposed by Peters and colleagues (Peters et al., 2006). To obtain the results of Peters' test, the log-transformed effect size is regressed on the inverse of the sample size:

$$\log\psi_k = \beta_0 + \beta_1\frac{1}{n_k} \tag{9.3}$$

In this formula, $\log \psi_k$ can stand for any log-transformed effect size based on binary outcome data (e.g. the odds ratio), and n_k is the total sample size of study k.

Importantly, when fitting the regression model, each study k is given a different weight w_k, depending on its sample size and event counts. This results in a *weighted* linear regression, which is similar (but not identical) to a meta-regression model (see Chapter 8.1.3). The formula for the weights w_k looks like this:

$$w_k = \frac{1}{\left(\frac{1}{a_k + c_k} + \frac{1}{b_k + d_k}\right)} \tag{9.4}$$

Where a_k is the number of events in the treatment group, c_k is the number of events in the control group; b_k and d_k are the number of non-events in the treatment and control group, respectively (see Chapter 3.3.2.1). In contrast to Eggers' regression test, Peters' test uses β_1 instead of the intercept to test for funnel plot asymmetry. When the statistical test reveals that $\beta_1 \neq 0$, we can assume that asymmetry exists in our data.

When we have calculated a meta-analysis based on binary outcome data using the `metabin` (Chapter 4.2.3.1) or `metaprop` (Chapter 4.2.6) function, the `metabias` function can be used to conduct Peters' test. We only have to provide a fitting meta-analysis object and use `"peters"` as the argument in `method.bias`. Let us check for funnel plot asymmetry in the `m.bin` object we created in Chapter 4.2.3.1. As you might remember, we used the risk ratio as the summary measure for this meta-analysis.

```
metabias(m.bin, method.bias = "peters")
```

```
## Review:     Depression and Mortality
##
## Linear regression test of funnel plot asymmetry
## (based on sample size)
##
## data: m.bin
## t = -0.081, df = 16, p-value = 0.9
## alternative hypothesis: asymmetry in funnel plot
## sample estimates:
##      bias   se.bias intercept
## -11.1728  138.6121    0.5731
```

We see that the structure of the output looks identical to the one of Eggers' test. The output tells us that the results are the ones of a regression test `based on sample size`, meaning that Peters' method has been used. The test is not significant (t = -0.08, p = 0.94), indicating no funnel plot asymmetry.

Statistical Power of Funnel Plot Asymmetry Tests

It is advisable to only test for funnel plot asymmetry when our meta-analysis includes a sufficient number of studies. When the number of studies is low, the statistical power of Eggers' or Peters' test may not be high enough to detect real asymmetry. It is generally recommended to only perform a test when $K \geq 10$ (Sterne et al., 2011). By default, `metabias` will throw an error when the number of studies in our meta-analysis is smaller than that. However, it is possible (although not advised) to prevent this by setting the `k.min` argument in the function to a lower number.

9.2.1.4 Duval & Tweedie Trim and Fill Method

We have now learned several ways to examine (and test for) small-study effects in our meta-analysis. While it is good to know that publication bias may exist in our data, what we are primarily interested in is the *magnitude* of that bias. We want to know if publication bias has only distorted our estimate slightly, or if it is massive enough to change the interpretation of our findings.

In short, we need a method which allows us to calculate a *bias-corrected* estimate of the true effect size. Yet, we already learned that publication bias cannot be measured directly. We can only use small-study effects as a proxy that may *point* to publication bias. We can therefore only adjust for small-study effects to attain a corrected effect estimate, not for publication bias *per se*. When effect size asymmetry was indeed caused by publication bias, correcting for this imbalance will yield an estimate that better represents the true effect when *all* evidence is considered.

One of the most common methods to adjust for funnel plot asymmetry is the *Duval & Tweedie trim and fill method* (Duval and Tweedie, 2000). The idea behind this method is simple: it imputes "missing" effects until the funnel plot is symmetric. The pooled effect size of the resulting "extended" data set then represents the estimate when correcting for small-study effects. This is achieved through a simple algorithm, which involves the "trimming" and "filling" of effects (Schwarzer et al., 2015, chapter 5.3.1):

- **Trimming**. First, the method identifies all the outlying studies in the funnel plot. In our example from before, these would be all small studies scattered around the right side of the plot. Once identified, these studies are *trimmed*: they are removed from the analysis, and the pooled effect is recalculated without them. This step is usually performed using a fixed-effect model.
- **Filling**. For the next step, the recalculated pooled effect is now assumed to be the center of all effect sizes. For each trimmed study, one additional effect size is added, mirroring its results on the other side of the funnel. For example, if the recalculated mean effect is 0.5 and a trimmed study has an effect of 0.8, the mirrored study will be given an effect of 0.2. After this is done for all trimmed studies, the funnel

plot will look roughly symmetric. Based on all data, including the trimmed and imputed effect sizes, the average effect is then recalculated again (typically using a random-effects model). The result is then used as the estimate of the corrected pooled effect size.

An important caveat pertaining to the trim-and-fill method is that it does not produce reliable results when the between-study heterogeneity is large (Peters et al., 2007; Terrin et al., 2003; Simonsohn et al., 2014a). When studies do not share one true effect, it is possible that even large studies deviate substantially from the average effect. This means that such studies are also trimmed and filled, even though it is unlikely that they are affected by publication bias. It is easy to see that this can lead to invalid results.

We can apply the trim and fill algorithm to our data using the `trimfill.meta` function in *{meta}*. The function has very sensible defaults, so it is sufficient to simply provide it with our meta-analysis object. In our example, we use our `m.gen` object again. However, before we start, let us first check the amount of I^2 heterogeneity we observed in this meta-analysis.

```
m.gen$I2
```

```
## [1] 0.6264
```

We see that, with I^2 = 63%, the heterogeneity in our analysis is substantial. In light of the trim and fill method's limitations in heterogeneous data sets, this could prove problematic. We will therefore conduct two trim and fill analyses: one with all studies, and a sensitivity analysis in which we exclude two outliers identified in chapter 5.4 (i.e. study 3 and 16). We save the results to `tf` and `tf.no.out`.

```
# Using all studies
tf <- trimfill.meta(m.gen)

# Analyze with outliers removed
tf.no.out <- trimfill.meta(update.meta(m.gen,
                                       subset = -c(3, 16)))
```

First, let us have a look at the first analysis, which includes all studies.

```
print(tf)
```

```
## Review:     Third Wave Psychotherapies
##                              SMD             95%-CI %W(random)
## [...]
## Filled: Warnecke et al.   0.0520 [-0.4360;  0.5401]       3.8
## Filled: Song & Lindquist  0.0395 [-0.4048;  0.4837]       4.0
## Filled: Frogeli et al.    0.0220 [-0.3621;  0.4062]       4.2
## Filled: Call et al.      -0.0571 [-0.5683;  0.4541]       3.8
```

```
## Filled: Gallego et al.   -0.0729 [-0.5132;  0.3675]        4.0
## Filled: Kang et al.      -0.6230 [-1.2839;  0.0379]        3.3
## Filled: Shapiro et al.   -0.8277 [-1.4456; -0.2098]        3.4
## Filled: DanitzOrsillo    -1.1391 [-1.8164; -0.4618]        3.3
##
## Number of studies combined: k = 26 (with 8 added studies)
##
##                         SMD            95%-CI    t p-value
## Random effects model 0.3428 [0.1015; 0.5841] 2.93  0.0072
##
## Quantifying heterogeneity:
##  tau^2 = 0.2557 [0.1456; 0.6642]; tau = 0.5056 [0.3816; 0.8150];
##  I^2 = 76.2% [65.4%; 83.7%]; H = 2.05 [1.70; 2.47]
##
## [...]
##
## Details on meta-analytical method:
## - Inverse variance method
## - Restricted maximum-likelihood estimator for tau^2
## - Q-profile method for confidence interval of tau^2 and tau
## - Hartung-Knapp adjustment for random effects model
## - Trim-and-fill method to adjust for funnel plot asymmetry
```

We see that the trim-and-fill procedure added a total of eight studies. Trimmed and filled studies include our detected outliers, but also a few other smaller studies with relatively high effects. We see that the imputed effect sizes are all very low, and some are even highly negative. The output also provides us with the estimate of the corrected effect, which is $g = 0.34$. This is still significant, but much lower than the effect of $g = 0.58$ we initially calculated for `m.gen`. Now, let us compare this to the results of the analysis in which outliers were removed.

```
print(tf.no.out)
```

```
## Review:     Third Wave Psychotherapies
## [...]
##
## Number of studies combined: k = 22 (with 6 added studies)
##
##                         SMD            95%-CI    t p-value
## Random effects model 0.3391 [0.1904; 0.4878] 4.74  0.0001
##
## Quantifying heterogeneity:
##  tau^2 = 0.0421 [0.0116; 0.2181]; tau = 0.2053 [0.1079; 0.4671];
##  I^2 = 50.5% [19.1%; 69.7%]; H = 1.42 [1.11; 1.82]
## [...]
```

With $g = 0.34$, the results are nearly identical. Overall, the trim-and-fill method indicates that the pooled effect of $g = 0.58$ in our meta-analysis is overestimated due to small-study effects. In reality, the effect may be considerably smaller. It is likely that this overestimation has been caused by publication bias, but this is not certain. Other explanations are possible too, and this could mean that the trim-and-fill estimate is invalid.

Lastly, it is also possible to create a funnel plot including the imputed studies. We only have to apply the `funnel.meta` function to the output of `trimfill.meta`. In the following code, we create contour-enhanced funnel plots for both trim and fill analyses (with and without outliers). Using the `par` function, we can print both plots side by side.

```
# Define fill colors for contour
contour <- c(0.9, 0.95, 0.99)
col.contour <- c("gray75", "gray85", "gray95")
ld <- c("p < 0.1", "p < 0.05", "p < 0.01")

# Use 'par' to create two plots in one row (row, columns)
par(mfrow=c(1,2))

# Contour-enhanced funnel plot (full data)
funnel.meta(tf,
            xlim = c(-1.5, 2),
            contour = contour,
            col.contour = col.contour)
legend(x = 1.1, y = 0.01,
       legend = ld,
       fill = col.contour)
title("Funnel Plot (Trim & Fill Method)")

# Contour-enhanced funnel plot (outliers removed)
funnel.meta(tf.no.out,
            xlim = c(-1.5, 2),
            contour = contour,
            col.contour = col.contour)
legend(x = 1.1, y = 0.01,
       legend = ld,
       fill = col.contour)
title("Funnel Plot (Trim & Fill Method) - Outliers Removed")
```

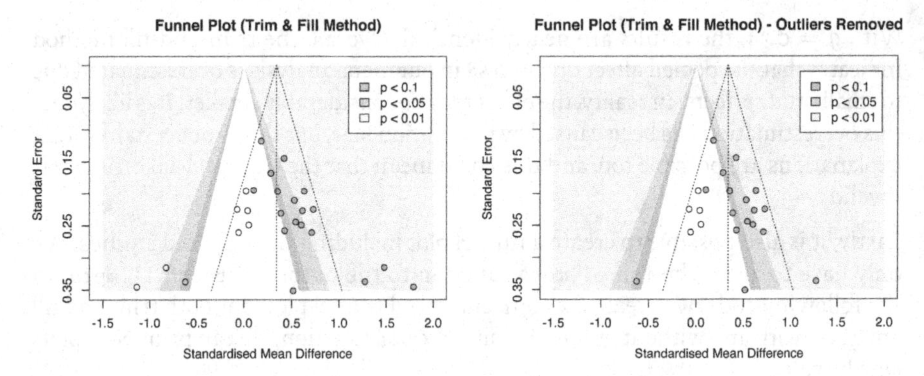

In these funnel plots, the imputed studies are represented by circles that have no fill color.

9.2.1.5 PET-PEESE

Duval & Tweedie's trim-and-fill method is relatively old, and arguably one of the most common methods to adjust for small-study effects. However, as we mentioned, it is an approach that is far from perfect, and not the only way to estimate a bias-corrected version of our pooled effect. In recent years, a method called *PET-PEESE* (Stanley and Doucouliagos, 2014; Stanley, 2008) has become increasingly popular; particularly in research fields where SMDs are frequently used as the outcome measure (for example, psychology or educational research). Like all previous techniques, PET-PEESE is aimed at small-study effects, which are seen as a potential indicator of publication bias.

PET-PEESE is actually a combination of two methods: the *precision-effect test* (PET) and the *precision-effect estimate with standard error* (PEESE). Let us begin with the former. The PET method is based on a simple regression model, in which we regress a study's effect size on its standard error:

$$\theta_k = \beta_0 + \beta_1 SE_{\theta_k} \tag{9.5}$$

Like in Peters' test, we use a weighted regression. The study weight w_k is calculated as the inverse of the variance–just like in a normal (fixed-effect) meta-analysis:

$$w_k = \frac{1}{s_k^2} \tag{9.6}$$

It is of note that the regression model used by the PET method is equivalent to the one of Eggers' test. The main difference is that in the PET formula, the β_1 coefficient quantifies funnel asymmetry, while in Eggers' test, this is indicated by the intercept.

When using the PET method, however, we are not interested in the funnel asymmetry measured by β_1, but in the intercept β_0. This is because, in the formula above, the intercept represents the so-called *limit effect*. This limit effect is the expected effect size of a study with a *standard error of zero*. This is the equivalent of an observed effect size measured without sampling error. All things being equal, we know that an effect size measured without sampling error ϵ_k will represent the true overall effect itself. The idea behind the PET method is to *control* for the effect of small studies by including the standard error as a predictor. In theory, this should lead to an intercept β_0 which represents the true effect in our meta-analysis after correction for all small-study effects:

$$\hat{\theta}_{\text{PET}} = \hat{\beta}_{0_{\text{PET}}} \tag{9.7}$$

The formula for the PEESE method is very similar. The only difference is that we use the *squared* standard error as the predictor (i.e. the effect size variance s^2_k):

$$\theta_k = \beta_0 + \beta_1 SE^2_{\theta_k} \tag{9.8}$$

While the formula for the study weights w_k remains the same. The idea behind squaring the standard error is that small studies are particularly prone to reporting *highly* over-estimated effects. This problem, it is assumed, is far less pronounced for studies with high statistical power.

While the PET method works best when the true effect captured by β_0 is *zero*, PEESE shows a better performance when the true effect is *not* zero. Stanley and Doucouliagos (2014) therefore proposed to combine both methods, in order to balance out their individual strengths. The resulting approach is the PET-PEESE method. PET-PEESE uses the intercept β_0 of either PET or PEESE as the estimate of the corrected true effect. Whether PET or PEESE is used depends on the size of the intercept calculated by the PET method. When $\beta_{0_{\text{PET}}}$ is significantly larger than zero in a one-sided test with $\alpha = 0.05$, we use the intercept of PEESE as the true effect size estimate. If PET's intercept is not significantly larger than zero, we remain with the PET estimate.

In most implementations of regression models in R, it is conventional to test the significance of coefficients using a two-sided test (i.e. we test if a β weight significantly differs from zero, no matter the direction). To assume a one-sided test with $\alpha = 0.05$, we already regard the intercept as significant when $p < 0.1$, and when the estimate of β_0 is larger than zero[2]. The rule to obtain the true effect size as estimated by PET-PEESE, therefore, looks like this:

$$\hat{\theta}_{\text{PET-PEESE}} = \begin{cases} \text{P}(\beta_{0_{\text{PET}}} = 0) < 0.1 \text{ and } \hat{\beta}_{0_{\text{PET}}} > 0: & \hat{\beta}_{0_{\text{PEESE}}} \\ \text{else}: & \hat{\beta}_{0_{\text{PET}}}. \end{cases} \tag{9.9}$$

[2]The latter condition ($\hat{\beta}_0 > 0$) only applies if positive effect sizes represent favorable outcomes (e.g. positive effect sizes mean that the intervention was effective). When negative effect sizes (e.g. SMD = -0.5) represent favorable outcomes, our one-side test should be in the other direction. This means that PEESE is used when the p value of PET's intercept is smaller than 0.1, and when the intercept estimate is *smaller* than zero.

It is somewhat difficult to wrap one's head around this if-else logic, but a hands-on example may help to clarify things. Using our `m.gen` meta-analysis object, let us see what PET-PEESE's estimate of the true effect size is.

There is currently no straightforward implementation of PET-PEESE in *{meta}*, so we write our own code using the linear model function `lm`. Before we can fit the PET and PEESE model, however, we first have to prepare all the variables we need in our data frame. We will call this data frame `dat.petpeese`. The most important variable, of course, is the standardized mean difference. No matter if we initially ran our meta-analysis using `metacont` or `metagen`, the calculated SMDs of each study will always be stored under `TE` in our meta-analysis object.

```
# Build data set, starting with the effect size
dat.petpeese <- data.frame(TE = m.gen$TE)
```

Next, we need the standard error of the effect size. For PET-PEESE, it is also advisable to use the modified standard error proposed by Pustejovsky and Rodgers (2019, see Chapter 9.2.1.2). Therefore, we use the adapted formula to calculate the corrected standard error `seTE_c`, so that it is not correlated with the effect size itself. We also save this variable to `dat.petpeese`. Furthermore, we add a variable `seTE_c2`, containing the *squared* standard error, since we need this as the predictor for PEESE.

```
# Experimental (n1) and control group (n2) sample size
n1 <- c(62, 72, 44, 135, 103, 71, 69, 68, 95,
        43, 79, 61, 62, 60, 43, 42, 64, 63)

n2 <- c(51, 78, 41, 115, 100, 79, 62, 72, 80,
        44, 72, 67, 59, 54, 41, 51, 66, 55)

# Calculate modified SE
dat.petpeese$seTE_c <- sqrt((n1+n2)/(n1*n2))

# Add squared modified SE (= variance)
dat.petpeese$seTE_c2 <- dat.petpeese$seTE_c^2
```

Lastly, we need to calculate the inverse-variance weights `w_k` for each study. Here, we also use the squared modified standard error to get an estimate of the variance.

```
dat.petpeese$w_k <- 1/dat.petpeese$seTE_c^2
```

Now, `dat.petpeese` contains all the variables we need to fit a weighted linear regression model for PET and PEESE. In the following code, we fit both models, and then directly print the estimated coefficients using the `summary` function. These are the results we get:

```
# PET
pet <- lm(TE ~ seTE_c, weights = w_k, data = dat.petpeese)
summary(pet)$coefficients
```

```
##             Estimate Std. Error t value Pr(>|t|)
## (Intercept)   -1.353     0.4432  -3.054 0.0075732
## seTE_c        11.190     2.5667   4.360 0.0004862
```

```
# PEESE
peese <- lm(TE ~ seTE_c2, weights = w_k, data = dat.petpeese)
summary(peese)$coefficients
```

```
##              Estimate Std. Error t value  Pr(>|t|)
## (Intercept)  -0.4366     0.2229  -1.959 0.0678222
## seTE_c2      33.3610     7.1784   4.647 0.0002683
```

To determine if PET or PEESE should be used, we first need to have a look at the results of the PET method. We see that the limit estimate is g = -1.35. This effect is significant (p < 0.10), but considerably smaller than zero, indicating that the PET estimate should be used.

However, with, g=-1.35, PET's estimate of the bias-corrected effect is not very credible. It indicates that in reality, the intervention type under study has a highly negative effect on the outcome of interest; that it is actually very harmful. That seems very unlikely. It may be possible that a "bona fide" intervention has no effect, but it is extremely uncommon to find interventions that are downright dangerous.

In fact, what we see in our results is a common limitation of PET-PEESE: it sometimes heavily *overcorrects* for biases in our data (Carter et al., 2019). This seems to be the case in our example: although all observed effect sizes have a positive sign, the corrected effect size is heavily negative. If we look at the second part of the output, we see that the same is also true for PEESE, even though its estimate is slightly less negative (g = -0.44).

When this happens, it is best not to interpret the intercept as a point estimate of the true effect size. We can simply say that PET-PEESE indicates, when correcting for small-sample effects, that the intervention type under study has *no effect*. This basically means that we set $\hat{\theta}_{\text{PET-PEESE}}$ to zero, instead of interpreting the negative effect size that was actually estimated.

Limitations of PET-PEESE

PET-PEESE can not only systematically over-correct the pooled effect size– it also sometimes *overestimates* the true effect, even when there is no publication bias at all. Overall, the PET-PEESE method has been found to perform badly when the number of included studies is small (i.e. $K < 20$), and the between-study heterogeneity is very high, i.e. $I^2 > 80\%$ (Stanley, 2017). Unfortunately, it is common to find meta-analyses with a small number of studies and high heterogeneity. This restricts the applicability of PET-PEESE, and we do not recommend its use as the *only* method to adjust for small-study effects. Yet, it is good to know that this method exists and how it can be applied since it has become increasingly common in some research fields.

9.2.1.6 Rücker's Limit Meta-Analysis Method

Another way to calculate an estimate of the adjusted effect size is to perform a *limit meta-analysis* as proposed by Rücker and colleagues (2011). This method is more sophisticated than PET-PEESE and involves more complex computations. Here, we therefore focus on understanding the general idea behind this method and let R do the heavy lifting after that.

The idea behind Rücker's method is to build a meta-analysis model which explicitly accounts for bias due to small-study effects. As a reminder, the formula of a (random-effects) meta-analysis can be defined like this:

$$\hat{\theta}_k = \mu + \epsilon_k + \zeta_k \tag{9.10}$$

Where $\hat{\theta}_k$ is the observed effect size of study k, μ is the true overall effect size, ϵ_k is the sampling error, and ζ_k quantifies the deviation due to between-study heterogeneity.

In a limit meta-analysis, we extend this model. We account for the fact that the effect sizes and standard errors of studies are not independent when there are small-study effects. This is assumed because we know that publication bias particularly affects small studies, and that small studies will therefore have a larger effect size than big studies. In Rücker's method, this bias is added to our model by introducing a new term θ_{Bias}. It is assumed that θ_{Bias} interacts with ϵ_k and ζ_k. It becomes larger as ϵ_k increases. The adapted formula looks like this:

$$\hat{\theta}_k = \mu_* + \theta_{\text{Bias}}(\epsilon_k + \zeta_k) \tag{9.11}$$

It is important to note that in this formula, μ_* does not represent the overall true effect size anymore, but a global mean that has no direct equivalent in a "standard" random-effects meta-analysis (unless $\theta_{\text{Bias}} = 0$).

The next step is similar to the idea behind PET-PEESE (see previous chapter). Using the formula above, we suppose that studies' effect size estimates become increasingly precise, meaning that their individual sampling error ϵ_k approaches zero. This means that ϵ_k ultimately drops out of the equation:

$$\text{E}(\hat{\theta}_k) \rightarrow \mu_* + \theta_{\text{Bias}}\zeta_k \quad \text{as} \quad \epsilon_k \rightarrow 0. \tag{9.12}$$

In this formula, $\text{E}(\hat{\theta}_k)$ stands for the *expected value* of $\hat{\theta}_k$ as ϵ_k approaches zero. The formula we just created is the one of a "limit meta-analysis". It provides us with an adjusted estimate of the effect when removing the distorting influence of studies with a large standard error. Since ζ_k is usually expressed by the between-study heterogeneity variance τ^2 (or its square root, the standard deviation τ), we can use it to replace ζ_k in the equation, which leaves us with this formula:

$$\hat{\theta}_* = \mu_* + \theta_{\text{Bias}}\tau \tag{9.13}$$

Where $\hat{\theta}_*$ stands for the estimate of the *pooled* effect size after adjusting for small-study effects. Rücker's method uses maximum likelihood to estimate the parameters in this formula, including the "shrunken" estimate of the true effect size $\hat{\theta}_*$. Furthermore, it is also possible to obtain a shrunken effect size estimate $\hat{\theta}_{*_k}$ for each individual study k, using this formula:

$$\hat{\theta}_{*_k} = \mu_* + \sqrt{\frac{\tau^2}{SE^2_k + \tau^2}}(\hat{\theta}_k - \mu_*) \tag{9.14}$$

in which SE^2_k stands for the squared standard error (i.e. the observed variance) of k, and with $\hat{\theta}_k$ being the originally observed effect size[3].

An advantage of Rücker's limit meta-analysis method, compared to PET-PEESE, is that the heterogeneity variance τ^2 is explicitly included in the model. Another more practical asset is that this method can be directly applied in R, using the `limitmeta` function. This function is included in the *{metasens}* package (Schwarzer et al., 2020). Since *{metasens}* and *{meta}* have been developed by the same group of researchers, they usually work together quite seamlessly. To conduct a limit meta-analysis of our `m.gen` meta-analysis, for example, we only need to provide it as the first argument in our call to `limitmeta`.

```
# Install 'metasens', then load from library
library(metasens)

# Run limit meta-analysis
limitmeta(m.gen)
```

[3]This formula can be derived from a less simplified version of equation 9.11. A technical description of how to do this can be found in Rücker and colleagues (2011), equations 2.4 to 2.6.

```
## Results for individual studies
## (left: original data; right: shrunken estimates)
##
##                               SMD            95%-CI      SMD            95%-CI
## Call et al.                  0.70 [ 0.19; 1.22]   -0.05 [-0.56; 0.45]
## Cavanagh et al.              0.35 [-0.03; 0.73]   -0.09 [-0.48; 0.28]
## DanitzOrsillo                1.79 [ 1.11; 2.46]    0.34 [-0.33; 1.01]
## de Vibe et al.               0.18 [-0.04; 0.41]    0.00 [-0.22; 0.23]
## Frazier et al.               0.42 [ 0.13; 0.70]    0.13 [-0.14; 0.42]
## Frogeli et al.               0.63 [ 0.24; 1.01]    0.13 [-0.25; 0.51]
## Gallego et al.               0.72 [ 0.28; 1.16]    0.09 [-0.34; 0.53]
## Hazlett-Stevens & Oren       0.52 [ 0.11; 0.94]   -0.00 [-0.41; 0.40]
## Hintz et al.                 0.28 [-0.04; 0.61]   -0.05 [-0.38; 0.26]
## Kang et al.                  1.27 [ 0.61; 1.93]    0.04 [-0.61; 0.70]
## Kuhlmann et al.              0.10 [-0.27; 0.48]   -0.29 [-0.67; 0.08]
## Lever Taylor et al.          0.38 [-0.06; 0.84]   -0.18 [-0.64; 0.26]
## Phang et al.                 0.54 [ 0.06; 1.01]   -0.11 [-0.59; 0.36]
## Rasanen et al.               0.42 [-0.07; 0.93]   -0.25 [-0.75; 0.25]
## Ratanasiripong               0.51 [-0.17; 1.20]   -0.48 [-1.17; 0.19]
## Shapiro et al.               1.47 [ 0.86; 2.09]    0.26 [-0.34; 0.88]
## Song & Lindquist             0.61 [ 0.16; 1.05]    0.00 [-0.44; 0.44]
## Warnecke et al.              0.60 [ 0.11; 1.08]   -0.09 [-0.57; 0.39]
##
## Result of limit meta-analysis:
##
##  Random effects model        SMD              95%-CI      z      pval
##      Adjusted estimate -0.0345 [-0.3630; 0.2940] -0.21    0.8367
##    Unadjusted estimate  0.5771 [ 0.3782; 0.7760] -0.21 < 0.0001
## [...]
```

The output first shows us the original (left) and shrunken estimates (right) of each study. We see that the adjusted effect sizes are considerably smaller than the observed ones–some are even negative now. In the second part of the output, we see the adjusted pooled effect estimate. It is g = -0.03, indicating that the overall effect is approximately zero when correcting for small-study effects. If the small-study effects are indeed caused by publication bias, this result would be discouraging. It would mean that our initial finding has been completely spurious and that selective publication has concealed the fact that the treatment is actually ineffective. Yet again, it is hard to prove that publication bias has been the only driving force behind the small-study effects in our data.

It is also possible to create funnel plots for the limit meta-analysis: we simply have to provide the results of `limitmeta` to the `funnel.limitmeta` function. This creates a funnel plot which looks exactly like the one produced by `funnel.meta`. The only difference is that a *gray curve* is added to the plot. This curve indicates the adjusted average effect size when the standard error on the y-axis is zero, but also symbolizes the increasing bias due to small-study effects as the standard error increases. When generating a funnel plot for `limitmeta` objects, it is also possible to include the shrunken study-level effect size estimates, by setting the `shrunken` argument to `TRUE`. Here is the code to produce these plots:

```
# Create limitmeta object
lmeta <- limitmeta(m.gen)

# Funnel with curve
funnel.limitmeta(lmeta, xlim = c(-0.5, 2))

# Funnel with curve and shrunken study estimates
funnel.limitmeta(lmeta, xlim = c(-0.5, 2), shrunken = TRUE)
```

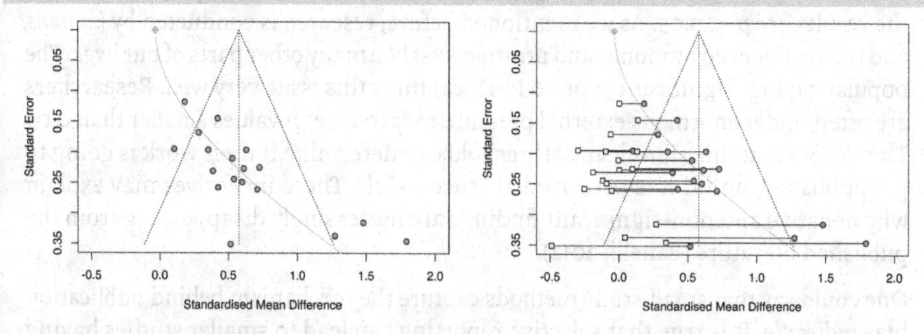

Note that `limitmeta` can not only be applied to meta-analyses which use the standardized mean difference–any kind of *{meta}* meta-analysis object can be used. To exemplify this, let us check the adjusted effect size of `m.bin`, which used the risk ratio as the summary measure.

```
limitmeta(m.bin)
```

```
## Result of limit meta-analysis:
##
## Random effects model        RR          95%-CI    z     pval
##    Adjusted estimate 2.2604 [1.8066; 2.8282] 7.13 < 0.0001
##  Unadjusted estimate 2.0217 [1.5786; 2.5892] 7.13 < 0.0001
```

We see that in this analysis, the original and adjusted estimate are largely identical. This is not very surprising, given that Peters' test (Chapter 9.2.1.3) already indicated that small-study effects seem to play a minor role in this meta-analysis.

9.2.2 P-Curve

Previously, we covered various approaches that assess the risk of publication bias by looking at small-study effects. Although their implementation differs, all of these methods are based on the idea that selective reporting causes a study's effect size to depend on its sample size. We assume that studies with a higher standard error (and thus a lower precision) have higher average effect sizes than large studies. This is because only small studies with a very high effect size are published, while others remain in the file drawer.

While this "theory" certainly sounds intuitive, one may also argue that it somewhat misses the point. Small-study methods assume that publication bias is driven by *effect sizes*. A more realistic stance, however, would be to say that it operates through *p-values*. In practice, research findings are only considered worth publishing when the results are $p < 0.05$. As we mentioned before, research is conducted by *humans*, and thus influenced by money and prestige–just like many other parts of our lives. The popular saying "significant p, or no PhD" captures this issue very well. Researchers are often under enormous external pressure to "produce" p-values smaller than 0.05. They know that this significance threshold can determine if their work is going to get published, and if it is perceived as "successful". These incentives may explain why negative and non-significant findings are increasingly disappearing from the published literature (Fanelli, 2012).

One could say that small-study methods capture the mechanism behind publication bias *indirectly*. It is true that selective reporting can lead to smaller studies having higher effects. Yet, this is only correct because very high effects increase the chance of obtaining a test statistic for which $p < 0.05$. For small-study effect methods, there is hardly a difference between a study in which $p = 0.049$, and a study with a p-value of 0.051. In practice, however, this tiny distinction can mean the world to researchers.

In the following, we will introduce a method called *p-curve*, which focuses on p-values as the main driver of publication bias (Simonsohn et al., 2014a,b, 2015). The special thing about this method is that it is restricted to *significant* effect sizes, and how their p-values are distributed. It allows to assess if there is a true effect behind our meta-analysis data, and can estimate how large it is. Importantly, it also explicitly controls for questionable research practices such as p-hacking, which small-study effect methods do not.

P-curve is a relatively novel method. It was developed in response to the "replication crisis" that affected the social sciences in recent years (Ioannidis, 2005; Open Science Collaboration et al., 2015; McNutt, 2014). This crisis was triggered by the observation that many seemingly well-established research findings are in fact spurious–they can not be systematically replicated. This has sparked renewed interest in methods to detect publication bias, since this may be a logical explanation for failed replications. Meta-analyses, by not adequately controlling for selective reporting, may have simply reproduced biases that already exist in the published literature. P-curve was also developed in response to deficiencies of standard publication bias methods, in particular the Duval & Tweedie trim-fill-method. Simonsohn and colleagues (2014b) found that the trim-and-fill approach usually only leads to a *small* downward correction, and often misses the fact that there is no true effect behind the analyzed data at all.

P-curve is, as it says in the name, based on a curve of p-values. A p curve is like a histogram, showing the number of studies in a meta-analysis for which $p < 0.05$, $p < 0.04$, $p < 0.03$, and so forth. The p-curve method is based on the idea that the shape of this histogram of p-values depends on the sample sizes of studies, and–more importantly–on the *true* effect size behind our data.

To illustrate this, we simulated the results of nine meta-analyses. To make patterns clearly visible, each of these imaginary meta-analyses contains the huge number of $K = 10^5$ studies. In each of the nine simulations, we assumed different sample sizes for each individual study (ranging from $n = 20$ to $n = 100$), and a different true effect size (ranging from $\theta = 0$ to 0.5). We assumed that all studies in a meta-analysis share one true effect size, meaning effects follow the fixed-effect model. Then, we took the p-value of all significant effect sizes in our simulations and created a histogram. The results can be seen in the plot below.

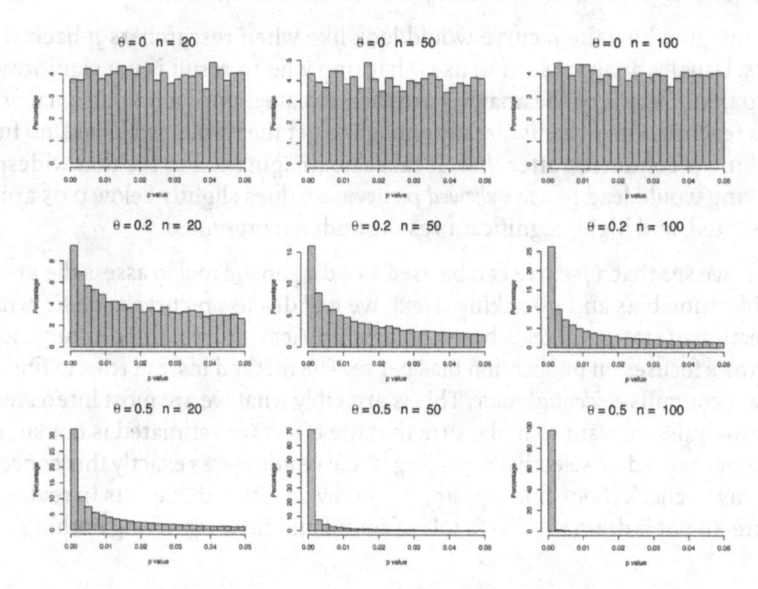

FIGURE 9.1: P-curves for varying study sample size and true effect.

The first row displays the distribution of significant p-values when there is no true effect. We see that the pattern is identical in all simulations, no matter how large the sample size of the individual studies. The p-values in all three examples seem to be evenly distributed: a barely significant value of $p = 0.04$ seems to be just as likely as $p = 0.01$. Such a flat p curve emerges when there is no underlying effect in our data, i.e. when the *null hypothesis* of $\theta = 0$ is true. When this is the case, p values are assumed to follow a *uniform* distribution: every p value is just as likely as the other.

When the null hypothesis ($\theta = 0$) is true, it is still possible to find significant effect sizes just by chance. This results in a *false positive*, or α error. But this is unlikely, and we know exactly *how unlikely*. Since they are uniformly distributed when the effect size is zero, 5% of all p values can be expected to be smaller than 0.05. This is exactly the significance threshold of $\alpha = 0.05$ that we commonly use in hypothesis testing to reject the null hypothesis.

The p curve looks completely different in the second and third row. In these examples, the null hypothesis is false, and a true effect exists in our data. This leads to a *right-skewed* distribution of p-values. When our data capture a true effect, highly significant (e.g. $p = 0.01$) effect sizes are more likely than effects that are barely significant (e.g. $p = 0.049$). This right-skew becomes more and more pronounced as the true effect size and study sample size increase. Yet, we see that a right-skewed p curve even emerges when the studies in our meta-analysis are drastically under-powered (i.e. containing only $n = 20$ participants while aiming to detect a small effect of $\theta = 0.2$). This makes it clear that p curves are very sensitive to changes in the true underlying effect size. When a true effect size exists, we will often be able to detect it just by looking at the distribution of p-values that are significant.

Now, imagine how the p curve would look like when researchers p-hacked their results. Usually, analysts start to use p-hacking when a result is not significant but *close* to that. Details of the analysis are then tweaked until a p value smaller than 0.05 is reached. Since that is already enough to get the results published, no further p-hacking is conducted after that. It takes no imagination to see that widespread p-hacking would lead to a *left-skewed* p curve: p values slightly below 0.05 are over-represented, and highly significant results under-represented.

In sum, we see that a p curve can be used as a diagnostic tool to assess the presence of publication bias and p-hacking. Next, we will discuss p-curve *analysis*, which is a collection of statistical tests based on an empirical p curve. Importantly, none of these tests focuses on publication bias per se. The method instead tries to find out if our data contains *evidential value*. This is arguably what we are most interested in a meta-analysis: we want to make sure that the effect we estimated is not spurious; an artifact caused by selective reporting. P-curve addresses exactly this concern. It allows us to check if our findings are driven by an effect that exists in reality, or if they are–to put it dramatically–"a tale of sound and fury, signifying nothing".

9.2.2.1 Testing for Evidential Value

To evaluate the presence of evidential value, p-curve uses two types of tests: a *test for right-skewness*, and a *test for 33% power* (the latter can be seen as a test for flatness of the p curve). We begin with the test for right-skewness. As we learned, the right-skewness of the p curve is a function of studies' sample sizes and their true underlying effect. Therefore, a test which allows us to confirm that the p curve of our meta-analysis is significantly right-skewed is very helpful. When we find a significant right-skew in our distribution of significant p values, this would indicate that our results are indeed driven by a true effect.

9.2.2.1.1 Test for Right-Skewness

To test for right-skewness, the p-curve method first uses a *binomial test*. These tests can be used for data that follows a binomial distribution. A binomial distribution can be assumed for data that can be divided into two categories (e.g. success/failure, head/tail, yes/no), where p indicates the probability of one of the outcomes, and $q = 1 - p$ is the probability of the other outcome. To use a binomial test, we have to split our p curve into two sections. We do this by counting the number of p-values that are <0.025, and then the number of significant p-values that are >0.025. Since values in our p curve can range from 0 to 0.05, we essentially use the middle of the x-axis as our cut-off. When the p curve is indeed right-skewed, we would expect that the number of p-values in the two groups differ. This is because the probability p of obtaining a result that is smaller than 0.025 is considerably higher than the probability q of getting values that are higher than 0.025.

Imagine that our p curve contains eight values, seven of which are below 0.025. We can use the `binom.test` function in R to test how likely it is to find such data under the null hypothesis that small and high p-values are equally likely[4]. Since we assume that small p-values are more frequent than high p-values, we can use a one-sided test by setting the `alternative` argument to `"greater"`.

```
k <- 7   # number of studies p<0.025
n <- 8   # total number of significant studies
p <- 0.5 # assumed probability of k (null hypothesis)

binom.test(k, n, p, alternative = "greater")$p.value
```

```
## [1] 0.03516
```

We see that the binomial test is significant ($p < 0.035$). This means there are significantly more high than low p-values in our example. Overall, this indicates that the p curve is right-skewed, and that there is a true effect.

A drawback of the binomial test is that is requires us to dichotomize our p-values, while they are in fact continuous. To avoid information loss, we need a test which does

[4]Under the null hypothesis, we assume that the true probability p of obtaining results that are <0.025 is $\pi = 0.5$. Our goal is to calculate the probability of getting $k = 7$ or more highly significant studies when there are $n = 8$ significant studies in total. This probability, the p value of a one-sided binomial test, can be obtained using this formula:

$$\text{P}(X \geq k) = \sum_{k=7}^{8} \frac{n!}{k!\,(n-k)!} p^k (1-p)^{n-k}$$

Where $p = 0.5$, and the exclamation mark stands for the factorial. The sum symbol tells us that we have to sum up everything to the right for values of $k = 7$ or higher, until $k = n$.

not require us to transform our data into bins. P-curve achieves this by calculating a p-value for each p-value, which results in a so-called pp-value of each study. The pp-value tells us how likely it is to get a value *at least as high* as p when the p curve is flat (i.e. when there is no true effect). It gives the probability of a p-value when *only* significant values are considered. Since p-values follow a uniform distribution when $\theta = 0$, pp-values are nothing but significant p-values which we project to the $[0, 1]$ range. For continuous outcomes measures, this is achieved through multiplying the p-value by 20, for example: $p = 0.023 \times 20 = 0.46 \rightarrow pp$.

Using the pp_k-value of each significant study k in our meta-analysis, we can test for right-skewness using *Fisher's method*. This method is an "archaic" type of meta-analysis developed by R. A. Fisher in the early 20th century (see Chapter 1.2). Fisher's method allows to aggregate p values from several studies, and to test if at least one of them measures a true effect (i.e. it tests if the distribution of submitted p-values is right-skewed). It entails log-transforming the pp-values, summing the result across all studies k, and then multiplying by -2. The resulting value is a test statistic which follows a χ^2 distribution (see Chapter 5.1.1) with $2 \times K$ degrees of freedom (where K is the total number of pp-values)[5]:

$$\chi^2_{2K} = -2 \sum_{k=1}^{K} \log(pp_k) \tag{9.15}$$

Let us try out Fisher's method in a brief example. Imagine that our p-curve contains five p values: p = 0.001, 0.002, 0.003, 0.004 and 0.03. To test for right-skewness, we first have to transform these p-values into the pp-value:

```
p <- c(0.001, 0.002, 0.003, 0.004, 0.03)
pp <- p*20

# Show pp values
pp
```

```
## [1] 0.02 0.04 0.06 0.08 0.60
```

Using equation 9.15, we can calculate the value of χ^2 using this code:

```
chi2 <- -2*sum(log(pp))
chi2
```

```
## [1] 25.96
```

This results in χ^2 = 25.96. Since five studies were included, the degrees of freedom are d.f. $= 2 \times 5 = 10$. We can use this information to check how likely our data are under the null hypothesis of no effect/no right-skewness. This can be done in R

[5] In newer versions, p-curve uses *Stouffer's method* instead of the one by Fisher to test for right-skewness (Simonsohn et al., 2015). Both methods are closely related, but Stouffer's method is based on z scores instead of p-values.

using the pchisq function, which we have to provide with our value of χ^2 as well as the number of d.f.:

```
pchisq(26.96, df = 10, lower.tail = FALSE)
```

```
## [1] 0.002643
```

This gives us a p-value of 0.0026. This means that the null hypothesis is very unlikely, and therefore rejected. The significant value of the χ^2 test tells us that, in this example, the p-values are indeed right-skewed. This can be seen as evidence for the assumption that there is evidential value behind our data.

9.2.2.1.2 Test for Flatness

We have seen that the right-skewness test can be used to determine if the distribution of significant p-values represents a true overall effect. The problem is that this test depends on the *statistical power* of our data. Therefore, when the right-skewness test is *not* significant, this does not automatically mean that there is no evidential value. Two things are possible: either there is indeed no true effect, or the number of values in our p curve is simply too small to render the χ^2 test significant–even if the data is in fact right-skewed.

Thus, we have to rule out lacking power as an explanation of a non-significant right-skewness test. The null hypothesis of the right-skewness test is that there is no evidential value. In the test, we essentially try to reject this null hypothesis by showing that our empirical p curve is *not* flat. Now, we have to turn this logic around to show that the p curve *is* flat. We can do this by changing the null hypothesis. Instead of no effect, our new null assumes that the p curve contains a *small* effect, and as a consequence is *slightly* right-skewed. In a test for flatness, the goal then becomes to show that our p curve is *not* slightly right-skewed. Or, to put it differently, we want to confirm that the p curve is significantly flatter than the one we expect for a very, very small effect. When this is the case, we can say that even a very small effect can be ruled out for the data at hand, and that there is likely no evidential value at all.

P-curve analysis achieves this through a test of 33% power. The idea is to construct the *expected* pp-value (i.e. the probability of p) of each significant study k when the true effect is very small. By very small, we mean an effect size that can be detected with a power of 33% using the study's sample size. This 33% threshold is, to some extent, arbitrary; it was chosen by the inventors of p-curve as a rough indicator of an effect that is on the verge of being practically negligible. We will spare you the statistical details behind how the 33% power pp-values are determined, but it is important to know that it involves the use of a *non-central* distribution, such as the non-central F, t and χ^2 distribution, depending on the outcome measure[6]. In the following section

[6]In the p-curve R function that we will use in this book, effect sizes are first transformed to z scores (using the formula $z = \hat\theta/SE_{\hat\theta}$). To calculate expected pp-values for the 33% power tests, a non-central

on effect size estimation using p-curve (Chapter 9.2.2.2), we will describe the concept behind a non-central distribution in greater detail.

In sum, the flatness test first involves calculating *pp*-values based on an effect detectable with 33% power, for each significant p value. If the 33% power estimate fits the distribution of our p-values well, the 33% power *pp*-values will be uniformly distributed–just as p-values follow a uniform distribution when the data fits the null hypothesis $\theta = 0$ well. Thus, we can apply the same methods we also used in the right-skewness test, but this time, we use the 33% power *pp*-values for our calculations. The only difference now is that we are not particularly keen on rejecting the null hypothesis: this would mean that we reject the notion that at least a small effect exists in our data. We would be forced to say that there is either an effect small enough to be negligible–or that there is no effect at all.

9.2.2.1.3 *Interpretation of P-Curve Results*

We have now covered several tests that allow to analyze an empirical p-curve. Do not worry too much if you found some of the statistical concepts difficult to understand. It takes some time to get a grasp of the methodology behind p-curve, and the following hands-on example will certainly be helpful in this respect. The most important part is to understand the *idea* behind the p-curve tests, and how their results can be interpreted. In this section, we will now focus on the latter.

When we interpret p-curve, we have to make sense out of four test results: the ones of the binomial right-skewness and flatness test, as well as the ones of the right-skewness and flatness test based on *pp*-values. To make things worse, p-curve analysis also involves two additional tests that we have not covered yet: a right-skewness and flatness test based on the *half p curve*. These tests are identical to the *pp*-value-based tests we covered before, but they are only applied to high p-values (i.e. $p < 0.025$).

The half p curve tests were introduced as a safeguard against *ambitious p-hacking* (Simonsohn et al., 2015). Although arguably less likely, it is possible that researchers may have p-hacked results until they become highly significant. This, however, may distort the shape of the p curve: it may not appear left-skewed, but slightly right-skewed, even when there is no true effect. A test based on the half p curve can control for this, because it becomes increasingly difficult, even for ambitious p-hackers, to obtain *very* high p-values (e.g. $p < 0.01$), unless there is a true effect. Since, by definition, the half p curve only contains values smaller than 0.025, no binomial test is performed.

When interpreting p-curve's results, we essentially try to answer two questions. The first one is: does our p-curve indicate the presence of evidential value? This can be evaluated using the right-skewness tests. In case we can not confirm the presence of evidential value, we turn to the second question: is evidential value absent or

χ^2 distribution with d.f. = 1 is used. This is possible because the distribution of χ^2_1 and z (or, more precisely: z^2) is equivalent.

inadequate? This can be assessed using the flatness tests. In practice, this guideline may be used (Simonsohn et al., 2015):

- **Evidential value present**: The right-skewness test is significant for the half p curve ($p < 0.05$) <u>or</u> the p-value of the right-skewness test is <0.1 for *both* the half and full curve.
- **Evidential value absent or inadequate**: The flatness test is significant with $p < 0.05$ for the full curve <u>or</u> the flatness test for the half curve *and* the binomial test are $p < 0.1$.

How to Interpret a "No-No" Case

Every p-curve analysis eventually ends up with one of three outcomes. When the right-skewness test is significant, we conclude that evidential value is present. When the right-skewness test is not significant, but the flatness test is, this indicates that evidential value is absent (or the effect is very, very small). The third and last outcome is the trickiest. For lack of a better word, we call it a "no-no" case. It arises when we can neither verify that evidential value is present, nor that it is absent (i.e. neither the right-skewness test, nor the flatness test is significant).

In terms of interpretation, a "no-no" case means that we can not confirm that a true effect is present, but that we are not able to rule out a relatively small effect either. This third outcome frequently occurs when the p curve only contains a few studies and is admittedly somewhat disappointing. This result essentially communicates that we do not know if a true effect exists when looking a the p curve, and that more evidence is needed to clarify things.

9.2.2.1.4 *P-Curve Analysis in R*

By now, we have learned a lot about the theory behind p-curve analysis, so it is high time we start applying the technique in a real-world example. Luckily, Simonsohn, Simmons and Nelson, the inventors of p-curve, have developed an application which automatically conducts all the tests we previously discussed, and returns the relevant results. This *p-curve app* can also be found online[7]. To use p-curve in R, we can resort to the `pcurve` function. This function emulates the behavior of the app and was designed specifically for meta-analysis objects created with the *{meta}* package.

[7] `http://p-curve.com/`

The "pcurve" Function

The pcurve function is included in the *{dmetar}* package. Once *{dmetar}* is installed and loaded on your computer, the function is ready to be used. If you did not install *{dmetar}*, follow these instructions:

1. Access the source code of the function online[a].
2. Let R "learn" the function by copying and pasting the source code in its entirety into the console (bottom left pane of R Studio), and then hit "Enter".
3. Make sure that the *{stringr}* and *{poibin}* package is installed and loaded.

[a]https://raw.githubusercontent.com/MathiasHarrer/dmetar/master/R/pcurve2.R

Using the pcurve function is easy. We only need to provide the function with a *{meta}* meta-analysis object that we created previously. In this example, we again use our m.gen meta-analysis object, which we created with the metagen function (Chapter 4.2.1). However, before we run the analysis, we remove the two outlying studies that we identified previously (study 3 and study 16, see Chapter 5.4). We will cover later why this is a good idea.

```
library(dmetar)
library(meta)

# Update m.gen to exclude outliers
m.gen_update <- update.meta(m.gen, subset = -c(3, 16))

# Run p-curve analysis
pcurve(m.gen_update)
```

```
## P-curve analysis
##  -----------------------
## - Total number of provided studies: k = 16
## - Total number of p<0.05 studies included into the
##   analysis: k = 9 (56.25%)
## - Total number of studies with p<0.025: k = 8 (50%)
##
## Results
##  -----------------------
##                     pBinomial  zFull pFull  zHalf pHalf
## Right-skewness test     0.020 -3.797 0.000 -2.743 0.003
## Flatness test           0.952  1.540 0.938  3.422 1.000
```

```
## Note: p-values of 0 or 1 correspond to p<0.001 and p>0.999,
## respectively.
##
## Power Estimate: 66% (31.1%-87.8%)
##
## Evidential value
## ------------------------
## - Evidential value present: yes
## - Evidential value absent/inadequate: no
```

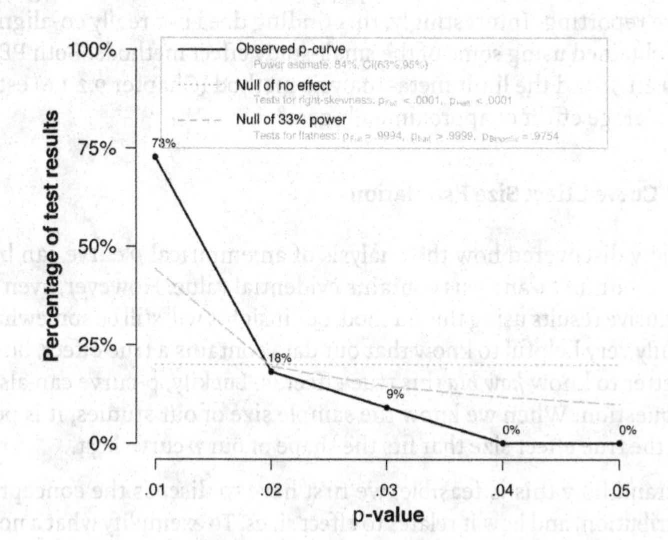

We see that running the pcurve function results in two outputs: our p-curve analysis results, and a plot of the observed p curve.

The output tells us that the meta-analysis contained $k = 9$ significant effects that were included in the p-curve. Most of these studies ($k = 8$) had highly significant results (i.e. $p < 0.025$). The `Results` section contains the main outcomes of the analysis. We see that all three right-skewness tests are significant: the binomial test (`pBinom`; $p = 0.02$), the full p curve pp-value test (`pFull`; $p < 0.001$) and the test based on the half p curve (`pHalf`; $p = 0.003$)[8]. Applying the criteria we set up in Chapter 9.2.2.1.3, this indicates that our data contains evidential value. In the next line, we see that the three flatness tests are *not* significant, with ps ranging from 0.938 to 1. This, quite logically, tells us that evidential value is neither absent nor inadequate. The same interpretation is provided for us in the `Evidential value` section of the output.

[8] In the output, we also see that the test statistic associated with the full and half p curve test is a z score. This is because the pcurve function uses Stouffer's instead Fisher's method to aggregate the results. As we mentioned in footnote 6 in this chapter, both methods are closely related.

The p curve plot produced by the function contains three lines: a solid one, representing the empirical p curve based on our data; a dashed line showing the expected p value distribution assuming 33% power; and a dotted line, which shows the uniform distribution we would expect when there is no effect. The solid line is visibly right-skewed, just as we would expect when the studies measure a true effect.

Overall, these results indicate the presence of evidential value and that there is a true non-zero effect. We can still not rule out that publication bias has affected the results of our meta-analysis. But, based on p-curve's results, we can conclude that the pooled effect we found is not completely spurious; it is not just a "mirage" produced by selective reporting. Interestingly, this finding does not really co-align with the results we obtained using some of the small-study effect methods. Both PET-PEESE (Chapter 9.2.1.5) and the limit meta-analysis method (Chapter 9.2.1.6) estimated a corrected average effect of approximately zero[9].

9.2.2.2 P-Curve Effect Size Estimation

We have now discovered how the analysis of an empirical p curve can be used to determine if our meta-analysis contains evidential value. However, even when we reach conclusive results using this method, our insights will still be somewhat limited. It is certainly very helpful to know that our data contains a true effect, but it would be even better to know *how big* this true effect is. Luckily, p-curve can also help us with this question. When we know the sample size of our studies, it is possible to search for the true effect size that fits the shape of our p curve best.

To understand how this is feasible, we first have to discuss the concept of a *non-central* distribution, and how it relates to effect sizes. To exemplify what a non-central distribution is, we start with the arguably most common statistical test there is: the two-sample t-test. A t-test is commonly used to examine if the means of two groups differ. The null hypothesis in a t-test is that both means μ_1 and μ_2 are identical, and that, therefore, their difference is zero. When the null hypothesis is true, we assume that the t statistic follows a *central* t distribution. The central t distribution looks similar to a standard normal distribution. Since the null hypothesis assumes that there is no difference, the t distribution is centered around zero.

Quite logically, this central t distribution will not represent the reality well when the null hypothesis is incorrect. When there is a true difference between the means, we do not expect the t-values to be centered around zero. Instead, the t statistics will follow a *non-central* t distribution. This non-central distribution is usually asymmetric, and tends to have a wider spread. Most importantly, however, its center is "shifted" away from zero. The magnitude of this shift is controlled by the *non-centrality parameter* δ.

---

[9]An attentive reader may have recognized that for the small-study effect methods, outliers were not removed, and that this may have caused the different results. We checked for this possibility by re-running PET-PEESE and a limit meta-analysis without the statistical outliers, which lead to largely the same results as before. If you want to, you can try to verify this yourself in R.

The higher δ, the further the peak of the non-central distribution will lie away from zero.

The graph below illustrates this behavior. On the left, we see a central t distribution, for which $\delta = 0$. To the right, a non-central t distribution with $\delta = 5$ is displayed. We see that the right curve is less symmetrical, and that it peaks around 5, whereas the center of the symmetrical, central t distribution is zero.

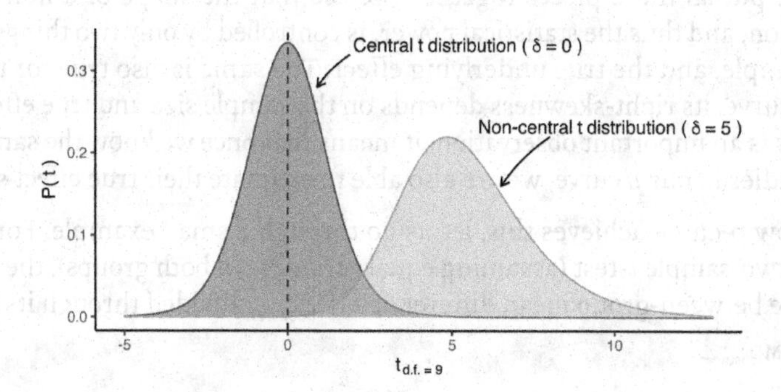

While the left central distribution shows the expected t-values when the null hypothesis is true, the right one shows the expected t-values when the *alternative hypothesis* is correct. Another way to think about this is to say that the left curve shows the t distribution when there is *no effect*, while the right one shows the distribution when there *is* an effect. The central distribution could, for example, represent an SMD of 0, while the non-central distribution represents the expected t-value distribution for an effect of SMD = 1.3 (this value is made up). The higher the effect size (i.e. the bigger the difference between the two samples), the higher the non-centrality parameter δ will be, and the non-central distribution will move further and further away from zero.

As we mentioned, the non-central t distribution can be used to model the alternative hypothesis in a t-test. The reason why it is so uncommon to see a non-central distribution in statistical textbooks, however, is because we usually do not *need* it. In statistical hypothesis testing, our alternative hypotheses are usually *non-specific*. When we calculate a two-sample t-test, we are only interested in the null hypothesis ("there is no difference between the groups"). When our data does not fit the null hypothesis well, we reject it and conclude that *some* effect exists. The alternative hypothesis in such a test is simply the opposite of the null hypothesis: that the mean difference between the two samples is *not* zero; not that the effect has this or that size.

A *specific* alternative hypothesis is usually only needed when we want to calculate the *power* of a statistical test. When setting up an experiment, it is conventional to plan for a sample size large enough to make sure that the probability of a *false negative* is 20% or lower. By using an adequate sample size, we want to make sure that a true effect can be detected, provided that it exists. This probability of a statistical test to uncover a true effect is its statistical power. It is defined as 1 minus the probability of a false positive, also known as β. To calculate the required sample size for an adequately powered t-test, we need the non-central t distribution, because it shows us the expected behavior of t-values when there is an effect. To calculate the sample size, we also need to assume a value for the true effect size, since it affects the non-centrality parameter δ, and thus the shape of the non-central distribution.

When we put all these pieces together, we see that the shape of a non-central t distribution, and thus the statistical power, is controlled by only two things: the size of our sample, and the true underlying effect. The same is also true for the shape of our p curve: its right-skewness depends on the sample size and true effect in our data. This is an important observation: it means that once we know the sample sizes of the studies in our p curve, we are also able to estimate their true effect size.

To see how p-curve achieves this, let us go through a small example. For an independent two-sample t-test (assuming equal variances in both groups), the value of t equals the between-group mean difference $\text{MD}_{\text{between}}$ divided through its standard error $SE_{\text{MD}_{\text{between}}}$:

$$t_{\text{d.f.}} = \frac{\text{MD}_{\text{between}}}{SE_{\text{MD}_{\text{between}}}} \tag{9.16}$$

When we insert the formula for the between-group mean difference and standard error (see equation 3.14 and 3.15 in Chapter 3.3.1.1), we get the following formula:

$$t_{n_1+n_2-2} = \frac{\hat{\mu}_1 - \hat{\mu}_2}{s_{\text{pooled}}\sqrt{\dfrac{1}{n_1}+\dfrac{1}{n_2}}} \tag{9.17}$$

We see in the equation that the degrees of freedom of t are defined as the combined sample size ($n_1 + n_2$) of both groups minus 2.

Using this formula, we can calculate the t-value based on data reported in a primary study. Imagine that a study contained $n_1 = 30$ participants in the experimental group, and $n_2 = 20$ participants in the control group. The study reports a mean of 13 and 10 for group 1 and 2, respectively, and that both groups had a standard deviation of 5. Based on this data, we can use the following code to calculate t:

```
# Calculate mean difference
md <- 13-10

# Calculate SE of mean difference
n1 <- 30
n2 <- 20
s1 <- s2 <- 5
s_pooled <- sqrt((((n1-1)*s1^2) + ((n2-1)*s2^2))/
```

```
                  ((n1-1)+(n2-1)))

se <- s_pooled*sqrt((n1+n2)/(n1*n2))

# Calculate t-value (equivalent to 2-sample t-test with equal variance)
md/se
```

```
## [1] 2.078
```

The result is $t_{48} = 2.078$. Does this result support the null hypothesis that both means are identical and that there is no effect? To answer this question, we can use the `pt` function. This function gives us the probability of finding a t-value greater than 2.078 when d.f. = 48 *and* provided that the null hypothesis is true. This probability is equal to the p-value of a one-sided t-test.

```
pt(2.078, df = 48, lower.tail = F)
```

```
## [1] 0.02154
```

The result is $p = 0.02$, which means that the test is significant. Therefore, we reject the null hypothesis that the effect of the experimental group is zero (or negative). As a consequence, we accept the alternative hypothesis: there is a positive effect favoring the experimental groups (assuming that higher scores represent better outcomes).

We now know that the central t distribution underlying the null hypothesis of the t test does not suit our empirical data well. We also know that a non-central t distribution fits our data better–but we do not know which one. For now, we can only guess which true effect size, and thus which non-centrality parameter δ, really represents the population from which our empirical t value was drawn.

As a first guess, we could assume that the true effect size behind our finding is a standardized mean difference of $\theta = 0.6$. This would mean that we found $t = 2.078$ because there was a medium-to-large effect in the experiment. Based on this value of θ, the non-centrality parameter is calculated using this formula:

$$\delta = \frac{\theta}{\sqrt{\frac{n_1+n_2}{n_1 n_2}}} \tag{9.18}$$

Where n_1 and n_2 are the sample sizes in both groups. In our example, we can calculate the non-centrality parameter δ using this code:

```
theta <- 0.6
delta <- theta/sqrt((n1+n2)/(n1*n2))
```

```
# Show delta
delta
```

```
## [1] 2.078
```

To see how a non-central t distribution with $\delta = 2.078$ looks like, let us do a little simulation. Using the `rt` function we draw one million random t-values with 48 degrees of freedom, twice: once when assuming a non-centrality parameter of zero (which is equal to the null of no effect), and once using the value of δ that we just calculated (meaning that the true effect is SMD = 0.6). Then, using a pipe and the `hist` function, we let R draw a histogram of both simulations.

```
# '1 with 6 zeros' can also be written as '1e6' in R
rt(n = 1e6, df = 48, ncp = 0) %>%
  hist(breaks = 100,
       col = "gray50",
       xlim = c(-4,8),
       ylim = c(0, 40000),
       xlab = "t-value",
       main = NULL)

rt(n = 1e6, df = 48, ncp = delta) %>%
  hist(breaks = 100,
       col = "gray95",
       xlim = c(-4,8),
       ylim = c(0, 40000),
       add = T)
```

This is the resulting plot:

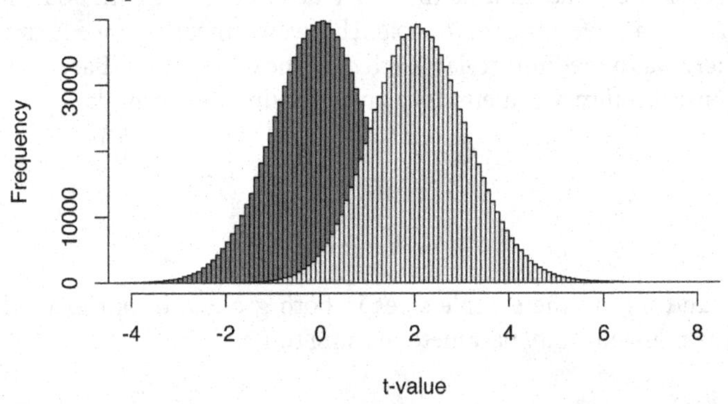

In it, we see the central t distribution (no effect) on the left, and the non-central distribution ($\theta = 0.6$) to the right. Because we already have a moderately sized sample ($N = 50$), the non-central distribution looks less right-skewed than in the previous visualization. Nevertheless, it is clearly visible that the distribution of our assumed alternative hypothesis is shifted to the right, and peaks at the value of δ.

The central question, of course, is: how likely is it to obtain a value greater than $t_{48} = 2.078$ when this alternative distribution is indeed the correct one, and when the true effect is in fact $\theta = 0.6$? To examine this question, we can use the pt function again, but this time also provide our assumed non-centrality parameter δ. This information can be added using the ncp argument. Let us check what result we get.

```
# Remember that t=2.078
pt(2.078, df = 48, ncp = delta, lower.tail = FALSE)
```

```
## [1] 0.5045
```

We see that the probability of obtaining a value greater than our result under the specified alternative hypothesis is roughly 50%. This means that about half of the values are expected to be higher, and the other half lower than the t-value we found. All in all, this indicates that a non-central t distribution assuming 48 degrees of freedom and a true effect of 0.6 approximates our finding very well. It seems very likely that the true population effect of the study is SMD = 0.6.

The steps we just made are, essentially, also the ones that p-curve uses to determine the true effect size. For every significant p-value in the p curve, it calculates the probability of getting a value greater than t, assuming:

1. a certain effect size/non-centrality parameter;
2. that the p-value is based on x degrees of freedom (which we can derive from the study's sample size); and
3. knowing that only significant values ($p < 0.05$) are included in the p-curve.

This results in a pp-value for each significant study k. Based on all that we just covered, the formula for the pp-value of a study k can be expressed like this:

$$pp(t_k) = \mathrm{P}(t > t_k \mid \delta, \text{d.f.}, p < 0.05) \tag{9.19}$$

Since the degrees of freedom of a study are usually known, the only unknown in the equation is δ, and thus the true effect size θ. However, it is possible to *find* this true effect size, because we know that the distribution of pp-values will be *uniform* when we assume the correct true effect size/δ-value. Just like p-values follow a uniform distribution when our findings conform with the null hypothesis, the pp-values are uniformly distributed when the results conform with the *correct* non-central distribution (i.e. the point alternative hypothesis). Therefore, we only have to try out many, many possible *candidate effect sizes*, plug the resulting δ-value into the equation above, and evaluate the skewness of the resulting pp-values. The candidate effect size that comes closest to a uniform distribution of pp-values then represents our estimate of the true effect. P-curve uses the D distance metric of a so-called Kolmogorov-Smirnov (KS) test to capture how much a *pp* distribution deviates from a uniform distribution.

P-curve's effect estimation method is also implemented in the `pcurve` function. To use it, we have to set the `effect.estimation` argument to `TRUE`. We also have to specify `N`, the sample size of each study. Lastly, we can control the search space for candidate effect sizes using `dmax` and `dmin`. Here, we tell `pcurve` to search for effect sizes between Cohen's $d = 0$ and 1. Please note that `dmin` must always be zero or greater–the minimum that p-curve can detect is no effect.

```
# Add experimental (n1) and control group (n2) sample size
# Sample sizes of study 3 and 16 removed
n1 <- c(62, 72, 135, 103, 71, 69, 68, 95,
        43, 79, 61, 62, 60, 43, 64, 63)

n2 <- c(51, 78, 115, 100, 79, 62, 72, 80,
        44, 72, 67, 59, 54, 41, 66, 55)

# Run p-curve analysis with effect estimation
pcurve(m.gen_update,
       effect.estimation = TRUE,
       N = n1+n2,
       dmin = 0,
       dmax = 1)
```

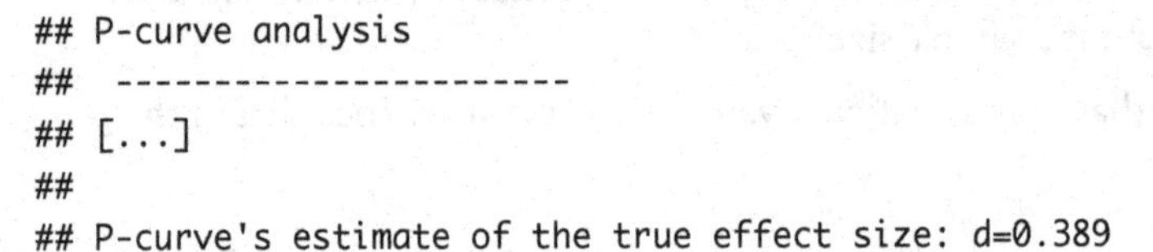

```
## P-curve analysis
##  -----------------------
## [...]
##
## P-curve's estimate of the true effect size: d=0.389
```

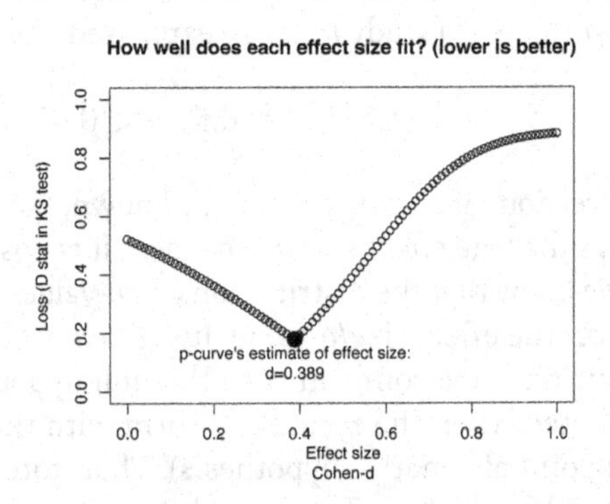

The output now contains two new elements: an estimate of the true effect size, as well as a plot showing results of the effect size search. In the plot, the candidate effect sizes form a smooth, V-shaped gradient, which peaks at an effect size of $d = 0.389$.

At this point, the difference between the calculated *pp* distribution and a uniform distribution (represented by the value of D on the y-axis) is minimal, which means that this it represents the best estimate of the true effect. Importantly, p-curve's estimate of the effect size is *only* trustworthy when the plot shows a V-shape similar to the one here. Other, more erratic shapes indicate that p-curve may not have found the minimum. A plot with a smooth, descending line might indicate that our search space is simply too narrow. In this case, it makes sense to re-run the analysis with a higher `dmax` value.

Overall, p-curve's estimate of $d = 0.389$ is somewhat lower than the pooled effect we found in the meta-analysis ($g = 0.45$ when outliers are removed). However, it is still large enough to be in line with our previous finding, that the studies contain evidential value.

Reporting P-Curve Analysis Results

When we report the results of a p-curve analysis, it is a good idea to at least include the p-values of the three right-skewness and flatness tests, as well as how these results are interpreted. When the true effect size was estimated, this should also be included. All of these results can be combined in a table that looks somewhat like this:

		Full Curve		Half Curve		Evidential Value		
	p_{Binomial}	z_{Full}	p_{Full}	z_{Half}	p_{Half}	present	absent	$\hat{d}$
Right-Skewness Test	0.020	-3.80	<0.001	-2.74	0.003	yes	no	0.39
Flatness Test	0.952	1.54	0.938	3.42	>0.999			

The developers of p-curve also highly recommend to create a *disclosure table* for each analysis, describing from which part of an article the result was extracted, and how it was originally reported. An example for such a disclosure table, along with several other practical guidelines, can be found in Simonsohn et al. (2014a).

P-Curve & Between-Study Heterogeneity

We still owe you an explanation of why we excluded outlying studies from the p-curve analysis. With outliers included, our meta-analysis had a between-study heterogeneity of $I^2 = 63\%$, which is quite substantial. This is problematic because it has been found that p-curve is not a robust method to estimate the true effect size when the between-study heterogeneity of our data is high.

Van Aert and colleagues (2016) have therefore proposed to only use p-curve when the heterogeneity is small to moderate. They proposed a threshold of $I^2 = 50\%$ as rule of thumb to determine if p-curve can be applied. When the between-study heterogeneity in our meta-analysis is higher than that, one workaround is to p-curve effect sizes without outliers, like we did in the example. An even better solution is to perform a separate analysis in sensible subgroups of studies, provided they exist.

9.2.3 Selection Models

The last type of publication bias method we cover are so-called *selection models*. Although selection models have been proposed to examine the impact of selective publication for some time (Hedges, 1992; Iyengar and Greenhouse, 1988; Hedges and Vevea, 1996; Hedges, 1984), interest in their application has particularly increased in the last few years (McShane et al., 2016; Carter et al., 2019).

All publication bias methods we covered previously are based on some kind of "theory", which is used to explain why and how selective publication affects the results of a meta-analysis. Small-study effect methods, for example, assume that a study's risk of non-publication is proportional to its sample and effect size. P-curve is based on the idea that a p-value of 0.05 serves as a "magic threshold", where results with $p \geq 0.05$ are generally much more likely to be missing in our data than statistically significant findings.

Selection models can be seen as a generalized version of these methods. They allow to model *any* kind of process through which we think that publication bias has affected our results. This makes them very versatile: selection models can be used to model our data based on very simple, or highly sophisticated hypotheses concerning the genesis of publication bias.

The idea behind all selection models is to specify a distribution which predicts, often in a highly idealized way, how likely it is that some study is published (i.e. "selected"), depending on its results. Usually, this result is the study's p-value, and a selection model can be seen like a function that returns the probability of publication for varying values of p. Once such a selection function has been defined, it can be used to "remove" the assumed bias due to selective publication, and derive a corrected estimate of the true effect size.

Yet, this corrected effect will only be appropriate when the selection model we defined is indeed correct. We always have to keep in mind that our model is just one of many ways to explain the selection process–even if our model seems to fit the data well. The exact processes through which publication bias has shaped our results will inevitably remain unknown. Nevertheless, selection models can be enormously helpful to broadly assess if, and more importantly, *how* publication may have influenced our data.

In this chapter, we will cover two types of (rather simple) selection models based on *step functions*. Therefore, let us first clarify what step functions are.

9.2.3.1 Step Function Selection Models

To perform any kind of selection model analysis, we need two ingredients: an *effect size model*, and the *selection model* itself. We can think of both of these models as *functions*, which use some input value x and then return the probability of that value.

The effect size model, described by the function $f(x_k)$, is identical to the random-effects model. It assumes that the observed effect sizes $\hat\theta_k$ are normally distributed around an average effect μ, and deviate from μ due to sampling error and between-study heterogeneity variance τ^2. Knowing μ, τ^2, a study's standard error, and that effect sizes are normally distributed, the function $f(x_k)$ predicts how likely it is to observe some effect size value x_k–assuming that there is no publication bias.

Yet, when there *is* publication bias, this effect size distribution, and thus $f(x)$ itself, is an incorrect representation of reality. Due to selective publication, some studies are over-represented in our data–presumably those with surprisingly high effect sizes and small samples. This means that we have, without knowing, given these studies a higher weight in our pooling model. We, therefore, need to derive a more "realistic" version of $f(x)$, which incorporates the fact that some results had a greater chance of being included than others; that they were given a higher "weight".

This is achieved through a *weight function* $w(p_k)$. The weight function tells us the selection probability of a study k, depending on its p-value. Based on this, we can define an adapted version of $f(x_k)$, which also incorporates the publication bias mechanism. This function $f^*(x_k)$ is symbolized by this formula (Vevea and Woods, 2005):

$$f^*(x_k) = \frac{w(p_k)f(x_k)}{\int w(p_k)f(x_k)dx_k} \tag{9.20}$$

Where the denominator in the fraction stands for the integral of $w(p_k)f(x_k)$. The weight function $w(p_k)$ in this equation represents our assumed selection model.

Although $w(p_k)$ can technically have any shape, it is often implemented as a *step function* (Hedges and Vevea, 1996). When $w(p_k)$ is a step function, this means that values p_k which fall into the same interval (e.g. all p-values smaller than 0.05) are selected with the same probability. This interval-specific selection probability is denoted with ω_i, and can differ from interval to interval. Essentially, we split up the range of possible p-values (0 to 1) into different segments, and give each segment its own selection probability $\omega_1, \omega_2, ..., \omega_c$. The size of the segments is determined by several *cut-points* (which we denote with a_i). The number of cut-points, as well as their exact value, can be chosen by us. For example, when $w(p_k)$ contains four segments (and thus four cut-points), we can define its inner workings like so:

$$w(p_k) = \begin{cases} \omega_1 & \text{if } 0 \leq p_k \leq a_1 \\ \omega_2 & \text{if } a_1 \leq p_k \leq a_2 \\ \omega_3 & \text{if } a_2 \leq p_k \leq a_3 \\ \omega_4 & \text{if } a_3 \leq p_k \leq a_4 \text{ (where } a_4 = 1). \end{cases} \tag{9.21}$$

We see that for any value of p, the function above returns a specific selection probability ω, based on the p-value interval into which our value falls. To make this more concrete, we now define a selection model with actual values filled in for the cut-points a_i and selection probabilities ω_i.

We could assume, for example, that the publication bias mechanism in our meta-analysis can be described with three cut-points. First, there is $a_1 = 0.025$. This value equals a one-sided p value of 0.025, and two-sided p-value of 0.05. Since this is the conventional significance threshold used in most studies, it makes sense to assume that all p-values smaller $a_1 = 0.025$ have a selection probability ω_1 of 100%. After all, there is no reason to put a study into the file drawer when its results were positive. The next cut-point we define is $a_2 = 0.05$. For results in this range, we assume a selection probability of 80%: still high, but lower compared to results that are clearly significant. Then, we specify a large interval, ranging from $a_2 = 0.05$ to $a_3 = 0.5$, in which the selection probability is 60%. Lastly, for studies with a very high p-value of ≥ 0.5, we define an even lower probability of $\omega_4 = 35\%$. This results in a selection model as depicted in Figure 9.2 below.

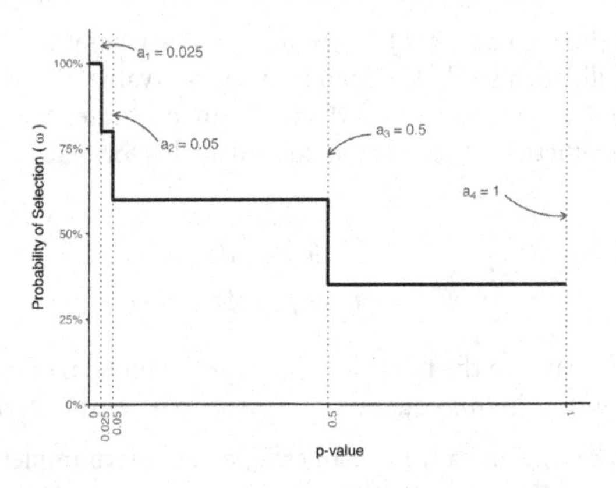

FIGURE 9.2: Selection model based on a step function.

When we define a selection model based on a step function, we usually only specify the cut-points a_i. These are the only fixed parameters in our model, while the selection probabilities $\omega = \omega_1, \omega_2, \dots, \omega_c$ are estimated from the data. Based on the formula in equation 9.20, the selection model can then be fitted to our data. This involves using maximum likelihood procedures to jointly estimate ω, as well as a corrected estimate of μ and τ^2 which takes the disparate selection probabilities ω into account. The resulting corrected estimate of μ then represents the true average effect size when controlling for the assumed publication bias mechanism.

Previously, we expressed the selection probabilities ω as percentages ranging from 0% to 100%. However, when we fit the selection model, ω_i is not estimated as an absolute selection probability, but in terms of its *relative likelihood* of selection. This entails giving the first interval in the step function a reference value of 1, while all other values of ω represent the likelihood of selection *in relation* to this reference group. If, for example, we estimate a value of ω_2=0.5 in the second interval, this means that studies in this segment were only half as likely to be selected compared to the first interval (for which $\omega_1 = 1$).

Of course, our corrected estimate of the true average effect μ will only be accurate when our selection model itself is appropriate. A rough indication of this is a significant likelihood ratio test (LRT) of the selection model parameters. The test is based on the null hypothesis that there is no selection, and that the relative selection likelihood is identical for all intervals (i.e. that $\omega_1 = \omega_2 = \dots = \omega_c$). It should be noted, however, that this significance test has been found to frequently produce anti-conservative results (Hedges and Vevea, 1996). This means that its results should be interpreted cautiously.

In theory, the number of cut-points a_i used in our selection model can be chosen *ad libitum*. Yet, with every additional cut-point, an additional value of ω_i has to be estimated. Depending on the size of our meta-analysis, this can soon lead to the problem that only few, if any, studies are available for each interval. This makes it increasingly difficult to estimate each ω_i properly. Complex selection models with many cut-points can therefore only be applied when the number of studies is large (i.e. $K \geq 100$).

Unfortunately, most meta-analyses only contain a small number of studies. This means that only simple selection models with very few cut-points can be applied. One variant of such a simple model the *three-parameter selection model*, which we will discuss next. This model has the advantage of being applicable even when the number of included studies is small (e.g. $K = 15$–20).

9.2.3.1.1 Three-Parameter Selection Model

The three-parameter model is a selection model with only one cut-point (McShane et al., 2016). It is called a *three-parameter* model because only three parameters need to be estimated: the true effect μ, the between-study heterogeneity variance τ^2, and the relative likelihood of the second interval ω_2[10].

In the three-parameter selection model, the single cut-point a_1 is set to 0.025, which is equal to a one-sided p-value of 0.05. This divides the range of p-values into two bins: those which can be considered statistically significant, and those which are not significant. Thus, ω_2 represents the probability that a non-significant result is selected for publication[11].

The `selmodel` function in the *{metafor}* package allows to fit various kinds of selection models in R[12]. It can also be used for three-parameter selection models, which we will try out now. As before, we will use the `ThirdWave` data set for our example.

The `selmodel` function only accepts meta-analysis objects created by *{metafor}*'s `rma` function (see Chapter 8.3.3). Therefore, we have to create such an object first. In our call to `rma`, we use settings identical to the ones we used for *{meta}*'s `metagen` function (Chapter 4.2.1).

```
library(metafor)

# We name the new object 'm.rma'
m.rma <- rma(yi = TE,
             sei = seTE,
             data = ThirdWave,
             slab = Author,
             method = "REML",
             test = "knha")
```

Using the `m.rma` object, we can now fit a three-parameter selection model using `selmodel`. To tell the function that we want to apply a step function, we have to set the `type` argument to `"stepfun"`. In the `steps` argument, we can specify the cut-point, which, in our model, is a_1=0.025. Let us have a look at the results:

---

[10]It is not necessary to estimate ω_1, since this interval serves as the reference category with a selection likelihood fixed at 1.

[11]P-curve (Chapter 9.2.2) can be seen as a special type of three-parameter selection model. It also uses a p-value of 0.05 as the cut-point for selection, but only focuses on the significant results (for which the selection probability is assumed to be 100%); and it assumes that τ^2 is zero. This means that for p-curve, only one parameter is actually estimated from the data: the true effect size μ.

[12]At the time we are writing this, the `selmodel` function is only available through the *development version* of *{metafor}*. The development version of *{metafor}* can be downloaded by first installing the *{remotes}* package, and then running this code: `remotes::install_github("wviechtb/metafor")`. After that, *{metafor}* can be called from the library and the `selmodel` function should be available. Please note that it is likely that, by the time you are reading this, the `selmodel` function has already been integrated into the standard version of *{metafor}*, which means that installation of the development version is not necessary anymore.

```
selmodel(m.rma,
         type = "stepfun",
         steps = 0.025)
```

```
## [...]
##
## Model Results:
##
## estimate      se    zval    pval   ci.lb   ci.ub
##   0.5893  0.1274  4.6260  <.0001  0.3396  0.8390  ***
##
## Test for Selection Model Parameters:
## LRT(df = 1) = 0.0337, p-val = 0.8544
##
## Selection Model Results:
##
##                      k  estimate      se    pval   ci.lb   ci.ub
## 0     < p <= 0.025  11    1.0000     ---     ---     ---     ---
## 0.025 < p <= 1       7    1.1500  0.8755  0.8639  0.0000  2.8660
##
## ---
## Signif. codes:  0 '***' 0.001 '**' 0.01 '*' 0.05 '.' 0.1 ' ' 1
```

Under `Model Results`, we can see that the selection model's estimate of the true average effect size is $g = 0.59$ (95%CI: 0.34-0.84). Interestingly, this estimate is nearly identical to the pooled effect size that we obtained previously ($g = 0.58$).

Overall, this does *not* indicate that our meta-analysis was substantially biased by a lower selection probability of non-significant results. This finding is corroborated by the `Test of Selection Model Parameters`, which is not significant ($\chi^2_1 = 0.034$, $p = 0.85$), and thus tells us that ω_1 and ω_2 do not differ significantly from each other.

Under `Selection Model Results`, we can see an estimate of the relative selection likelihood in both bins. We see that, with $\omega_2 = 1.15$, the selection probability in the second segment is actually slightly higher than in the first one. In case of substantial publication bias, we would expect just the opposite: that the relative selection likelihood of a non-significant result is considerably *lower* compared to significant findings.

As a sensitivity analysis, we can change a_1 from 0.025 to 0.05, and then re-run the analysis. Setting the cut-point to 0.05 means that we assume that a two-sided p value between 0.05 and 0.10 is just as "publishable" as one below 0.05. It could be, for example, that results which are significant on a "trend level" are still likely to be selected–or that some of the original studies used a one-sided test to evaluate if the study groups differ. Let us see if this altered cut-point changes the results:

```
selmodel(m.rma,
         type = "stepfun",
         steps = 0.05)
```

```
## [...]
##
## Model Results:
##
## estimate      se    zval    pval   ci.lb   ci.ub
##   0.3661  0.1755  2.0863  0.0370  0.0222  0.7100  *
##
## Test for Selection Model Parameters:
## LRT(df = 1) = 3.9970, p-val = 0.0456
##
## Selection Model Results:
##
##                      k  estimate     se    pval   ci.lb   ci.ub
## 0    < p <= 0.05    15    1.0000    ---     ---     ---     ---
## 0.05 < p <= 1        3    0.1786 0.1665 <.0001  0.0000  0.5050  ***
##
## ---
## Signif. codes:  0 '***' 0.001 '**' 0.01 '*' 0.05 '.' 0.1 ' ' 1
```

Interestingly enough, we now see a different pattern. The new average effect size estimate is $g = 0.37$, which is smaller than before. Furthermore, the likelihood test is significant, indicating that the intervals differ. We can see that, with $\omega_2 = 0.18$, the selection likelihood of values $p > 0.1$ (two-sided) is much lower than the one of (marginally) significant p-values. This indicates that our pooled effect may have been slightly distorted by selective reporting–particularly because studies with *clearly* non-significant results landed in the file-drawer.

9.2.3.1.2 *Fixed Weights Selection Model*

In the three-parameter selection model we just discussed, only a single cut-point a_1 is specified, while the selection likelihood is freely estimated by the model. As we mentioned, the fact that three-parameter models only use one cut-point makes them applicable even to meta-analyses with relatively few studies. This is because there is a lower chance that the model will "run out of studies" in some of the bins.

However, the large sample size requirement of selection models may be avoided if the selection likelihoods ω_i do not have to be estimated from the data. We can simply provide a fixed value of ω_i for each interval, and then check what the estimate of μ is under the imposed model. While this approach allows to fit more complex selection models (i.e. models with more cut-points), it also comes with a drawback. When imposing such a *fixed weights selection model*, we simply assume that all pre-specified ω_i values are correct. Yet, when the model is not appropriate, this means that the estimate of the average effect will not be trustworthy. A fixed weights model should therefore be seen as a way to check how the true effect size *would* look like, if the assumed selection process applies.

Vevea and Woods (2005) provide a few examples of how such multi-cutpoint, fixed weights selection models can look like. The plot below shows two illustrative examples of a step function representing moderate and severe selection:

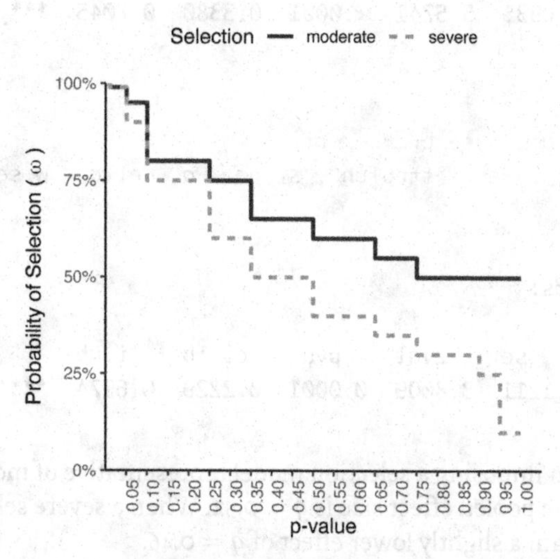

As a sensitivity analysis, we could check how the estimate of μ changes when we assume that these selection models are appropriate for our meta-analysis. To do this, we have to define all the cut-points used in the models displayed above, as well as the likelihood ω_i given to each interval.

```
# Define the cut-points
a <- c(0.005, 0.01, 0.05, 0.10, 0.25, 0.35, 0.50,
       0.65, 0.75, 0.90, 0.95, 0.99, 0.995)

# Define the selection likelihood for each interval
# (moderate/severe selection)
w.moderate <- c(1, 0.99, 0.95, 0.80, 0.75, 0.65, 0.60,
                0.55, 0.50, 0.50, 0.50, 0.50, 0.50, 0.50)
w.severe <- c(1, 0.99, 0.90, 0.75, 0.60, 0.50, 0.40, 0.35,
              0.30, 0.25, 0.10, 0.10, 0.10, 0.10)
```

Once these parameters are defined, we can use them in our call to `selmodel`. The new cut-points must be provided to the `steps` arguments, and the fixed likelihoods to `delta`:

```
# Fit model assuming moderate selection
selmodel(m.rma, type = "stepfun", steps = a, delta = w.moderate)
```

```
## [...]
##
## Model Results:
##
## estimate      se    zval    pval   ci.lb   ci.ub
##   0.5212  0.0935  5.5741  <.0001  0.3380  0.7045  ***
##
## [...]
```

```
# Fit model assuming severe selection
selmodel(m.rma, type = "stepfun", steps = a, delta = w.severe)
```

```
## [...]
## Model Results:
##
## estimate      se    zval    pval   ci.lb   ci.ub
##   0.4601  0.1211  3.8009  0.0001  0.2229  0.6974  ***
## [...]
```

We see that, when imposing a selection model representative of moderate selection, the estimate of the pooled effect size is $g = 0.52$. When a severe selection process is assumed, we obtain a slightly lower effect of $g = 0.46$.

Both of these results indicate that our observed effect is quite robust, even when controlling for selective publication. Importantly, however, these estimates are valid if–and only if–the selection model we specified is representative of the reality.

Other Selection Model Functions

In the examples, we only discussed step functions as the basis of selection models. It should noted, however, that this is not the only type of function one can use to model the selection process. The `selmodel` function also contains several functions based on *continuous* distributions, for example, the half-normal, logistic, or negative-exponential selection model. These models can be chosen by changing the specification of the type argument. It is beyond the scope of this guide to discuss all of these models, but the documentation of the `selmodel` function provides an excellent introduction. You can access the documentation by running `?selmodel` in R once *{metafor}* is loaded.

9.3 Which Method Should I Use?

This concludes our discussion of statistical methods for publication bias. This chapter has been quite long, and one might ask why we discussed so many different approaches. Is it not enough to simply choose one method, assess the risk of publication bias with it, and then move on?

The short answer is no. Publication bias methods remain a highly active research topic, and many studies have evaluated the performance of different approaches over the years (e.g. Simonsohn et al., 2014b; Stanley, 2017; van Aert et al., 2016; McShane et al., 2016; Rücker et al., 2011; Terrin et al., 2003; Peters et al., 2007). Alas, no clear winner has yet become apparent. On the contrary, there is evidence that no publication bias method consistently outperforms all the others (Carter et al., 2019).

It is possible, and in fact quite common, that different publication bias methods yield wildly different results. Our own hands-on exercises in this chapter are a very good example. Although we used the same data set each time, estimates of the true bias-corrected effect ranged from practically zero to $g = 0.59$. This underlines that the choice of method can have a profound impact on the results, and thus on our conclusions. While some methods indicated that our pooled effect completely disappears once we control for small-study effects, others largely corroborated our initial finding.

To address this issue, we recommend to always use *several* methods when evaluating publication bias. It is often difficult, if not impossible, to know which approach is suited best for our data and if its results are trustworthy. As we mentioned before, the exact extent to which selective reporting has affected our results will always be unknown. However, by applying several publication bias techniques, we can produce something similar to a *range* of credible true effects.

The size of this range can be used to guide our interpretation. If, for example, both PET-PEESE, p-curve and the three-parameter selection model arrive at estimates that are close to our initial pooled effect, this boosts our confidence in the robustness of our finding. Less so if we discover that the methods disagree. This means that the impact of publication bias and small-study effects is much more uncertain, as is the trustworthiness of our pooled effect.

In any case, results of publication bias methods should always be interpreted with caution. There are instances where the results of publication bias analyses have led to contentious debates–for example, in the "ego-depletion" literature (Friese et al., 2019). It is important to keep in mind that the best way to control for publication bias is to perform an adequate search for unpublished evidence, and to change publication practices altogether. Every statistical "proof" of publication bias that we as meta-analysts can come up with is weak at best.

To make it easier for you to decide which method may be applied when we created a brief overview of the advantages and disadvantages of each approach (see Table 9.1).

TABLE 9.1: Methods to estimate the true effect size corrected for publication bias: Overview of advantages and disadvantages.

	Advantages	**Disadvantages**
Duval & Tweedie Trim-and-Fill	Very heavily used in practice. Can be interpreted by many researchers.	Often fails to correct the effect size enough, for example, when the true effect is zero. Not robust when the heterogeneity is very large; often outperformed by other methods.
PET-PEESE	Based on a simple and intuitive model. Easy to implement and interpret.	Sometimes massively over- or underestimates the effect. Weak performance for meta-analyses with few studies, low sample sizes, and high heterogeneity.
Limit Meta-Analysis	Similar approach as PET-PEESE, but explicitly models between-study heterogeneity.	Performance is less well studied than the one of other methods. May fail when the number of studies is very low (<10) and heterogeneity very high.
P-Curve	Has been shown to outperform other methods (particularly trim-and-fill) when its assumptions are met.	Works under the assumption of no heterogeneity, which is unlikely in practice. Requires a minimum number of significant effect sizes. Less easy to interpret and communicate.
Selection Models	Can potentially model any kind of assumed selection process. The three-parameter selection model has shown good performance in simulation studies.	Only valid when the selection model describes the publication bias process adequately. Assume that other small-study effects are not relevant. Can be difficult to interpret and requires background knowledge.

The table contains both statistical and practical considerations and should be seen as neither comprehensive nor final. Publication bias methods are an ongoing field of investigation, and it is likely that things will look differently once more evidence has been established.

Lastly, it is of note that there is currently no method providing acceptable results when the between-study heterogeneity is high (van Aert et al., 2016, i.e. $I^2 \approx 75\%$). This means that publication bias analyses of meta-analyses with very high heterogeneity should at best be avoided altogether. Analyses without outliers or in more homogeneous subgroups can often be used as a practical workaround, but do not solve the general problem.

□

9.4 Questions & Answers

Test your knowledge!

1. How can the term "publication bias" be defined? Why is publication bias problematic in meta-analyses?
2. What other reporting biases are there? Name and explain at least three.
3. Name two questionable research practices (QRPs) and explain how they can threaten the validity of our meta-analysis.
4. Explain the core assumptions behind small-study effect methods.
5. When we find out that our data displays small-study effects, does this automatically mean that there is publication bias?
6. What does p-curve estimate: the true effect of all studies included in our meta-analysis, or just the true effect of all *significant* effect sizes?
7. Which publication bias method has the best performance?

Answers to these questions are listed in Appendix A at the end of this book.

9.5 Summary

- Publication bias occurs when some studies are systematically missing in the published literature, and thus in our meta-analysis. Strictly defined, publication bias exists when the probability of a study to get published depends on its results. However, there is also a range of other *reporting biases*. These reporting biases also influence how likely it is that a finding will end up in our meta-analysis. Examples are citation bias, language bias, or outcome reporting bias.

- It is also possible that *published* evidence is biased, for example, due to questionable research practices (QRPs). Two common QRPs are p-hacking and HARKing, and both can increase the risk of overestimating effects in a meta-analysis.
- Many publication bias methods are based on the idea of *small-study effects*. These approaches assume that only small studies with a surprisingly high effect size obtain significant results and are therefore selected for publication. This leads to an asymmetric funnel plot, which can be a sign of publication bias. But it does not have to be. Various "benign" causes of small-study effects are also possible.
- A relatively novel method, p-curve, is based on the idea that we can control for evidential value just by looking at the pattern of significant ($p < 0.05$) effects in our data. It can be used to test for both the presence and absence of a true effect, and can estimate its magnitude.
- Selection models are a very versatile method and can be used to model different publication bias processes. However, they only provide valid results when the assumed model is adequate and often require a very large number of studies. A very simple selection model, the three-parameter model, can also be used for smaller data sets.
- No publication bias method consistently outperforms all the others. It is therefore advisable to always apply several techniques and interpret the corrected effect size cautiously. Thorough searches for unpublished evidence mitigate the risk of publication bias in a much better way than current statistical approaches.

Part III

Advanced Methods

10

"Multilevel" Meta-Analysis

Welcome to the advanced methods section. In the previous part of the guide, we took a deep dive into topics that we consider highly relevant for almost every meta-analysis. With this background, we can now proceed to somewhat more advanced techniques.

We consider the following methods "advanced" because their mathematical underpinnings are more involved, or because of their implementation in R. However, if you have worked yourself through the previous chapters of the guide, you should be more than well equipped to understand and implement the contents that are about to follow. Many of the following topics merit books of their own, and what we cover here should only be considered as a brief introduction. Where useful, we will therefore also provide literature for further reading.

This first chapter deals with the topic of "multilevel" meta-analyses. You probably wonder why we put the word "multilevel" into quotation marks. Describing a study as a "multilevel" meta-analysis insinuates that this is something special or extraordinary compared to "standard" meta-analyses. Yet, that is not true. Every meta-analytic model presupposes a multilevel structure of our data to pool results (Pastor and Lazowski, 2018). In the chapters before, we have already fitted a multilevel (meta-analysis) model several times–without even knowing.

When people talk about multilevel meta-analysis, what they think of are *three-level meta-analysis models*. Such models are indeed somewhat different to the fixed-effect and random-effects model we already know. In this chapter, we will therefore first describe why meta-analysis naturally implies a multilevel structure of our data, and how we can extend a conventional meta-analysis to a three-level model. As always, we will also have a look at how such models can be fitted in R using a hands-on example.

10.1 The Multilevel Nature of Meta-Analysis

To see why meta-analysis has multiple levels by default, let us go back to the formula of the random-effects model that we discussed in Chapter 4.1.2:

$$\hat\theta_k = \mu + \epsilon_k + \zeta_k \tag{10.1}$$

DOI: 10.1201/9781003107347-10

We discussed that the terms ϵ_k and ζ_k are introduced in a random-effects model because we assume that there are two sources of variability. The first one is caused by the sampling error (ϵ_k) of individual studies, which leads effect size estimates to deviate from the true effect size θ_k. The second one, ζ_k, represents the between-study heterogeneity. This heterogeneity is caused by the fact that the true effect size of some study k is again only part of an overarching *distribution of true effect sizes*. This distribution is from where the individual true effect size θ_k was drawn. Therefore, our aim in the random-effects model is to estimate the mean of the distribution of true effect sizes, denoted with μ.

The two error terms ϵ_k and ζ_k correspond with the two levels in our meta-analysis data: the "participant" level (level 1) and the "study" level (level 2). Figure 10.1 below symbolizes this structure.

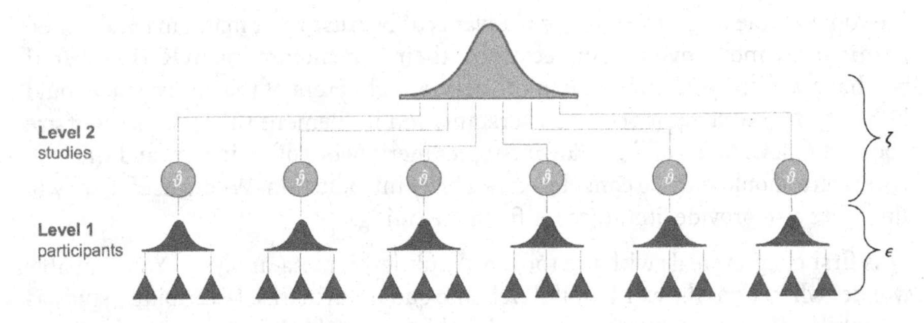

FIGURE 10.1: Multilevel structure of the conventional random-effects model.

At the lowest level (level 1) we have the participants (or patients, specimens, etc., depending on the research field). These participants are part of larger units: the studies included in our meta-analysis. This overlying layer of studies constitutes our second level.

When we conduct a meta-analysis, data on level 1 usually already reaches us in a "pooled" form (e.g. the authors of the paper provide us with the mean and standard deviation of their studied sample instead of the raw data). Pooling on level 2, the study level, however, has to be performed as part of the meta-analysis. Traditionally, such type of data is called *nested*: one can say that participants are "nested" within studies.

Let us go back to the random-effects model formula in equation (10.1). Implicitly, this formula already describes the multilevel structure of our meta-analysis data. To make this more obvious, we have to split the equation into two formulas, where each corresponds to one of the two levels. If we do this, we get the following result:

Level 1 (participants) model:

$$\hat{\theta}_k = \theta_k + \epsilon_k \tag{10.2}$$

Level 2 (studies) model:

$$\theta_k = \mu + \zeta_k \tag{10.3}$$

You might have already detected that we can substitute θ_k in the first equation with its definition in the second equation. What we then obtain is a formula exactly identical to the one of the random-effects model from before. The fixed-effects model can also be written in this way–we only have to set ζ_k to zero.

Evidently, our plain old meta-analysis model already has multilevel properties "built in". It exhibits this property because we assume that participants are nested within studies in our data.

This makes it clear that meta-analysis naturally possesses a multilevel structure. It is possible to expand this structure even further in order to better capture certain mechanisms that generated our data. This is where *three-level models* (Cheung, 2014; Assink et al., 2016) come into play.

Statistical independence is one of the core assumptions when we pool effect sizes in a meta-analysis. If there is a dependency between effect sizes (i.e. effect sizes are correlated), this can artificially reduce heterogeneity and thus lead to false-positive results. This issue is known as the *unit-of-analysis error*, which we already covered before (see Chapter 3.5.2). Effect size dependence can stem from different sources (Cheung, 2014):

- *Dependence introduced by the authors of the individual studies.* For example, scientists conducting the study may have collected data from multiple sites, compared multiple interventions to one single control group, or used different questionnaires to measure the same outcome. In all of these scenarios, we can assume that some kind of dependency is introduced within the reported data.
- *Dependence introduced by the meta-analyst herself.* As an example, think of a meta-analysis that focuses on some psychological mechanism. This meta-analysis includes studies which were conducted in different cultural regions of the world (e.g. East Asian and Western European societies). Depending on the type of psychological mechanism, it could be that results of studies conducted in the same cultural region are more similar compared to those conducted in a different culture.

We can take such dependencies into account by integrating a third layer into the structure of our meta-analysis model. For example, one could model that effect sizes based on different questionnaires are nested within studies. Or one could create a model in which studies are nested within cultural regions.

This creates a three-level meta-analysis model, as illustrated by the next figure.

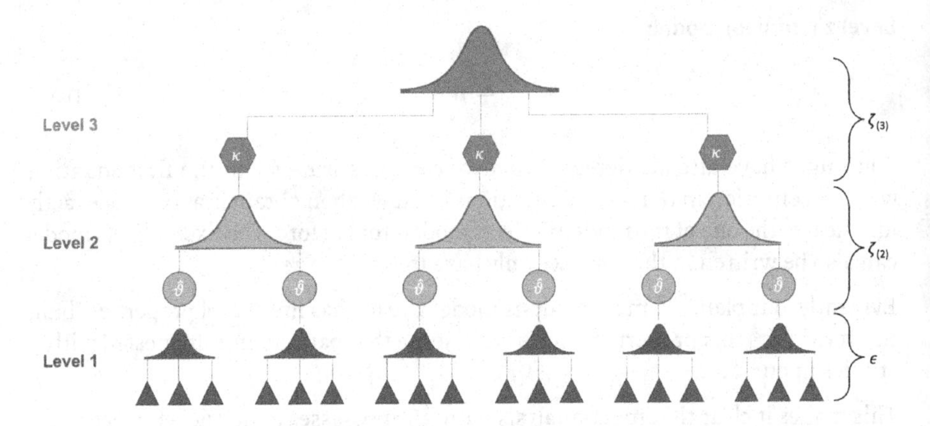

We see that a three-level model contains three pooling steps. First, researchers themselves "pool" the results of individual participants in their primary studies, and report the aggregated effect size. Then, on level 2, these effect sizes are nested within several *clusters*, denoted by κ. These cluster can either be individual studies (i.e. many effect sizes are nested in one study) or subgroups of studies (i.e. many studies are nested in one subgroup, where each study contributes only one effect size).

Lastly, pooling the aggregated cluster effects leads to the overall true effect size μ. Conceptually, this average effect is very close to the pooled true effect μ in a fixed- or random-effects model. The difference, however, is that it is based on a model in which we explicitly account for dependent effect sizes in our data.

It is possible to write down the formula of the three-level model using the same level notation we used before. The greatest distinction is that now, we need to define three formulas instead of two:

Level 1 model:

$$\hat{\theta}_{ij} = \theta_{ij} + \epsilon_{ij} \tag{10.4}$$

Level 2 model:

$$\theta_{ij} = \kappa_j + \zeta_{(2)ij} \tag{10.5}$$

Level 3 model:

$$\kappa_j = \mu + \zeta_{(3)j} \tag{10.6}$$

Where $\hat{\theta}_{ij}$ is an estimate of the true effect size θ_{ij}. The term ij can be read as "some effect size i nested in cluster j". Parameter κ_j is the average effect size in cluster j, and μ the overall average population effect. Like before, we can piece these formulas together and thus reduce the formula to one line:

$$\hat{\theta}_{ij} = \mu + \zeta_{(2)ij} + \zeta_{(3)j} + \epsilon_{ij} \tag{10.7}$$

We see that, in contrast to the random-effects model, this formula now contains *two* heterogeneity terms. One is $\zeta_{(2)ij}$, which stands for the *within-cluster* heterogeneity on level 2 (i.e. the *true* effect sizes within cluster j follow a distribution with mean κ_j). The other is $\zeta_{(3)j}$, the *between-cluster* heterogeneity on level 3. Consequentially, fitting a three-level meta-analysis model does not only involve the estimation of one heterogeneity variance parameter τ^2. We have to estimate two τ^2 values: one for level 2 and the other for level 3.

The *{metafor}* package is particularly well suited for fitting meta-analytic three-level models. It uses (restricted) maximum likelihood procedures to do so. Previously, we primarily used functions of the *{meta}* package to run meta-analyses. We did this because this package is a little less technical, and thus better suited for beginners. Yet, the *{metafor}* package, as we have seen in Chapter 8.3.3, is also fairly easy to use once the data is prepared correctly. How exactly one can use *{metafor}* to fit three-level models in R will be the topic of the next section.

10.2 Fitting Three-Level Meta-Analysis Models in R

As mentioned before, we need the *{metafor}* package to fit three-level meta-analysis models. Therefore, we need to load it from our library first.

```
library(metafor)
```

In our hands-on example, we will use the `Chernobyl` data set. This data set is loosely based on a real meta-analysis which examined the correlation between ionizing radiation ("nuclear fallout") and mutation rates in humans, caused by the devastating 1986 Chernobyl reactor disaster (Møller and Mousseau, 2015).

The "Chernobyl" Data Set

The `Chernobyl` data set is part of the *{dmetar}* package. If you have installed *{dmetar}*, and loaded it from your library, running `data(Chernobyl)` automatically saves the data set in your R environment. The data set is then ready to be used.

If you do not have *{dmetar}* installed, you can download the data set as an *.rda* file from the Internet[a], save it in your working directory, and then click on it in your R Studio window to import it.

[a]https://www.protectlab.org/meta-analysis-in-r/data/Chernobyl.rda

```
# Load data set from 'dmetar'
library(dmetar)
data("Chernobyl")
```

To see the general structure of the data, we can use the head function. This prints the first six rows of the data frame that we just loaded into our global environment.

```
head(Chernobyl)
```

```
##                      author  cor   n    z se.z var.z radiation es.id
## 1 Aghajanyan & Suskov (2009) 0.20  91 0.20 0.10  0.01       low  id_1
## 2 Aghajanyan & Suskov (2009) 0.26  91 0.27 0.10  0.01       low  id_2
## 3 Aghajanyan & Suskov (2009) 0.20  92 0.20 0.10  0.01       low  id_3
## 4 Aghajanyan & Suskov (2009) 0.26  92 0.27 0.10  0.01       low  id_4
## 5      Alexanin et al. (2010) 0.93 559 1.67 0.04  0.00       low  id_5
## 6      Alexanin et al. (2010) 0.44 559 0.47 0.04  0.00       low  id_6
```

The data set contains eight columns. The first one, author, displays the name of the study. The cor column shows the (un-transformed) correlation between radiation exposure and mutation rates, while n stands for the sample size. The columns z, se.z, and var.z are the Fisher-z transformed correlations (Chapter 3.2.3.1), as well their standard error and variance. The radiation column serves as a moderator, dividing effect sizes into subgroups with low and high overall radiation exposure. The es.id column simply contains a unique ID for each effect size (i.e. each row in our data frame).

A peculiar thing about this data set is that it contains repeated entries in author. This is because most studies in this meta-analysis contributed more than one observed effect size. Some studies used several methods to measure mutations or several types of index persons (e.g. exposed parents versus their offspring), all of which leads to multiple effects per study.

Looking at this structure, it is quite obvious that effect sizes in our data set are not independent. They follow a nested structure, where various effect sizes are nested in one study. Thus, it might be a good idea to fit a three-level meta-analysis in order to adequately model these dependencies in our data.

10.2.1 Model Fitting

A three-level meta-analysis model can be fitted using the `rma.mv` function in *{metafor}*. Here is a list of the most important arguments for this function, and how they should be specified:

- `yi`. The name of the column in our data set which contains the calculated effect sizes. In our example, this is `z`, since Fisher-z transformed correlations have better mathematical properties than "untransformed" correlations.
- `V`. The name of the column in our data set which contains the *variance* of the calculated effect sizes. In our case, this is `var.z`. It is also possible to use the *squared* standard error of the effect size, since $SE_k^2 = v_k$.
- `slab`. The name of the column in our data set which contains the study labels, similar to `studlab` in *{meta}*.
- `data`. The name of the data set.
- `test`. The test we want to apply for our regression coefficients. We can choose from `"z"` (default) and `"t"` (recommended; uses a test similar to the Knapp-Hartung method).
- `method`. The method used to estimate the model parameters. Both `"REML"` (recommended; restricted maximum-likelihood) and `"ML"` (maximum likelihood) are possible. Please note that other types of between-study heterogeneity estimators (e.g. Paule-Mandel) are not applicable here.

The most important argument, however, is `random`. Arguably, it is also the trickiest one. In this argument, we specify a formula which defines the (nested) random effects. For a three-level model, the formula always starts with `~ 1`, followed by a vertical bar `|`. Behind the vertical bar, we assign a *random effect* to a grouping variable (such as studies, measures, regions, etc.). This grouping variable is often called a *random intercept* because it tells our model to assume different effects (i.e. intercepts) for each group.

In a three-level model, there are two grouping variables: one on level 2 and another on level 3. We assume that these grouping variables are nested: several effects on level 2 together make up a larger cluster on level 3.

There is a special way through which we can tell `rma.mv` to assume such nested random effects. We do this using a slash (`/`) to separate the higher- and lower-level grouping variable. To the left of `/`, we put in the level 3 (cluster) variable. To the right, we insert the lower-order variable nested in the larger cluster. Therefore, the general structure of the formula looks like this: `~ 1 | cluster/effects_within_cluster`.

In our example, we assume that individual effect sizes (level 2; defined by `es.id`) are nested within studies (level 3; defined by `author`). This results in the following formula: `~ 1 | author/es.id`. The complete `rma.mv` function call looks like this:

```
full.model <- rma.mv(yi = z,
                     V = var.z,
                     slab = author,
                     data = Chernobyl,
                     random = ~ 1 | author/es.id,
                     test = "t",
                     method = "REML")
```

We gave the output the name `full.model`. To print an overview of the results, we can use the `summary` function.

```
summary(full.model)
```

```
## Multivariate Meta-Analysis Model (k = 33; method: REML)
## [...]
## Variance Components:
##
##                 estim    sqrt  nlvls  fixed        factor
## sigma^2.1   0.1788  0.4229     14     no        author
## sigma^2.2   0.1194  0.3455     33     no  author/es.id
##
## Test for Heterogeneity:
## Q(df = 32) = 4195.8268, p-val < .0001
##
## Model Results:
##
## estimate      se    tval    pval   ci.lb   ci.ub
##   0.5231  0.1341  3.9008  0.0005  0.2500  0.7963  ***
## [...]
```

First, have a look at the `Variance Components`. Here, we see the random-effects variances calculated for each level of our model. The first one, `sigma^2.1`, shows the level 3 *between-cluster* variance. In our example, this is equivalent to the between-study heterogeneity variance τ^2 in a conventional meta-analysis (since clusters represent studies in our model). The second variance component `sigma^2.2` shows the variance *within* clusters (level 2). In the `nlvls` column, we see the number of groups on each level. Level 3 has 14 groups, equal to the $K = 14$ included studies. Together, these 14 studies contain 33 effect sizes, as shown in the second row.

Under `Model Results`, we see the estimate of our pooled effect, which is $z = 0.52$ (95%CI: 0.25–0.80). To facilitate the interpretation, it is advisable to transform the effect back to a normal correlation. This can be done using the `convert_z2r` function in the *{esc}* package:

```
library(esc)
convert_z2r(0.52)
```

```
## [1] 0.4777
```

We see that this leads to a correlation of approximately $r \approx 0.48$. This can be considered large. There seems to be a substantial association between mutation rates and exposure to radiation from Chernobyl.

The `Test for Heterogeneity` in the output points at true effect size differences in our data ($p < 0.001$). This result, however, is not very informative. We are more interested in the precise amount of heterogeneity variance captured by each level in our model. It would be good to know how much of the heterogeneity is due to differences *within* studies (level 2), and how much is caused by *between*-study differences (level 3).

10.2.2 Distribution of Variance across Levels

We can answer this question by calculating a multilevel version of I^2 (Cheung, 2014). In conventional meta-analyses, I^2 represents the amount of variation not attributable to sampling error (i.e. the between-study heterogeneity). In three-level models, this heterogeneity variance is split into two parts: one attributable to true effect size differences *within* clusters, and the other to *between*-cluster variation. Thus, there are two I^2 values, quantifying the percentage of total variation associated with either level 2 or level 3.

The "var.comp" Function

The `var.comp` function in *{dmetar}* can be used to calculate multilevel I^2 values. Once *{dmetar}* is installed and loaded on your computer, the function is ready to be used. If you did not install *{dmetar}*, follow these instructions:

1. Access the source code of the function online[a].
2. Let R "learn" the function by copying and pasting the source code in its entirety into the console (bottom left pane of R Studio), and then hit "Enter".
3. Make sure that the *{ggplot2}* package is installed and loaded.

[a] `https://raw.githubusercontent.com/MathiasHarrer/dmetar/master/R/mlm.variance.distribution.R`

The `var.comp` function only needs a fitted `rma.mv` model as input. We save the output in `i2` and then use the `summary` function to print the results.

```
i2 <- var.comp(full.model)
summary(i2)
```

```
##            % of total variance    I2
## Level 1                  1.255   ---
## Level 2                 39.525 39.53
## Level 3                 59.220 59.22
## Total I2: 98.75%
```

In the output, we see the percentage of total variance attributable to each of the three levels. The sampling error variance on level 1 is very small, making up only roughly 1%. The value of $I^2_{\text{Level 2}}$, the amount of heterogeneity variance within clusters, is much higher, totaling roughly 40%. The largest share, however, falls to level 3. Between-cluster (here: between-study) heterogeneity makes up $I^2_{\text{Level 3}} = 59\%$ of the total variation in our data.

Overall, this indicates that there is substantial between-study heterogeneity on the third level. Yet, we also see that a large proportion of the total variance, more than one third, can be explained by differences *within studies*.

It is also possible to visualize this distribution of the total variance. We only have to plug the `var.comp` output into the `plot` function.

```
plot(i2, greyscale = TRUE)
```

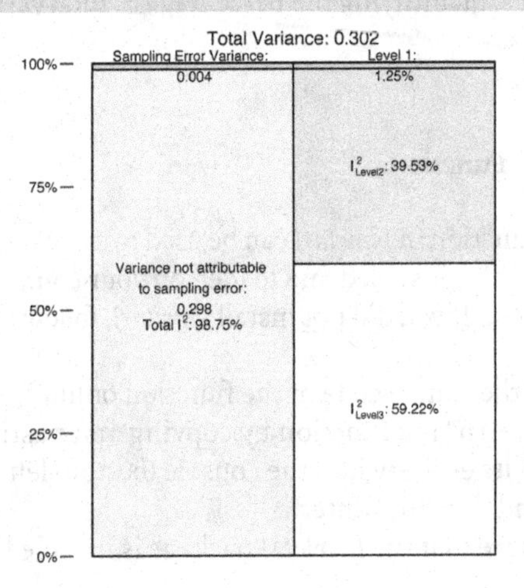

10.2.3 Comparing Models

Fitting a three-level model only makes sense when it represents the variability in our data better than a two-level model. When we find that a two-level model provides a fit comparable to a three-level model, *Occam's razor* should be applied: we favor the two-level model over the three-level model, since it is less complex, but explains our data just as well.

Fortunately, the *{metafor}* package makes it possible to compare our three-level model to one in which a level is removed. To do this, we use the `rma.mv` function again; but this time, set the variance component of one level to zero. This can be done by specifying the `sigma2` parameter. We have to provide a vector with the generic form `c(level 3, level 2)`. In this vector, we fill in `0` when a variance component should be set to zero, while using `NA` to indicate that a parameter should be estimated from the data.

In our example, it makes sense to check if nesting individual effect sizes in studies has improved our model. Thus, we fit a model in which the level 3 variance, representing the between-study heterogeneity, is set to zero. This is equal to fitting a simple random-effects model in which we assume that all effect sizes are independent (which we know they are not). Since level 3 is held constant at zero, the input for `sigma2` is `c(0, NA)`. This results in the following call to `rma.mv`, the output of which we save under the name `l3.removed`.

```
l3.removed <- rma.mv(yi = z,
                     V = var.z,
                     slab = author,
                     data = Chernobyl,
                     random = ~ 1 | author/es.id,
                     test = "t",
                     method = "REML",
                     sigma2 =  c(0, NA))

summary(l3.removed)
```

```
## [...]
## Variance Components:
##
##             estim    sqrt  nlvls  fixed        factor
## sigma^2.1  0.0000  0.0000     14    yes        author
## sigma^2.2  0.3550  0.5959     33     no  author/es.id
##
## Test for Heterogeneity:
## Q(df = 32) = 4195.8268, p-val < .0001
##
## Model Results:
##
## estimate      se    tval    pval   ci.lb   ci.ub
##   0.5985  0.1051  5.6938  <.0001  0.3844  0.8126  ***
## [...]
```

In the output, we see that `sigma^2.1` has been set to zero–just as intended. The overall effect has also changed. But is this result better than the one of the three-level model? To assess this, we can use the *anova* function to compare both models.

```
anova(full.model, l3.removed)
```

```
##          df   AIC   BIC  AICc logLik   LRT   pval      QE
## Full      3 48.24 52.64 49.10 -21.12              4195.82
## Reduced   2 62.34 65.27 62.76 -29.17 16.10 <.0001 4195.82
```

We see that the `Full` (three-level) model, compared to the `Reduced` one with two levels, does indeed show a better fit. The Akaike (AIC) and Bayesian Information Criterion (BIC) are lower for this model, which indicates favorable performance. The likelihood ratio test (`LRT`) comparing both models is significant ($\chi^2_1 = 16.1, p < 0.001$), and thus points in the same direction.

We can say that, although the three-level model introduces one additional parameter (i.e. it has 3 degrees of freedom instead of 2), this added complexity seems to be justified. Modeling of the nested data structure was probably a good idea, and has improved our estimate of the pooled effect.

However, please note that there are often good reasons to stick with a three-level structure–even when it does *not* provide a significantly better fit. In particular, it makes sense to keep a three-level model when we think that it is based on a solid theoretical rationale.

When our data contains studies with multiple effect sizes, for example, we *know* that these effects can not be independent. It thus makes sense to keep the nested model, since it more adequately represents how the data were "generated". If the results of *anova* in our example had favored a two-level solution, we would have concluded that effects within studies were *largely* homogeneous. But we likely would have reported results of the three-level model anyway. This is because we know that a three-level model represents the data-generating process better.

The situation is somewhat different when the importance of the cluster variable is unclear. Imagine, for example, that clusters on level 3 represent different cultural regions in a three-level model. When we find that the phenomenon under study shows no variation between cultures, it is perfectly fine to drop the third level and use a two-level model instead.

10.3 Subgroup Analyses in Three-Level Models

Once our three-level model is set, it is also possible to assess putative moderators of the overall effect. Previously in this guide, we discovered that subgroup analyses can be expressed as a meta-regression model with a dummy-coded predictor (Chapter 8.1). In a similar vein, we can add regression terms to a "multilevel" model, which leads to a *three-level mixed-effects model*:

$$\hat{\theta}_{ij} = \theta + \beta x_i + \zeta_{(2)ij} + \zeta_{(3)j} + \epsilon_{ij} \tag{10.8}$$

Where θ is the intercept and β is the regression weight of a predictor variable x. When we replace x_i with a dummy (Chapter 8.1), we get a model that can be used for subgroup analyses. When x is continuous, the formula above represents a three-level meta-regression model.

Categorical or continuous predictors can be specified in `rma.mv` using the `mods` argument. The argument requires a formula, starting with a tilde (~), and then the name of the predictor. Multiple meta-regression is also possible by providing more than one predictor (e.g. `~ var1 + var2`).

In our `Chernobyl` example, we want to check if correlations differ depending on the overall amount of radiation in the studied sample (low, medium, or high). This information is provided in the `radiation` column in our data set. We can fit a three-level moderator model using this code:

```
mod.model <- rma.mv(yi = z, V = var.z,
                    slab = author, data = Chernobyl,
                    random = ~ 1 | author/es.id,
                    test = "t", method = "REML",
                    mods = ~ radiation)

summary(mod.model)
```

```
## [...]
## Test of Moderators (coefficients 2:3):
## F(df1 = 2, df2 = 28) = 0.4512, p-val = 0.6414
##
## Model Results:
##                   estimate   se   tval  pval  ci.lb ci.ub
## intrcpt               0.58 0.36   1.63  0.11  -0.14  1.32
## radiationlow         -0.19 0.40  -0.48  0.63  -1.03  0.63
## radiationmedium       0.20 0.54   0.37  0.70  -0.90  1.31
## [...]
```

The first important output is the `Test of Moderators`. We see that $F_{2,28} = 0.45$, with $p = 0.64$. This means that there is no significant difference between the subgroups.

The `Model Results` are printed within a meta-regression framework. This means that we cannot directly extract the estimates in order to obtain the pooled effect sizes within subgroups.

The first value, the intercept (`intrcpt`), shows the z value when the overall radiation exposure was high ($z = 0.58$). The effect in the low and medium group can be obtained by adding their `estimate` to the one of the intercept. Thus, the effect in the low radiation group is $z = 0.58 - 0.19 = 0.39$, and the one in the medium exposure group is $z = 0.58 + 0.20 = 0.78$.

Reporting the Results of Three-Level (Moderator) Models

When we report the results of a three-level model, we should at least mention the estimated variance components alongside the pooled effect. The `rma.mv` function denotes the random-effects variance on level 3 and 2 with σ_1^2 and σ_2^2, respectively. When we report the estimated variance, however, using $\tau^2_{\text{Level 3}}$ and $\tau^2_{\text{Level 2}}$ may be preferable since this makes it clear that we are dealing with variances of *true (study) effects* (i.e. heterogeneity variance). Adding the multilevel I^2 values also makes sense, since they are easier for others to interpret–provided we first explain what they represent.

When you conducted a model comparison using `anova`, you may at least report the results of the likelihood ratio test. Results of moderator analyses can be reported in a table such as the one presented in Chapter 7.3. Here is one way to report the results in our example:

"The pooled correlation based on the three-level meta-analytic model was $r = 0.48$ (95%CI: 0.25-0.66; $p < 0.001$). The estimated variance components were $\tau^2_{\text{Level 3}} = 0.179$ and $\tau^2_{\text{Level 2}} = 0.119$. This means that $I^2_{\text{Level 3}} = 58.22\%$ of the total variation can be attributed to between-cluster, and $I^2_{\text{Level 2}} = 31.86\%$ to within-cluster heterogeneity. We found that the three-level model provided a significantly better fit compared to a two-level model with level 3 heterogeneity constrained to zero ($\chi^2_1 = 16.10$; $p < 0.001$)."

□

10.4 Questions & Answers

Test your knowledge!

1. Why is it more accurate to speak of "three-level" instead of "multilevel" models?
2. When are three-level meta-analysis models useful?
3. Name two common causes of effect size dependency.
4. How can the multilevel I^2 statistic be interpreted?
5. How can a three-level model be expanded to incorporate the effect of moderator variables?

Answers to these questions are listed in Appendix A at the end of this book.

10.5 Summary

- All random-effects meta-analyses are based on a multilevel model. When a third layer is added, we speak of a three-level meta-analysis model. Such models are well suited to handle *clustered* effect size data.
- Three-level models can be used for dependent effect sizes. When a study contributes more than one effect size, for example, we typically cannot assume that these results are independent. Three-level model control for this problem by assuming that effect sizes are *nested* in larger clusters (e.g. studies).
- In contrast to a conventional meta-analysis, three-level models estimate two heterogeneity variances: the random-effects variance *within* clusters and the *between*-cluster heterogeneity variance.
- It is also possible to test categorical or continuous predictors using a three-level model. This results in a three-level mixed-effects model.

   a. [illegible]

11

Structural Equation Modeling Meta-Analysis

In the last chapter, we showed that meta-analytic models have an inherent multilevel structure. This quality can be used, for example, to extend conventional meta-analysis to three-level models.

A peculiar thing about statistical methods is that they are often put into separate "boxes". They are treated as unrelated in research and practice, when in fact they are not. For many social science students, for example, it often comes as a surprise to hear that an *analysis of variance* (ANOVA) and a dummy-coded regression are doing essentially the same thing. This often happens because these two methods are traditionally used in different contexts, and taught as separate entities.

In a similar vein, it has been only fairly recently that researchers have started seeing multilevel models as a special form of a *structural equation model*, or SEM (Mehta and Neale, 2005; Bauer, 2003). As we learned, every meta-analysis is based on a multilevel model. As consequence, we can also treat meta-analyses as structural equation models in which the pooled effect size is treated as a latent (or unobserved) variable (Cheung, 2015a, chapter 4.6). In short: meta-analyses are multilevel models; therefore, they can be expressed as structural equation models too.

This does not only mean that we can conceptualize previously covered types of meta-analyses from a structural equation modeling perspective. It also allows us to use SEM to build more complex meta-analysis models. Using *meta-analytic* SEM, we can test *factor analytic* models, or perform *multivariate meta-analyses* which include more than one outcome of interest (to name just a few applications).

Meta-analytic SEM can be helpful when we want to evaluate if certain models in the literature actually hold up once we consider all available evidence. Conversely, it can also be used to check if a theory is not backed by the evidence; or, even more interestingly, if it only applies to a subgroup of individuals or entities.

Application of meta-analytic SEM techniques, of course, presupposes a basic familiarity with structural equation modeling. In the next section, we therefore briefly discuss the general idea behind structural equation modeling, as well as its meta-analytic extension.

DOI: 10.1201/9781003107347-11

11.1 What Is Meta-Analytic Structural Equation Modeling?

Structural equation modeling is a statistical technique used to test hypotheses about the relationship of *manifest* (observed) and *latent* variables (Kline, 2015, chapter 1). Latent variables are either not observed or *observable*. Personality, for example, is a construct which can only be measured indirectly, for instance through different items in a questionnaire. In SEM, an assumed relationship between manifest and latent variables (a "structure") is modeled using the manifest, measured variables, while taking their measurement error into account.

SEM analysis is somewhat different to "conventional" statistical hypothesis tests (such as t-tests, for example). Usually, statistical tests involve testing against a *null hypothesis*, such as $H_0 : \mu_1 = \mu_2$ (where μ_1 and μ_2 are the means of two groups). In such a test, the researcher "aims" to *reject* the null hypothesis, since this allows to conclude that the two groups differ. Yet in SEM, a specific structural model is proposed beforehand, and the researcher instead "aims" to *accept* this model if the goodness of fit is sufficient (Cheung, 2015a, chapter 2.4.6).

11.1.1 Model Specification

Typically, SEM are specified and represented mathematically through a series of *matrices*. You can think of a matrix as a simple table containing rows and columns, much like a `data.frame` object in R (in fact, most data frames can be easily converted to a matrix using the `as.matrix` function). Visually, SEM can be represented as *path diagrams*. Such path diagrams are usually very intuitive, and straightforward in their interpretation. Thus, we will start by specifying a SEM *visually* first, and then move on to the matrix notation.

11.1.1.1 Path Diagrams

Path diagrams represent our SEM graphically. There is no full consensus on how path diagrams should be drawn, yet there are a few conventions. Here are the main components of path diagrams, and what they represent.

Symbol	Name	Description
□	Rectangle	Manifest/observed variables.
○	Circle	Latent/unobserved variables.
△	Triangle	Intercept (fixed vector of 1s).

(continued)

Symbol	Name	Description
$\rightarrow$	Arrow	Prediction. The variable at the start of the arrow predicts the variable at the end of the arrow: Predictor $\rightarrow$ Target.
$\leftrightarrow$	Double Arrow	(Co-)Variance. If a double arrow connects two variables (rectangles/circles), it signifies the covariance/ correlation between the two variables. If a double arrow forms a loop on top of one single variable, it signifies the variance of that variable.

As an illustration, let us create a path diagram for a simple linear ("non-meta-analytic") regression model, in which we want to predict y with x. The model formula looks like this:

$$y_i = \beta_0 + \beta_1 x_i + e_i \tag{11.1}$$

Now, let us "deconstruct" this formula. In the model, x_i and y_i are the observed variables. There are no unobserved (latent) variables. The true population mean of y is the regression intercept β_0, while μ_x denotes the population mean of x. The variance of our observed predictor x is denoted with σ^2_x. Provided that x is not a perfect predictor of y, there will be some amount of residual error variance $\sigma^2_{e_y}$ associated with y. There are two regression coefficients: β_0, the intercept, and β_1, the slope coefficient of x.

Using these components, we can build a path diagram for our linear regression model, as seen below.

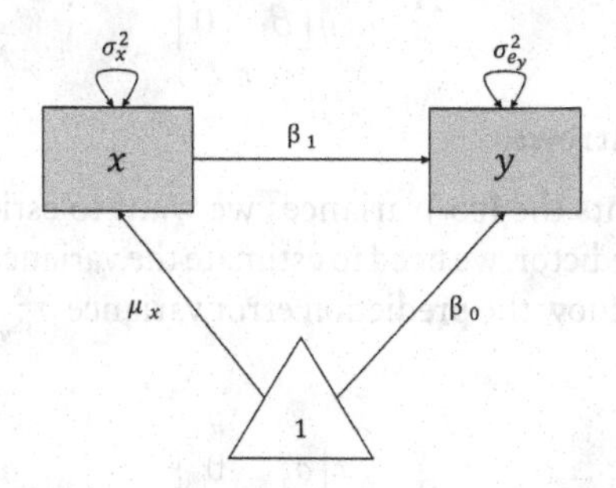

We can also use this graphical model as a starting point to reassemble the regression model equation. From the model, we can infer that y is influenced by two components: $x \times \beta_1$ and $1 \times \beta_0$. If we add these two parts together, we again arrive at the formula for y from before.

11.1.1.2 Matrix Representation

There are several ways to represent SEM through matrices (Jöreskog and Sörbom, 2006; Muthén and Muthén, 2012; McArdle and McDonald, 1984). Here, we will focus on the *Reticular Action Model* formulation, or RAM (McArdle and McDonald, 1984). We do this because this formulation is used by the *{metaSEM}* package which we will be introducing later on. RAM uses four matrices: $\boldsymbol{F}$, $\boldsymbol{A}$, $\boldsymbol{S}$, and $\boldsymbol{M}$. Because the $\boldsymbol{M}$ matrix is not necessary to fit the meta-analytic SEM we cover, we omit it here (see Cheung, 2015a, for a more extensive introduction).

We will now specify the remaining $\boldsymbol{A}$, $\boldsymbol{F}$ and $\boldsymbol{S}$ matrices for our linear regression model from before. The three matrices all have the same number of rows and columns, corresponding with the variables we have in our model: x and y. Therefore, the generic matrix structure of our regression model always looks like this:

$$\begin{array}{c} \\ x \\ y \end{array}\begin{array}{c} \begin{array}{cc} x & y \end{array} \\ \begin{bmatrix} i_{x,x} & i_{x,y} \\ i_{y,x} & i_{y,y} \end{bmatrix} \end{array}$$

The A Matrix: Single Arrows

The $\boldsymbol{A}$ matrix represents the asymmetrical (single) arrows in our path model. We can fill this matrix by searching for the *column* entry of the variable in which the arrow starts (x), and then for the matrix *row* entry of the variable in which the arrow ends (y). The value of our arrow, β_1, is put where the selected column and row intersect in the matrix ($i_{y,x}$). Given that there are no other paths between variables in our model, we fill the remaining fields with 0. Thus, the $\boldsymbol{A}$ matrix for our example looks like this:

$$A = \begin{array}{c} \\ x \\ y \end{array}\begin{array}{c} \begin{array}{cc} x & y \end{array} \\ \begin{bmatrix} 0 & 0 \\ \beta_1 & 0 \end{bmatrix} \end{array}$$

The S Matrix: Single Arrows

The $\boldsymbol{S}$ matrix represents the (co-)variances we want to estimate for the included variables. For x, our predictor, we need to estimate the variance σ^2_x. For our predicted variable y, we want to know the prediction error variance $\sigma^2_{e_y}$. Therefore, we specify $\boldsymbol{S}$ like this:

$$S = \begin{array}{c} \\ x \\ y \end{array}\begin{array}{c} \begin{array}{cc} x & y \end{array} \\ \begin{bmatrix} \sigma^2_x & 0 \\ 0 & \sigma^2_{e_y} \end{bmatrix} \end{array}$$

The F Matrix: Single Arrows

The $\boldsymbol{F}$ matrix allows us to specify the *observed* variables in our model. To specify that a variable has been observed, we simply insert 1 in the respective diagonal field of the matrix. Given that both x and y are observed in our model, we insert 1 into both diagonal fields:

$$F = \begin{array}{c} \\ x \\ y \end{array}\!\!\begin{array}{c} \begin{array}{cc} x & y \end{array} \\ \begin{bmatrix} 1 & 0 \\ 0 & 1 \end{bmatrix} \end{array}$$

Once these matrices are set, it is possible to estimate the parameters in our SEM, and to evaluate how well the specified model fits the data. This involves some matrix algebra and parameter estimation through maximum likelihood estimation, the mathematical minutiae of which we will omit here. If you are interested in understanding the details behind this step, you can have a look at Cheung (2015a), chapter 4.3.

11.1.2 Meta-Analysis from a SEM Perspective

We will now combine our knowledge about meta-analysis models and SEM to formulate meta-analysis as a structural equation model (Cheung, 2008).

To begin, let us return to the formula of the random-effects model. Previously, we already described that the meta-analysis model follows a multilevel structure (see Chapter 10.1), which looks like this:

Level 1

$$\hat{\theta}_k = \theta_k + \epsilon_k \tag{11.2}$$

Level 2

$$\theta_k = \mu + \zeta_k \tag{11.3}$$

On the first level, we assume that the effect size $\hat{\theta}_k$ reported in study k is an estimator of the true effect size θ_k. The observed effect size deviates from the true effect because of sampling error ϵ_k, represented by the variance $\widehat{\text{Var}}(\hat{\theta}_k) = v_k$.

In a random-effects model, we assume that even the true effect size of each study is only drawn from a population of true effect sizes at level 2. The mean of this true effect size population, μ, is what we want to estimate, since it represents the pooled effect size. To do this, we also need to estimate the variance of the true effect sizes $\widehat{\text{Var}}(\theta) = \tau^2$ (i.e. the between-study heterogeneity). The fixed-effect model is a special case of the random-effects model in which τ^2 is assumed to be zero.

It is quite straightforward to represent this model as a SEM graph. We use the parameters on level 1 as latent variables to "explain" how the effect sizes we observe came into being (Cheung, 2015a, chapter 4.6.2):

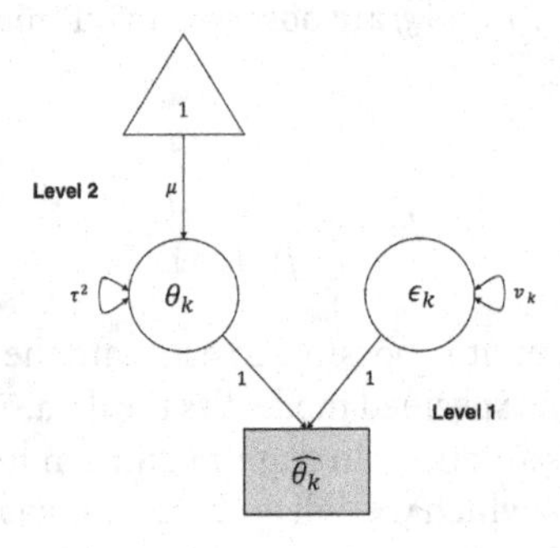

In the graphical model, we see that the observed effect size $\hat{\theta}_k$ of some study k is "influenced" by two arms: the sampling error ϵ_k with variance v_k, and the true effect size θ_k with variance τ^2.

11.1.3 The Two-Stage Meta-Analytic SEM Approach

Above, we defined the (random-effects) meta-analysis model from a SEM perspective. Although this is interesting from a theoretical standpoint, the model above is not more or less capable than the meta-analysis techniques we covered before: it describes that effect sizes are pooled assuming a random-effects model.

To really exploit the versatility of meta-analytic SEM, a two-stepped approach is required (Tang and Cheung, 2016; Cheung, 2015a, chapter 7). In *Two-Stage Structural Equation Modeling* (TSSEM), we first pool the effect sizes of each study. Usually, these effect sizes are correlations between several variables that we want to use for modeling. For each study k, we have a selection of correlations, represented by the vector $\boldsymbol{r_k} = (r_1, r_2, \dots, r_p)$, where p is the total number of (unique) correlations. Like in a normal random-effects model, we assume that each observed correlation in study k deviates from the true average correlation ρ due to sampling error ϵ_k and between-study heterogeneity ζ_k. When we take into account that $\boldsymbol{r_k}$ stands for *several* correlations contained in one study, we get the following equation for the random-effects model:

$$\boldsymbol{r_k} = \boldsymbol{\rho} + \boldsymbol{\zeta_k} + \boldsymbol{\epsilon_k}$$
$$\begin{bmatrix} r_1 \\ r_2 \\ \vdots \\ r_p \end{bmatrix} = \begin{bmatrix} \rho_1 \\ \rho_2 \\ \vdots \\ \rho_p \end{bmatrix} + \begin{bmatrix} \zeta_1 \\ \zeta_2 \\ \vdots \\ \zeta_p \end{bmatrix} + \begin{bmatrix} \epsilon_1 \\ \epsilon_2 \\ \vdots \\ \epsilon_p \end{bmatrix} \tag{11.4}$$

Using this model, we can calculate a vector of *pooled* correlations, $\boldsymbol{r}$. This first pooling step allows to evaluate the heterogeneity of effects between studies, and if a random-effects model or subgroup analyses should be used. Thanks to the maximum likelihood-based approach used by the *{metaSEM}* package, even studies with partially missing data can be included in this step.

In the second step, we then use *weighted least squares* (see Chapter 8.1.3) to fit the structural equation model we specified. The function for the specified model $\rho(\hat{\theta})$ is (Cheung and Chan, 2009; Cheung, 2015a, chapter 7.4.2):

$$F_{\text{WLS}}(\hat{\theta}) = (\boldsymbol{r} - \rho(\hat{\theta}))^{\top} \boldsymbol{V}^{-1} (\boldsymbol{r} - \rho(\hat{\theta})) \tag{11.5}$$

Where $\boldsymbol{r}$ is the pooled correlation vector. The important part of this formula is $\boldsymbol{V}^{-1}$, which is an inverted matrix containing the covariances of $\boldsymbol{r}$. This matrix is used for weighting. Importantly, the formula in this second step is the same no matter if we assume a random or fixed-effect model, because the between study-heterogeneity, if existent, is already taken care of in step 1.

11.2 Multivariate Meta-Analysis

Time to delve into our first worked meta-analytic SEM example. We will begin by using the SEM approach for a *multivariate meta-analysis*, which is something we have not covered yet. In multivariate meta-analyses, we try to estimate more than just one effect at the same time. Such types of meta-analyses are helpful in cases where we are studying a research topic for which there are several main outcomes, not just one.

Imagine that we are examining the effects of some type of treatment. For this treatment, there could be two types of outcomes which are deemed as important by most experts and are thus assessed in most studies. Multivariate meta-analyses can address this, by estimating the effect sizes for both outcomes *jointly* in one model. This multivariate approach also allows us to take the correlation between the two outcomes into account. This can be used to determine if studies with a high effect size on one outcome also have higher effect sizes on the other outcome. Alternatively, we might also find out that there is a negative relationship between the two outcomes or no association at all.

It is of note that multivariate meta-analysis can also be performed outside a SEM framework (Schwarzer et al., 2015, chapter 7; Gasparrini et al., 2012). Here, however, we will to show you how to perform them from a SEM perspective. In this and the following examples, we will work with *{metaSEM}*, a magnificent package for meta-analytic SEM developed by Mike Cheung (2015b). As always, we first have to install the *{metaSEM}* package and load it from your library.

```
library(metaSEM)
```

In our example, we will use *{dmetar}*'s `ThirdWave` data set again (see Chapter 4.2.1). By default, this data set only contains effects on one outcome, perceived stress. Now, imagine that most studies in this meta-analysis also measured effects on *anxiety*, which is another important mental health-related outcome. We can therefore use a multivariate meta-analysis to jointly estimate effects on stress and anxiety, and how they relate to each other.

To proceed, we therefore have to create a new data frame first, in which data for both outcomes is included. First, we define a vector containing the effects on anxiety (expressed as Hedges' g) as reported in each study, as well as their standard error. We also need to define a vector which contains the *covariance* between stress and anxiety reported in each study. One study did not assess anxiety outcomes, so we use `NA` in the three vectors to indicate that the information is missing.

```
# Define vector with effects on anxiety (Hedges g)
Anxiety <- c(0.224,0.389,0.913,0.255,0.615,-0.021,0.201,
             0.665,0.373,1.118,0.158,0.252,0.142,NA,
             0.410,1.139,-0.002,1.084)

# Standard error of anxiety effects
Anxiety_SE <- c(0.193,0.194,0.314,0.165,0.270,0.233,0.159,
                0.298,0.153,0.388,0.206,0.256,0.256,NA,
                0.431,0.242,0.274,0.250)

# Covariance between stress and anxiety outcomes
Covariance <- c(0.023,0.028,0.065,0.008,0.018,0.032,0.026,
                0.046,0.020,0.063,0.017,0.043,0.037,NA,
                0.079,0.046,0.040,0.041)
```

Then, we use this data along with information from `ThirdWave` to create a new data frame called `ThirdWaveMV`. In this data set, we include the effect size *variances* `Stress_var` and `Anxiety_var`, which can be obtained by squaring the standard error.

```
ThirdWaveMV <- data.frame(Author = ThirdWave$Author,
                          Stress = ThirdWave$TE,
                          Stress_var = ThirdWave$seTE^2,
                          Anxiety = Anxiety,
                          Anxiety_var = Anxiety_SE^2,
                          Covariance = Covariance)

format(head(ThirdWaveMV), digits = 2)
```

```
##              Author Stress Stress_var Anxiety Anxiety_var Covariance
## 1      Call et al.   0.71      0.068   0.224       0.037      0.023
## 2 Cavanagh et al.   0.35      0.039   0.389       0.038      0.028
## 3   DanitzOrsillo   1.79      0.119   0.913       0.099      0.065
```

```
## 4  de Vibe et al.   0.18    0.014  0.255    0.027    0.008
## 5  Frazier et al.   0.42    0.021  0.615    0.073    0.018
## 6  Frogeli et al.   0.63    0.038 -0.021    0.054    0.032
```

As we can see, the new data set contains the effect sizes for both stress and anxiety, along with the respective sampling variances. The `Covariance` column stores the covariance between stress and anxiety as measured in each study.

A common problem in practice is that the covariance (or correlation) between two outcomes is not reported in original studies. If this is the case, we have to *estimate* the covariance, based on a reasonable assumption concerning the correlation between the outcomes.

Imagine that we do not know the covariance in each study yet. How can we estimate it? A good way is to look for previous literature which assessed the correlation between the two outcomes, optimally in the same kind of context we are dealing with right now. Let us say we found in the literature that stress and anxiety are very highly correlated in post-tests of clinical trials, with $r_{\text{S,A}} \approx 0.6$. Based on this assumed correlation, we can approximate the co-variance of some study k using the following formula (Schwarzer et al., 2015, chapter 7):

$$\widehat{\text{Cov}}(\theta_1, \theta_2) = SE_{\theta_1} \times SE_{\theta_2} \times \hat{\rho}_{1,2} \tag{11.6}$$

Using our example data and assuming $r_{\text{S,A}} \approx 0.6$, this formula can implemented in R like so:

```
# We use the square root of the variance since SE = sqrt(var)
cov.est <- with(ThirdWaveMV,
                sqrt(Stress_var) * sqrt(Anxiety_var) * 0.6)
```

Please note that, when we calculate covariances this way, the choice of the assumed correlation can have a profound impact on the results. Therefore, it is highly advised to (1) always report the assumed correlation coefficient, and (2) conduct sensitivity analyses, where we inspect how results change depending on the correlation we choose.

11.2.1 Specifying the Model

To specify a multivariate meta-analysis model, we do not have to follow the TSSEM procedure (see previous chapter) programmatically, nor do we have to specify any RAM matrices. For such a relatively simple model, we can use the `meta` function in *{metaSEM}* to fit a meta-analytic SEM in just one step. To use `meta`, we only have to specify three essential arguments:

- `y`. The columns of our data set which contain the effect size data. In a multivariate meta-analysis, we have to combine the effect size columns we want to include using `cbind`.

- v. The columns of our data set which contain the effect size variances. In a multivariate meta-analysis, we have to combine the variance columns we want to include using cbind. We also have to include the column containing the covariance between the effect sizes. The structure of the argument should be cbind(variance_1, covariance, variance_2).
- data. The data set in which the effect sizes and variances are stored.

We save our fitted model under the name m.mv. Importantly, before running meta, please make sure that the *{meta}* package is *not* loaded. Some functions in *{meta}* and *{metaSEM}* have the same name, and this can lead to errors when running the code in R. It is possible to "unload" packages using the detach function. Therefore, we first make sure that *{meta}* is unloaded and then fit the model. The resulting m.mv object can be inspected using summary.

```
detach(package:meta, unload = TRUE)

m.mv <- meta(y = cbind(Stress, Anxiety),
             v = cbind(Stress_var, Covariance, Anxiety_var),
             data = ThirdWaveMV)

summary(m.mv)
```

```
## [...]
## Coefficients:
##              Estimate Std.Error lbound ubound z value Pr(>|z|)
## Intercept1      0.570     0.087  0.399  0.740  6.5455  5.9e-13 ***
## Intercept2      0.407     0.083  0.244  0.570  4.9006  9.5e-09 ***
## Tau2_1_1        0.073     0.049 -0.023  0.169  1.4861   0.1372
## Tau2_2_1        0.028     0.035 -0.041  0.099  0.8040   0.4214
## Tau2_2_2        0.057     0.042 -0.025  0.140  1.3643   0.1725
## ---
## Signif. codes:  0 '***' 0.001 '**' 0.01 '*' 0.05 '.' 0.1 ' ' 1
## [...]
##
## Heterogeneity indices (based on the estimated Tau2):
##                               Estimate
## Intercept1: I2 (Q statistic)    0.6203
## Intercept2: I2 (Q statistic)    0.5292
##
## Number of studies (or clusters): 18
## [...]
## OpenMx status1: 0 ("0" or "1": The optimization is considered fine.
## Other values may indicate problems.)
```

11.2.2 Evaluating the Results

Given that the SEM model is fitted using the maximum likelihood algorithm, the first thing we always do is check the `OpenMx status` right at the end of the output. Maximum likelihood is an optimization procedure, in which parameters are changed iteratively until the optimal solution for the data at hand is found. However, especially with more complex models, it can happen that this optimum is not reached even after many iterations; the maximum likelihood algorithm will then stop and output the parameter values it has approximated so far. Yet, those values for our model components will very likely be incorrect and should not be trusted.

The `OpenMx status` for our model is `0`, which indicates that the maximum likelihood estimation worked fine. If the status would have been anything other than 0 or 1, it would have been necessary to rerun the model, using this code:

```
rerun(m.mv)
```

In the output, the two pooled effect sizes are shown as `Intercept1` and `Intercept2`. The effect sizes are numbered in the order in which we inserted them into our call to `meta`. We can see that the pooled effect sizes are $g_{\text{Stress}} = 0.57$ and $g_{\text{Anxiety}} = 0.41$. Both effect sizes are significant. Under `Heterogeneity indices`, we can also see the values of I^2, which are $I^2_{\text{Stress}} = 62\%$ and $I^2_{\text{Anxiety}} = 53\%$, indicating substantial between-study heterogeneity in both outcomes.

The direct estimates of the between-study heterogeneity variance τ^2 are also provided. We see that there are not only two estimates, but three. To understand what this means, we can extract the "random" values from the `m.mv` object.

```
tau.coefs <- coef(m.mv, select = "random")
```

Then, we use the `vec2symMat` function to create a matrix of the coefficients. We give the matrix rows and columns the names of our variables: `Stress` and `Anxiety`.

```
# Create matrix
tc.mat <- vec2symMat(tau.coefs)

# Label rows and columns
dimnames(tc.mat)[[1]] <- dimnames(tc.mat)[[2]] <- c("Stress",
                                                     "Anxiety")

tc.mat
```

```
##          Stress Anxiety
## Stress  0.07331 0.02894
## Anxiety 0.02894 0.05753
```

We now understand better what the three τ^2 values mean: they represent the between-study variance (heterogeneity) in the diagonal of the matrix. In the other two fields, the matrix shows the estimated covariance between stress and anxiety. Given that the covariance is just an unstandardized version of a correlation, we can transform these values into correlations using the cov2cor function.

```
cov2cor(tc.mat)
```

```
##         Stress Anxiety
## Stress  1.0000  0.4457
## Anxiety 0.4457  1.0000
```

We see that, quite logically, the correlations in the diagonal elements of the matrix are 1. The correlation between effects on stress and anxiety is $r_{\text{S,A}} = 0.45$. This is an interesting finding: it shows that there is a positive association between a treatment's effect on perceived stress and its effect on anxiety. We can say that treatments which have high effects on stress seem to have higher effects on anxiety too.

It is of note that the confidence intervals presented in the summary of `m.mv` are Wald-type intervals (see Chapter 4.1.2.2). Such Wald-type intervals can sometimes be inaccurate, especially in small samples (DiCiccio and Efron, 1996). It may thus be valuable to construct confidence intervals in another way, by using *likelihood-based* confidence intervals. We can get these CIs by re-running the `meta` function and additionally specifying `intervals.type = "LB"`.

```
m.mv <- meta(y = cbind(Stress, Anxiety),
             v = cbind(Stress_var, Covariance, Anxiety_var),
             data = ThirdWaveMV,
             intervals.type = "LB")
```

We have already seen that the output for our m.mv contains non-zero estimates of the between-study heterogeneity τ^2. We can therefore conclude that the model we just fitted is a random-effects model. The `meta` function uses a random-effects model automatically. Considering the I^2 values in our output, we can conclude that this is indeed adequate. However, if we want to fit a fixed-effect model anyway, we can do so by re-running the analysis, and adding the parameter `RE.constraints = matrix(0, nrow=2, ncol=2)`. This creates a matrix of 0s which constrains all τ^2 values to zero:

```
m.mv <- meta(y = cbind(Stress, Anxiety),
             v = cbind(Stress_var, Covariance, Anxiety_var),
             data = ThirdWaveMV,
             RE.constraints = matrix(0, nrow=2, ncol=2))
```

11.2.3 Visualizing the Results

To plot the multivariate meta-analysis model, we can use the `plot` function. We also make some additional specifications to change the appearance of the plot. If you want to see all styling options, you can paste `?metaSEM::plot.meta` into the console and then hit Enter.

```
plot(m.mv,
     axis.labels = c("Perceived Stress", "Anxiety"),
     randeff.ellipse.col = "black",
     univariate.arrows.col = "black",
     univariate.polygon.col = "gray40",
     estimate.ellipse.col = "gray40",
     estimate.col = "black")
```

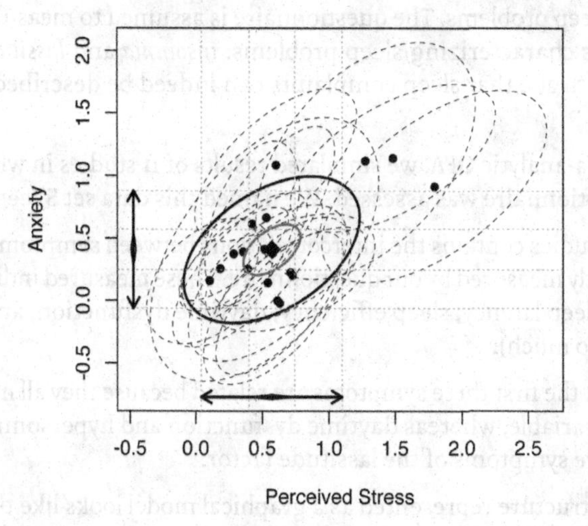

Let us go through what we see. The plot has two axes: an x-axis displaying the effects on stress, and a y-axis, which displays the effects on anxiety. We also see the pooled effect and its 95% confidence interval for both outcomes, symbolized by the black diamond.

In the middle of the plot, the pooled effect of both variables is shown as a black diamond. The smaller gray ellipse represents the 95% confidence interval of our pooled effect; while the larger black ellipse depicts the 95% *prediction* interval (Chapter 5.2)[1]. Lastly, the black dots show the individual studies, where the ellipses with dashed lines represent the 95% confidence intervals.

[1]These prediction intervals (or "plausible value intervals") are based on a different formula ($\hat{\mu} \pm 1.96 \times \hat{\tau}$, Raudenbush, 2009) than the one used by *{meta}* and *{metafor}* (equation 5.7 in Chapter 5.2), resulting in a slightly narrower interval.

11.3 Confirmatory Factor Analysis

Confirmatory Factor Analysis (CFA) is a popular SEM method in which one specifies how observed variables relate to assumed latent variables (Thompson, 2004, chapter 1). CFA is often used to evaluate the psychometric properties of questionnaires or other types of assessments. It allows researchers to determine if assessed variables indeed measure the latent variables they are intended to measure, and how several latent variables relate to each other.

For frequently used questionnaires, there are usually many empirical studies which report the correlations between the different questionnaire items. Such data can be used for meta-analytic SEM. This allows us to evaluate which latent factor structure is the most appropriate based on all available evidence.

In this example, we want to confirm the latent factor structure of a (fictitious) questionnaire for sleep problems. The questionnaire is assumed to measure two distinct latent variables characterizing sleep problems: *insomnia* and *lassitude*. Koffel and Watson (2009) argue that sleep complaints can indeed be described by these two latent factors.

To practice meta-analytic CFA, we simulated results of 11 studies in which our imaginary sleep questionnaire was assessed. We named this data set `SleepProblems`.

Each of these studies contains the intercorrelations between symptoms of sleep complaints as directly measured by our questionnaire. These measured indicators include sleep quality, sleep latency, sleep efficiency, daytime dysfunction, and *hypersomnia* (i.e. sleeping too much).

We assume that the first three symptoms are related because they all measure insomnia as a latent variable, whereas daytime dysfunction and hypersomnia are related because they are symptoms of the lassitude factor.

The proposed structure represented as a graphical model looks like this[2]:

[2]Please note that the labels in the path diagram are somewhat "idiosyncratic" to make identifying the relevant components of the model easier later on.

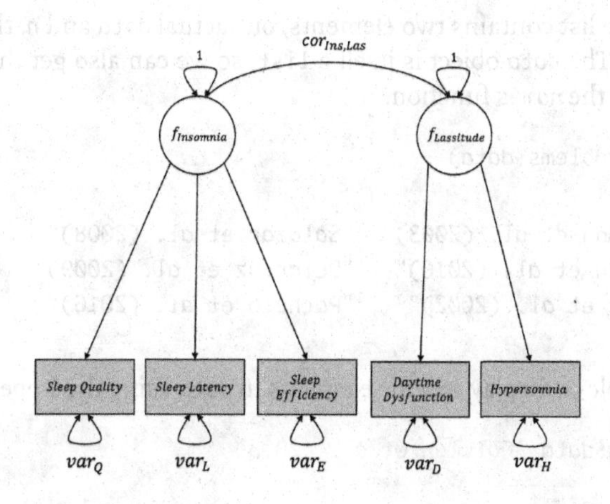

11.3.1 Data Preparation

Let us first have a look at the `SleepProblems` data we want to use for our model. This data set has a special structure: it is a `list` object, containing (1) another `list` of matrices and (2) a numeric vector. Lists are very versatile R objects and allow to bind together different elements in one big object. Lists can be accessed like data frames using the `$` operator. The `names` function can be used to print the names of objects in the list.

The "SleepProblems" Data Set

The `SleepProblems` data set is part of the *{dmetar}* package. If you have installed *{dmetar}*, and loaded it from your library, running `data(SleepProblems)` automatically saves the data set in your R environment. The data set is then ready to be used.

If you have not installed *{dmetar}*, you can download the data set as an *.rda* file from the Internet[a], save it in your working directory, and then click on it in your R Studio window to import it.

[a] `https://www.protectlab.org/meta-analysis-in-r/data/SleepProblems.rda`

```
data(SleepProblems)
names(SleepProblems)
```

```
## [1] "data" "n"
```

We see that the list contains two elements, our actual data and n, the sample size of each study. The data object is itself a list, so we can also get the names of its contents using the names function.

```
names(SleepProblems$data)
```

```
## [1] "Coleman et al. (2003)"  "Salazar et al. (2008)"
## [3] "Newman et al. (2016)"   "Delacruz et al. (2009)"
## [5] "Wyatt et al. (2002)"    "Pacheco et al. (2016)"
## [...]
```

It is also possible to display specific elements in data using the $ operator.

```
SleepProblems$data$`Coleman et al. (2003)`
```

```
##             Quality Latency Efficiency DTDysf HypSomnia
## Quality        1.00    0.39       0.53  -0.30     -0.05
## Latency        0.39    1.00       0.59   0.07      0.44
## Efficiency     0.53    0.59       1.00   0.09      0.22
## DTDysf        -0.30    0.07       0.09   1.00      0.45
## HypSomnia     -0.05    0.44       0.22   0.45      1.00
```

The data list contains 11 elements, one for each of the 11 included studies. A closer look at the Coleman et al. (2003) study reveals that the data are stored as correlation matrices with five variables. Each row and column in the matrix corresponds with one of the sleep complaint symptoms assessed by our questionnaire.

The Coleman et al. (2003) study contains reported correlations for each symptom combination. However, it is also possible to use studies which have missing values (coded as NA) in some of the fields. This is because meta-analytic SEM can handle missing data–at least to some degree.

Before we proceed, let us quickly show how you can construct such a list yourself. Let us assume that we have extracted correlation matrices of two studies, which we imported as data frames into R. Assuming that these data frames are called df1 and df2, we can use the following "recipe" to create a list object that is suitable for further analysis.

```
# Convert both data.frames to matrices
mat1 <- as.matrix(df1)
mat2 <- as.matrix(df2)

# Define the row labels
dimnames(mat1)[[1]] <- c("Variable 1", "Variable 2", "Variable 3")
dimnames(mat2)[[1]] <- c("Variable 1", "Variable 2", "Variable 3")

# Bind the correlation matrices together in a list
```

```
data <- list(mat1, mat2)
names(data) <- c("Study1", "Study2")

# Define sample size of both studies
n <- c(205, # N of study 1
       830) # N of study 2

# Bind matrices and sample size together
cfa.data <- list(data, n)
```

11.3.2 Model Specification

To specify our CFA model, we have to use the RAM specification and two-stage meta-analytic SEM procedure we mentioned before. The *{metaSEM}* package contains separate functions for each of the two stages, `tssem1` and `tssem2`. The first function pools our correlation matrices across all studies, and the second fits the proposed model to the data.

11.3.2.1 Stage 1

At the first stage, we pool our correlation matrices using the `tssem1` function. There are four important arguments we have to specify in the function.

- `Cov`. A `list` of correlation matrices we want to pool. Note that all correlation matrices in the list need to have an identical structure.
- `n`. A numeric vector containing the sample sizes of each study, in the same order as the matrices included in `Cov`.
- `method`. Specifies if we want to use a fixed-effect model (`"FEM"`) or random-effects model (`"REM"`).
- `RE.type`. When a random-effects model is used, this specifies how the random effects should be estimated. The default is `"Symm"`, which estimates all τ^2 values, including the covariances between two variables. When set to `"Diag"`, only the diagonal elements of the random-effects matrix are estimated. This means that we assume that the random effects are independent. Although `"Diag"` results in a strongly simplified model, it is often preferable, because less parameters have to be estimated. This particularly makes sense when the number of variables is high, or the number of studies is low.

In our example, we assume a random-effects model, and use `RE.type = "Diag"`. I will save the model as `cfa1`, and then call the `summary` function to retrieve the output.

```
cfa1 <- tssem1(SleepProblems$data,
               SleepProblems$n,
               method="REM",
               RE.type = "Diag")

summary(cfa1)
```

```
[...]
Coefficients:
            Estimate Std.Error lbound  ubound z value Pr(>|z|)
Intercept1     0.444     0.057  0.331   0.557   7.733  < 0.001 ***
Intercept2     0.478     0.042  0.394   0.561  11.249  < 0.001 ***
Intercept3     0.032     0.071 -0.106   0.172   0.459    0.645
Intercept4     0.132     0.048  0.038   0.227   2.756    0.005 **
Intercept5     0.509     0.036  0.438   0.581  13.965  < 0.001 ***
Intercept6     0.120     0.040  0.040   0.201   2.954    0.003 **
Intercept7     0.192     0.060  0.073   0.311   3.170    0.001 **
Intercept8     0.221     0.039  0.143   0.298   5.586  < 0.001 ***
Intercept9     0.189     0.045  0.100   0.279   4.163  < 0.001 ***
Intercept10    0.509     0.023  0.462   0.556  21.231  < 0.001 ***
Tau2_1_1       0.032     0.015  0.002   0.061   2.153    0.031 *
Tau2_2_2       0.016     0.008  0.000   0.032   1.963    0.049 *
Tau2_3_3       0.049     0.023  0.003   0.096   2.091    0.036 *
Tau2_4_4       0.019     0.010  0.000   0.039   1.975    0.048 *
Tau2_5_5       0.010     0.006 -0.001   0.022   1.787    0.073 .
Tau2_6_6       0.012     0.007 -0.002   0.027   1.605    0.108
Tau2_7_7       0.034     0.016  0.001   0.067   2.070    0.038 *
Tau2_8_8       0.012     0.006 -0.000   0.025   1.849    0.064 .
Tau2_9_9       0.017     0.009 -0.001   0.036   1.849    0.064 .
Tau2_10_10     0.003     0.002 -0.001   0.008   1.390    0.164
---
Signif. codes:  0 '***' 0.001 '**' 0.01 '*' 0.05 '.' 0.1 ' ' 1
[...]

Heterogeneity indices (based on the estimated Tau2):
                              Estimate
Intercept1: I2 (Q statistic)    0.9316
Intercept2: I2 (Q statistic)    0.8837
Intercept3: I2 (Q statistic)    0.9336
Intercept4: I2 (Q statistic)    0.8547
Intercept5: I2 (Q statistic)    0.8315
Intercept6: I2 (Q statistic)    0.7800
Intercept7: I2 (Q statistic)    0.9093
Intercept8: I2 (Q statistic)    0.7958
```

```
Intercept9: I2 (Q statistic)    0.8366
Intercept10: I2 (Q statistic)   0.6486

[...]
OpenMx status1: 0 ("0" or "1": The optimization is considered fine.
Other values may indicate problems.)
```

A look at the OpenMx status confirms that the model estimates are trustworthy. To make the results more easily digestible, we can extract the fixed effects (our estimated pooled correlations) using the coef function. We then make a symmetrical matrix out of the coefficients using vec2symMat and add the dimension names for easier interpretation.

```
# Extract the fixed coefficients (correlations)
fixed.coefs <- coef(cfa1, "fixed")

# Make a symmetric matrix
fc.mat <- vec2symMat(fixed.coefs, diag = FALSE)

# Label rows and columns
dimnames(fc.mat)[[1]] <- c("Quality", "Latency",
                           "Efficiency", "DTDysf", "HypSomnia")
dimnames(fc.mat)[[2]] <- c("Quality", "Latency",
                           "Efficiency", "DTDysf", "HypSomnia")

# Print correlation matrix (3 digits)
round(fc.mat, 3)
```

```
##            Quality Latency Efficiency DTDysf HypSomnia
## Quality      1.000   0.444      0.478  0.033     0.133
## Latency      0.444   1.000      0.510  0.121     0.193
## Efficiency   0.478   0.510      1.000  0.221     0.190
## DTDysf       0.033   0.121      0.221  1.000     0.509
## HypSomnia    0.133   0.193      0.190  0.509     1.000
```

We can now see the pooled correlation matrix for our variables. Looking back at the model output, we also see that all correlation coefficients are significant ($p < 0.05$), except one: the correlation between sleep quality and daytime dysfunction was not significant. From the perspective of our assumed model, this makes sense, because we expect these variables to load on different factors. We also see that the I^2 values of the different estimates are very large (65-93%).

11.3.2.2 Stage 2

After pooling the correlation matrices, it is now time to determine if our proposed factor model fits the data well. To specify our model, we have to use the RAM formulation this time, and specify the $\boldsymbol{A}$, $\boldsymbol{S}$ and $\boldsymbol{F}$ matrices. To fill the fields in each of these matrices, it is often best to construct an empty matrix first. Structure-wise, all matrices we define do not only contain the observed variables but also the assumed latent variables, `f_Insomnia` and `f_Lassitude`. Here is how we can create a zero matrix as a starting point:

```
# Create vector of column/row names
dims <- c("Quality", "Latency", "Efficiency",
          "DTDysf", "HypSomnia", "f_Insomnia", "f_Lassitude")

# Create 7x7 matrix of zeros
mat <- matrix(rep(0, 7*7), nrow = 7, ncol = 7)

# Label the rows and columns
dimnames(mat)[[1]] <- dimnames(mat)[[2]] <- dims
mat
```

```
##             Qlty Ltncy Effcncy DTDysf HypSmn f_Insmn f_Lsstd
## Quality        0     0       0      0      0       0       0
## Latency        0     0       0      0      0       0       0
## Efficiency     0     0       0      0      0       0       0
## DTDysf         0     0       0      0      0       0       0
## HypSomnia      0     0       0      0      0       0       0
## f_Insomnia     0     0       0      0      0       0       0
## f_Lassitude    0     0       0      0      0       0       0
```

$\boldsymbol{A}$ Matrix

In the $\boldsymbol{A}$ matrix, we specify the asymmetrical (i.e. single) arrows in our model. Each single arrow starts at the column variable and ends where the column intersects with the entry of the row variable. All other fields which do not represent arrows are filled with `0`.

We specify that an arrow has to be "estimated" by adding a character string to the $\boldsymbol{A}$ matrix. This character string begins with a starting value for the optimization procedure (usually somewhere between 0.1 and 0.3) followed by `*`. After the `*` symbol, we specify a label for the value. If two fields in the $\boldsymbol{A}$ matrix have the same label, this means that we assume that the fields have the same value.

In our example, we use a starting value of 0.3 for all estimated arrows, and label the fields according to the path diagram we presented before.

```
A <- matrix(c(0, 0, 0, 0, 0, "0.3*Ins_Q", 0          ,
              0, 0, 0, 0, 0, "0.3*Ins_L", 0          ,
              0, 0, 0, 0, 0, "0.3*Ins_E", 0          ,
              0, 0, 0, 0, 0, 0          , "0.3*Las_D",
              0, 0, 0, 0, 0, 0          , "0.3*Las_H",
              0, 0, 0, 0, 0, 0          , 0          ,
              0, 0, 0, 0, 0, 0          , 0
              ), nrow = 7, ncol = 7, byrow=TRUE)

# Label columns and rows
dimnames(A)[[1]] <- dimnames(A)[[2]] <- dims
```

The last step is to plug the A matrix into the `as.mxMatrix` function to make it usable for the stage 2 model.

```
A <- as.mxMatrix(A)
```

S Matrix

In the S matrix, we specify the variances we want to estimate. In our example, these are the variances of all observed variables, as well as the correlation between our two latent factors. First, we set the correlation of our latent factors with themselves to 1. Furthermore, we use a starting value of 0.2 for the variances in the observed variables, and 0.3 for the correlations. All of this can be specified using this code:

```
# Make a diagonal matrix for the variances
Vars <- Diag(c("0.2*var_Q", "0.2*var_L",
               "0.2*var_E", "0.2*var_D", "0.2*var_H"))

# Make the matrix for the latent variables
Cors <- matrix(c(1, "0.3*cor_InsLas",
                 "0.3*cor_InsLas", 1),
               nrow=2, ncol=2)

# Combine
S <- bdiagMat(list(Vars, Cors))

# Label columns and rows
dimnames(S)[[1]] <- dimnames(S)[[2]] <- dims
```

And again, we transform the matrix using `as.mxMatrix`.

```
S <- as.mxMatrix(S)
```

F Matrix

The $\boldsymbol{F}$ matrix, lastly, is easy to specify. In the diagonal elements of observed variables, we fill in 1. Everywhere else, we use 0. Furthermore, we only select the rows of the matrix in which at least on element is not zero (i.e. the last two rows are dropped since they only contain zeros).

```
# Construct diagonal matrix
F <- Diag(c(1, 1, 1, 1, 1, 0, 0))

# Only select non-null rows
F <- F[1:5,]

# Specify row and column labels
dimnames(F)[[1]] <- dims[1:5]
dimnames(F)[[2]] <- dims

F <- as.mxMatrix(F)
```

11.3.3 Model Fitting

Now, it is time to fit our proposed model to the pooled data. To do this, we use the `tssem2` function. We only have to provide the stage 1 model `cfa1`, the three matrices, and specify `diag.constraints=FALSE` (because we are not fitting a mediation model). We save the resulting object as `cfa2` and then access it using `summary`.

```
cfa2 <- tssem2(cfa1,
               Amatrix = A,
               Smatrix = S,
               Fmatrix = F,
               diag.constraints = FALSE)
summary(cfa2)
```

```
## [...]
## Coefficients:
##            Estimate Std.Error lbound ubound z value Pr(>|z|)
## Las_D         0.688     0.081  0.527  0.848   8.409  < 0.001 ***
## Ins_E         0.789     0.060  0.670  0.908  13.026  < 0.001 ***
## Las_H         0.741     0.088  0.568  0.914   8.384  < 0.001 ***
```

```
## Ins_L         0.658     0.053  0.553  0.763  12.275  < 0.001 ***
## Ins_Q         0.613     0.051  0.512  0.714  11.941  < 0.001 ***
## cor_InsLas    0.330     0.045  0.240  0.419   7.241  < 0.001 ***
## ---
## Signif. codes:  0 '***' 0.001 '**' 0.01 '*' 0.05 '.' 0.1 ' ' 1
##
## Goodness-of-fit indices:
##                                                  Value
## Sample size                                  3272.0000
## Chi-square of target model                      5.2640
## DF of target model                              4.0000
## p value of target model                         0.2613
## [...]
## RMSEA                                           0.0098
## RMSEA lower 95% CI                              0.0000
## RMSEA upper 95% CI                              0.0297
## [...]
## OpenMx status1: 0 ("0" or "1": The optimization is considered fine.
## Other values indicate problems.)
```

We see that the `OpenMx status` is `0`, meaning that the optimization worked fine. In the output, we are provided with estimates for the paths between the two latent factors and the observed symptoms, such as 0.69 for Lassitude → Daytime Dysfunction (`Las_D`). We also see that, according to the model, there is a significant correlation between the two latent factors: $r_{\text{Ins,Las}} = 0.33$.

Most importantly, however, we need to check how well the assumed model fits our data. This can be achieved by having a look at the `Goodness-of-fit indices`. We see that the goodness of fit test is *not* significant, with $\chi^2_4 = 5.26, p = 0.26$. Contrary to other statistical tests, this outcome is desired, since it means that we accept the null hypothesis that our model fits the data *well*.

Furthermore, we see that the *Root Mean Square Error of Approximation* (RMSEA) value is 0.0098. As a rule of thumb, a model can be considered to fit the data well when its RSMEA value is below 0.05, with smaller values indicating a better fit (Browne and Cudeck, 1993). Thus, this goodness of fit index also indicates that the model fits our data well.

> **Alternative Models**
>
> Please be aware that a common problem in SEM studies is that researchers often only focus on their own proposed model, and if it fits the data well. If it is found that the assumed model shows a close fit to the data,

many researchers often directly conclude that the data prove their theory. This is problematic because more than one model can fit well to the same data. Therefore, it is necessary to also check for alternative model hypotheses and structures. If the alternative model also fits the data well, it becomes less clear if our proposed structure is really the "correct" one.

11.3.4 Path Diagrams

After the model has been fitted, *{metaSEM}* makes it quite easy for us to visualize it graphically. However, to draw a path diagram, we first have to install and load the *{semPlot}* package (Epskamp, 2019).

```
library(semPlot)
```

To plot the model, we have to convert it into a format that *{semPlot}* can use. This can be done using the `meta2semPlot` function.

```
cfa.plot <- meta2semPlot(cfa2)
```

We can then use the `semPaths` function in *{semPlot}* to generate the graph. This function has many parameters, which can be accessed by typing `?semPaths` into the console, and then hitting Enter. Here is our code, and the resulting plot:

```
# Create Plot labels (left to right, bottom to top)
labels <- c("Sleep\nQuality",
            "Sleep\nLatency",
            "Sleep\nEfficiency",
            "Daytime\nDysfunction",
            "Hyper-\nsomnia","Insomnia",
            "Lassitude")

# Plot
semPaths(cfa.plot,
         whatLabels = "est",
         edge.color = "black",
         nodeLabels = labels,
         sizeMan = 10,
         sizeLat = 10,
         edge.label.cex = 1)
```

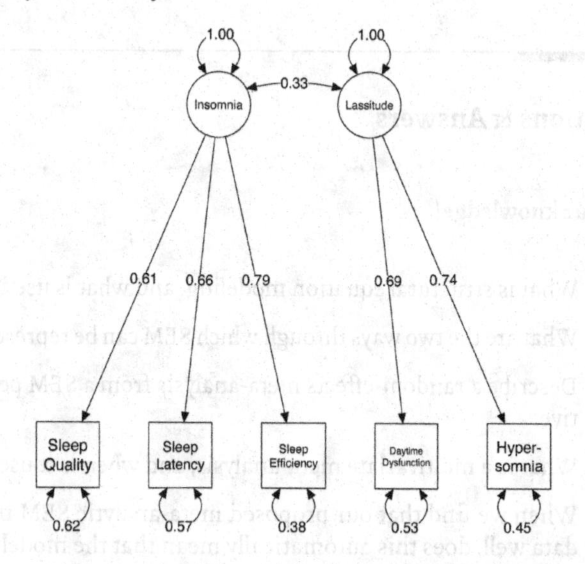

Further Reading

What we covered in this chapter should at best be seen as a rudimentary introduction to meta-analytic SEM. A much more elaborated discussion of this method can be found in Mike Cheung's definitive book *Meta-Analysis: A Structural Equation Modeling Approach* (2015a). This book also describes various other kinds of meta-analytic structural equation models that we have not covered and describes how they can be implemented using R.

If you are looking for a shorter (and openly accessible) resource, you can have a look at the *{metaSEM}* package vignette. The vignette provides a brief discussion of the theory behind meta-analytic SEM and includes several illustrations with R. After *{metaSEM}* is loaded, the vignette can be downloaded from the Internet by running `vignette("metaSEM")` in the console.

□

11.4 Questions & Answers

Test your knowledge!

1. What is structural equation modeling, and what is used for?
2. What are the two ways through which SEM can be represented?
3. Describe a random-effects meta-analysis from a SEM perspective.
4. What is a multivariate meta-analysis, and when is it useful?
5. When we find that our proposed meta-analytic SEM fits the data well, does this automatically mean that the model is the "correct" one?

Answers to these questions are listed in Appendix A at the end of this book.

11.5 Summary

- Structural equation modeling (SEM) is a statistical technique which can be used to test complex relationships between observed (i.e. manifest) and unobserved (i.e. latent) variables.
- Meta-analysis is based on a multilevel model and can therefore also be formulated from a SEM perspective. This can be used to "replicate" random-effects meta-analyses as structural equation models. More importantly, however, this allows us to perform meta-analyses which model more complex relationships between observed effect sizes.
- Meta-analytic SEM can be applied, for example, to perform multivariate meta-analyses. In multivariate meta-analyses, two or more outcomes are estimated jointly, while taking the correlation between both outcome measures into account.
- Another application of meta-analytic SEM is confirmatory factor analysis. To test the fit of a proposed factor model across all included studies, a two-step procedures must be used. At the first stage, correlation matrices of individual studies are pooled. Then, this pooled correlation matrix is used to fit the assumed SEM.

12

Network Meta-Analysis

When we perform meta-analyses of clinical trials or other types of intervention studies, we usually estimate the true effect size of *one* specific treatment. We include studies in which the same type of intervention was compared to similar control groups, for example a placebo. All else being equal, this allows to assess if a *specific* type of treatment is effective.

Yet, in many research areas, there is not only one "definitive" type of treatment–there are several ones. Migraine, for example, can be treated with various kinds of medications, and non-pharmaceutical therapy options also exist. Especially in "matured" research fields, it is often less relevant to show that some kind of treatment is beneficial. Instead, we want to find out which treatment is the *most* effective for some specific indication.

This leads to new problems. To assess the comparative effectiveness of several treatments in a conventional meta-analysis, sufficient head-to-head comparisons between two treatments need to be available. Alas, this is often not the case. In many research fields, it is common to find that only few–if any–trials have compared the effects of two treatments *directly*, in lieu of "weaker" control groups. This often means that traditional meta-analyses can not be used to establish solid evidence on the *relative* effectiveness of several treatments.

However, while direct comparisons between two or more treatments may not exist, *indirect* evidence is typically available. Different treatments may have been evaluated in *separate* trials, but all of these trials may have used the *same* control group. For example, it is possible that two medications were never compared directly, but that the effect of both medications compared to a pill placebo has been studied extensively.

Network meta-analysis can be used to incorporate such indirect comparisons, and thus allows us to compare the effects of several interventions simultaneously (Dias et al., 2013). Network meta-analysis is also known as *mixed-treatment comparison meta-analysis* (van Valkenhoef et al., 2012). This is because it integrates multiple direct and indirect treatment comparisons into one model, which can be formalized as a "network" of comparisons.

Network meta-analysis is a "hot" research topic. In the last decade, it has been increasingly picked up by applied researchers in the bio-medical field, and other disciplines. However, this method also comes with additional challenges and pitfalls, particularly with respect to heterogeneity and so-called *network inconsistency* (Salanti et al., 2014).

DOI: 10.1201/9781003107347-12

Therefore, it is important to first discuss the core components and assumptions of network meta-analysis models. The underpinnings of network meta-analysis can be a little abstract at times. We will therefore go through the essential details in small steps, in order to get a better understanding of this method.

12.1 What Are Network Meta-Analyses?

12.1.1 Direct & Indirect Evidence

First, we have to understand what we mean by a "network" of treatments. Imagine that we have extracted data from some randomized controlled trial i, which compared the effect of treatment A to another condition B (e.g. a wait-list control group). We can illustrate this comparison graphically:

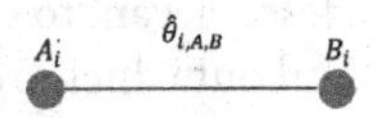

This visual representation of a treatment comparison is called a *graph*. Graphs are structures used to model how different objects relate to each other, and there is an entire sub-field of mathematics, *graph theory*, which is devoted to this topic.

Our graph has two core components. The first one are two circles (so-called *nodes*), which represent the two conditions A and B in trial i. The second component is the line connecting these two nodes. This line is called an *edge*. The edge represents how A and B relate to each other. In our case, the interpretation of the line is quite easy. We can describe the relationship between A and B in terms of the effect size $\hat\theta_{i,\text{A,B}}$ we observe when we compare A and B. This effect size can be expressed as, for example, an SMD or odds ratio, depending on the outcome measure.

Now, imagine that we have also obtained data from another study j. This trial also used the control condition B. But instead of administering A, this study used another treatment C. In study j, treatment C was also compared to B. We can add this information to our graph:

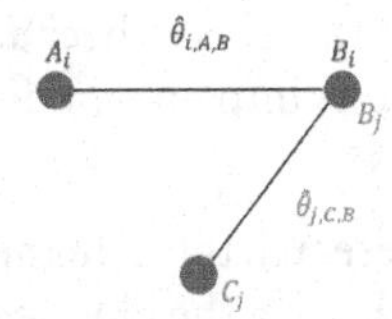

This creates our first small network. It is clearly visible that the graph now contains two effect size estimates: $\hat\theta_{i,\text{A,B}}$, comparing A to B, and $\hat\theta_{j,\text{C,B}}$, the comparison between

C and B. Since both of these effect sizes were directly observed in "real" trials, we call such information *direct evidence*. Therefore, we denote these effect sizes with $\hat{\theta}^{\text{direct}}_{\text{B,A}}$ and $\hat{\theta}^{\text{direct}}_{\text{B,C}}$. Condition B comes first in this notation because we determined it to be our *reference* group. We chose B as the reference condition because both trials used it as the control group.

In the new graph, all nodes (conditions) are either *directly* or *indirectly* connected. The B condition (our control group) is directly connected to all other nodes. It takes only one "step" in the graph to get from B to the two other nodes A and C: B → A, B → C. In contrast, A and C only have one direct connection, and they both connect to B: A → B and C → B.

However, there is an indirect connection between A and C. This connection exists because B serves as the link, or *bridge*, between the two conditions: A → B → C. As a result, there is *indirect evidence* for the relationship between A and C, which can be derived from the structure of the network:

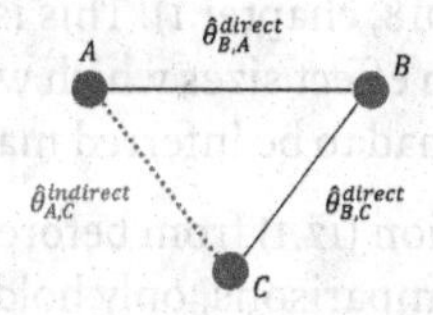

Using information from the directly observed edges, we can calculate the effect of the *indirectly* observed comparison between A and C. We denote this non-observed, indirect effect size with $\hat{\theta}^{\text{indirect}}_{\text{A,C}}$. The effect estimate can be derived using this formula (Dias et al., 2018, chapter 1):

$$\hat{\theta}^{\text{indirect}}_{\text{A,C}} = \hat{\theta}^{\text{direct}}_{\text{B,A}} - \hat{\theta}^{\text{direct}}_{\text{B,C}} \tag{12.1}$$

This step is a crucial component of network meta-analysis. The equation above lets us estimate the effect size of a comparison, even if it was never directly assessed in a trial.

Network meta-analysis involves combining both direct and indirect evidence in one model. Based on this information, we can estimate the (relative) effect of each included treatment. By adding indirect evidence, we also increase the precision of an effect size estimate, even when there is direct evidence for that specific comparison. Overall, network meta-analysis comes with several benefits:

- It allows us to pool all available information from a set of related studies in one analysis. Think of how we would usually deal in conventional meta-analyses with trials comparing different treatments to, say, a placebo. We would have to pool each comparison (e.g. treatment A compared to placebo, treatment B compared to placebo, treatment A compared to treatment B, etc.) in a separate meta-analysis.

- Network meta-analysis can incorporate indirect evidence in a network, which is not possible in conventional meta-analysis. In pairwise meta-analyses, we can only pool direct evidence from comparisons which were actually included in a trial.
- If all assumptions are met, and when the results are sufficiently conclusive, network meta-analyses allow us to infer which type of treatment may be preferable for the target population under study.

All of this sounds intriguing, but there are some important limitations we have to consider. First, look at how the variance of the indirect effect size estimate is calculated:

$$\text{Var}\left(\hat{\theta}_{\text{A,C}}^{\text{indirect}}\right) = \text{Var}\left(\hat{\theta}_{\text{B,A}}^{\text{direct}}\right) + \text{Var}\left(\hat{\theta}_{\text{B,C}}^{\text{direct}}\right) \tag{12.2}$$

To calculate the variance of the indirect comparison, we *add up* the variances of the direct comparisons. This means that effect sizes estimated from indirect evidence will always have a greater variance, and thus a lower precision, than the ones based on direct evidence (Dias et al., 2018, chapter 1). This is nothing but logical. We can have a much higher confidence in effect sizes which were estimated from observed data, compared to results which had to be inferred mathematically.

There is yet another issue. Equation (12.1) from before, which allows us to estimate indirect evidence from direct comparisons, only holds if a crucial pre-requisite is met: the assumption of *transitivity*. From a statistical standpoint, this assumption translates to network *consistency* (Efthimiou et al., 2016). In the following, we explain what both of these terms mean, and why they are important.

12.1.2 Transitivity & Consistency

Network meta-analyses are certainly a valuable extension of standard meta-analytic methods. Their validity, however, has not remained uncontested. Most of the criticism of network meta-analysis revolves around, as you might have guessed, the use of indirect evidence (Edwards et al., 2009; Ioannidis, 2006). This especially involves cases where direct evidence is actually available for a comparison.

The key issue is that, while participants in (randomized) trials are allocated to one of the treatment conditions (e.g., A and B) *by chance*, the trial conditions themselves were not randomly selected in our network. This is of course all but logical. It is usually no problem to randomize participants into one of several conditions of a trial. Yet, it is difficult to imagine a researcher determining treatment conditions to be used in a trial via, say, a dice roll, before rolling out her study. The composition of selected trial conditions will hardly ever follow a random pattern in a network meta-analysis.

This does not constitute a problem for network meta-analytic models *per se* (Dias et al., 2018, chapter 1). Our network meta-analysis model will only be biased when the selection, or non-selection, of a specific comparison within a trial depends on the true effect of that comparison (Dias et al., 2013). This statement is quite abstract, so let us elaborate on it a little.

The requirement we just mentioned is derived from the *transitivity* assumption of network meta-analyses. There is disagreement in the literature about whether this is an assumption unique to network meta-analysis, or simply an extension of the assumptions in conventional pairwise meta-analysis. The disagreement may also be partly caused by an inconsistent usage of terms in the literature (Dias et al., 2018; Efthimiou et al., 2016; Song et al., 2009; Lu and Ades, 2009).

The core tenet of the transitivity assumption is that we can combine direct evidence (e.g. from comparisons A — B and C — B) to create indirect evidence about a related comparison (e.g. A — C), as we have done before using formula (12.1) (Efthimiou et al., 2016).

The assumption of transitivity pertains to the concept of *exchangeability*. We already described this prerequisite in Chapter 4.1.2, where we discussed the random-effects model. The exchangeability assumption says that each true effect size θ_i of some comparison i is the result of a random, *independent* draw from an "overarching" distribution of true effect sizes.

To translate this assumption to our scenario, think of network meta-analysis as a set of K trials. Now, we pretend that each trial in our model contains *all possible* treatment comparisons in our network, denoted with M (e.g. A — B, A — C, B — C, and so forth). However, some of the treatment comparisons have been *"deleted"*, and are thus *"missing"* in some trials. The reason for this is that, in practice, studies cannot assess all possible treatment options (Dias et al., 2013). The key assumption is that the effect of a comparison, e.g. A — B, is *exchangeable* between trials–no matter if a trial actually assessed this comparison, or if it is is "missing". In network meta-analyses, exchangeability is fulfilled when the effect $\hat{\theta}_i$ of some comparison i is based on a random, independent draw from the overarching distribution of true effects, no matter if this effect size is derived through direct or indirect evidence.

The assumption of transitivity can be violated when covariates or other effect modifiers (such as the age group of the studied populations, or the treatment intensity) are not evenly distributed across trials assessing, for example, condition A versus B, and C versus B (Song et al., 2009). Transitivity as such can not be tested statistically, but the risk for violating this assumption can be attenuated by only including studies for which the population, methodology, and target condition are as similar as possible (Salanti et al., 2014).

The statistical manifestation of transitivity is called *consistency*, and a lack thereof is known as *inconsistency* (Efthimiou et al., 2016; Cipriani et al., 2013). Consistency means that the relative effect of a comparison (e.g. A — B) based on direct evidence does not differ from the one based on indirect evidence (Schwarzer et al., 2015, chapter 8):

$$\theta_{\text{A,B}}^{\text{indirect}} = \theta_{\text{A,B}}^{\text{direct}} \tag{12.3}$$

Several methods have been proposed to diagnose inconsistency in network meta-analysis models, including *net heat plots* (Krahn et al., 2013) and the *node splitting* method (Dias et al., 2010). We will describe these methods in greater detail in the following sections.

12.1.3 Network Meta-Analysis Models

This concludes our description of the basic theory and assumptions of network meta-analysis models. Before, we used a simple network with three nodes and edges as an illustration. In practice, however, the number of treatments included in a network meta-analysis is usually much higher. This quickly results in considerably more complex networks, for example, one which looks like this:

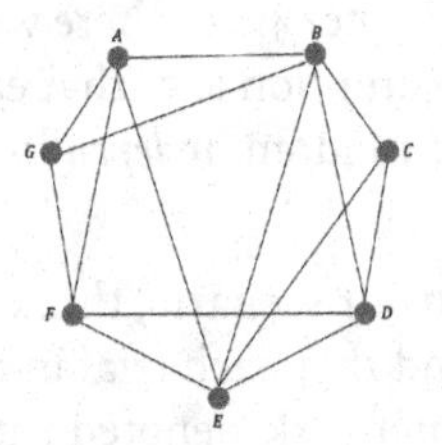

Yet, with an increasing number of treatments S in our network, the number of (direct and indirect) pairwise comparisons C we have to estimate skyrockets:

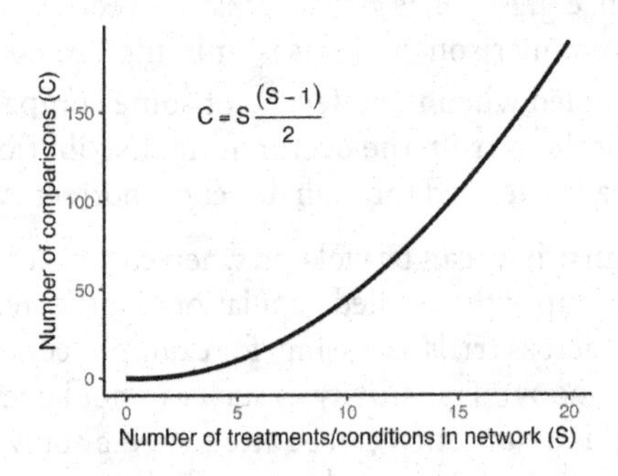

Therefore, we need a computational model which allows us to pool all available network data in an efficient and internally consistent manner. Several statistical approaches have been developed for network meta-analysis (Efthimiou et al., 2016). In the following chapters, we will discuss a *frequentist* as well as a *Bayesian hierarchical model*, and how they can be implemented in R.

Which Modeling Approach Should I Use?

While network meta-analysis models may differ in their statistical approach, the good thing is that all should produce the same results when the sample size is sufficient (Shim et al., 2019). In general, no network meta-analysis method is more or less valid than the other. You may therefore safely choose one or the other approach, depending on which one you find more intuitive, or based on the functionality of the R package which implements it (Efthimiou et al., 2016).

In most disciplines, methods based on frequentist inference are (still) much more common than Bayesian approaches. This means that some people might understand the kind of results produced by a frequentist model more easily. A disadvantage is that the implementation of frequentist network meta-analysis in R (which we will cover next) does not yet support meta-regression, while this is possible using a Bayesian model.

In practice, a useful strategy is to choose one approach for the main analysis, and then employ the other approach in a sensitivity analysis. If the two methods come to the same conclusion, this increases our confidence that the findings are trustworthy.

12.2 Frequentist Network Meta-Analysis

In the following, we will describe how to perform a network meta-analysis using the *{netmeta}* package (Rücker et al., 2020). This package allows to estimate network meta-analysis models within a *frequentist* framework. The method used by *{netmeta}* is derived from graph theoretical techniques, which were originally developed for electrical networks (Rücker, 2012).

The Frequentist Interpretation of Probability

Frequentism is a common theoretical approach to interpret the probability of some event E. Frequentist approaches define the probability of E in terms of how often E is expected to occur if we repeat some process (e.g., an experiment) *many, many times* (Aronow and Miller, 2019, chapter 1.1.1). Frequentist ideas are at the core of many statistical procedures that quantitative researchers use on a daily basis, for example, significance testing, calculation of confidence intervals, or p-values.

12.2.1 The Graph Theoretical Model

Let us now describe how the network meta-analysis model implemented in the *{netmeta}* package can be formulated. Imagine that we have collected effect size data from several trials. Then, we go through all K trials and count the total number of treatment comparisons contained in the studies. This number of pairwise comparisons is denoted with M.

We then calculate the effect size $\hat{\theta}_m$ for each comparison m, and collect all effect sizes in a vector $\hat{\boldsymbol{\theta}} = (\hat{\theta}_1, \hat{\theta}_2, \dots, \hat{\theta}_M)$. To run a network meta-analysis, we now need a model which describes how this vector of observed effect sizes $\hat{\boldsymbol{\theta}}$ was generated. In *{netmeta}*, the following model is used (Schwarzer et al., 2015, chapter 8):

$$\hat{\boldsymbol{\theta}} = \boldsymbol{X}\boldsymbol{\theta}_{\text{treat}} + \boldsymbol{\epsilon} \tag{12.4}$$

We assume that the vector of observed effects sizes $\hat{\boldsymbol{\theta}}$ was generated by the right side of the equation–our model. The first part, $\boldsymbol{X}$, is a $m \times n$ *design matrix*, in which the columns represent the different treatments n, and the rows represent the treatment comparisons m. In the matrix, a treatment comparison is defined by a 1 and -1 in the same row, where the column positions correspond with the treatments that are being compared.

The most important part of the formula is the vector $\boldsymbol{\theta}_{\text{treat}}$. This vector contains the *true* effects of the n unique treatments in our network. This vector is what our network meta-analysis model needs to estimate, since it allows us to determine which treatments in our network are the most effective ones.

The parameter $\boldsymbol{\epsilon}$ is a vector containing the sampling errors ϵ_m of all the comparisons. The sampling error of each comparison is assumed to be a random draw from a Gaussian normal distribution with a mean of zero and variance s^2_m:

$$\epsilon_m \sim \mathcal{N}(0, s^2_m) \tag{12.5}$$

To illustrate the model formula (see Schwarzer et al., 2015, p. 189), imagine that our network meta-analysis consists of $K = 5$ studies. Each study contains a unique treatment comparison (i.e. $K = M$). These comparisons are A − B, A − C, A − D, B − C, and B − D. This results in a vector of (observed) comparisons $\hat{\boldsymbol{\theta}} = (\hat{\theta}_{1,\text{A,B}}, \hat{\theta}_{2,\text{A,C}}, \hat{\theta}_{4,\text{A,D}}, \hat{\theta}_{4,\text{B,C}}, \hat{\theta}_{5,\text{B,D}})^\top$. Our aim is to estimate the true effect size of all four conditions included in our network, $\boldsymbol{\theta}_{\text{treat}} = (\theta_{\text{A}}, \theta_{\text{B}}, \theta_{\text{C}}, \theta_{\text{D}})^\top$. If we plug these parameters into our model formula, we get the following equation:

$$\hat{\boldsymbol{\theta}} = \boldsymbol{X}\boldsymbol{\theta}_{\text{treat}} + \boldsymbol{\epsilon}$$

$$\begin{bmatrix} \hat{\theta}_{1,\text{A,B}} \\ \hat{\theta}_{2,\text{A,C}} \\ \hat{\theta}_{3,\text{A,D}} \\ \hat{\theta}_{4,\text{B,C}} \\ \hat{\theta}_{5,\text{B,D}} \end{bmatrix} = \begin{bmatrix} 1 & -1 & 0 & 0 \\ 1 & 0 & -1 & 0 \\ 1 & 0 & 0 & -1 \\ 0 & 1 & -1 & 0 \\ 0 & 1 & 0 & -1 \end{bmatrix} \begin{bmatrix} \theta_{\text{A}} \\ \theta_{\text{B}} \\ \theta_{\text{C}} \\ \theta_{\text{D}} \end{bmatrix} + \begin{bmatrix} \epsilon_1 \\ \epsilon_2 \\ \epsilon_3 \\ \epsilon_4 \\ \epsilon_5 \end{bmatrix} \tag{12.6}$$

Of note is that in its current form, this model formula is problematic from a mathematical standpoint. Right now, the model is *overparameterized*. There are too many parameters in our model to be estimated based on the data at hand. This means that the design matrix $\boldsymbol{X}$ is not of *full rank*. A matrix is not of full rank when its rows are not all independent. Because we are dealing with a network of *comparisons*, it is clear that not all comparisons (rows) will be independent of each other in our matrix. For example, the row for comparison B − C above can be described as a *linear combination* of comparisons A − B and A − C[1].

The fact that the $\boldsymbol{X}$ matrix is not of full rank means that we can not invert it. This makes it impossible to directly estimate $\boldsymbol{\theta}_{\text{treat}}$ using a weighted least squares approach. While there can only be a maximum of $n-1$ *independent* treatment comparisons, our model always has to estimate the true effect of n treatments in $\boldsymbol{\theta}_{\text{treat}}$.

This is where the *graph theoretical* approach implemented in the *{netmeta}* provides a solution. We will spare you the tedious mathematical details behind this approach, particularly since that the *{netmeta}* package will do the heavy lifting for us anyway. Let us only mention that this approach involves constructing a so-called *Moore-Penrose pseudoinverse matrix*, which then allows for calculating the fitted values of our network model using a weighted least squares approach.

The procedure also takes care of *multi-arm* studies, which contribute more than one pairwise comparison (i.e. studies in which more than two conditions were compared). Multi-arm comparisons are *correlated* because at least one condition is compared more than once (Chapter 3.5.2). This means that the precision of multi-arm study comparisons is artificially increased–unless this is accounted for in our model.

The model also allows us to incorporate estimates of between-study heterogeneity. Like in the "conventional" random-effects model (Chapter 4.1.2), this is achieved by adding the estimated heterogeneity variance $\hat{\tau}^2$ to the variance of a comparison m: $s^2_m + \hat{\tau}^2$. In the *{netmeta}* package, the τ^2 values are estimated using an adaptation of the DerSimonian-Laird estimator (Jackson et al., 2013, see also Chapter 4.1.2.1).

An equivalent of I^2 can also be calculated, which now represents the amount of *inconsistency* in our network. Like in Higgins and Thompson's formula (see Chapter 5.1.2), this I^2 version is derived from Q. In network meta-analyses, however, Q

[1]When we multiply the first row in $\boldsymbol{X}$ with -1, and then add the values in the second row of the matrix, we get the values in the fourth row, which represents $\hat{\theta}_{4,\text{B,C}}$.

translates to the total heterogeneity in the *network* (also denoted with Q_{total}). Thus, the following formula is used:

$$I^2 = \max\left(\frac{Q_{\text{total}} - \text{d.f.}}{Q_{\text{total}}}, 0\right) \tag{12.7}$$

Where the degrees of freedom in our network are:

$$\text{d.f.} = \left(\sum_{k=1}^{K} p_k - 1\right) - (n - 1) \tag{12.8}$$

with K being the total number of studies, p is the number of conditions in some study k, and n is the total number of treatments in our network model.

12.2.2 Frequentist Network Meta-Analysis in R

After all this input, it is time for a hands-on example. In the following, we will use *{netmeta}* to conduct our own network meta-analysis. As always, we first install the package and then load it from the library.

```
library(netmeta)
```

12.2.2.1 Data Preparation

In this illustration, we use the `TherapyFormats` data. This data set is modeled after a real network meta-analysis assessing the effectiveness of different delivery formats of cognitive behavioral therapy for depression (Cuijpers et al., 2019b). All included studies are randomized controlled trials in which the effect on depressive symptoms was measured at post-test. Effect sizes of included comparisons are expressed as the standardized mean difference (SMD) between the two analyzed conditions.

The "TherapyFormats" Data Set

The `TherapyFormats` data set is part of the *{dmetar}* package. If you have installed *{dmetar}*, and loaded it from your library, running `data(TherapyFormats)` automatically saves the data set in your R environment. The data set is then ready to be used.

If you have not installed *{dmetar}*, you can download the data set as an *.rda* file from the Internet[a], save it in your working directory, and then click on it in your R Studio window to import it.

[a]`https://www.protectlab.org/meta-analysis-in-r/data/TherapyFormats.rda`

Let us have a look at the data.

```
library(dmetar)
data(TherapyFormats)

head(TherapyFormats[1:5])
```

```
##          author     TE  seTE treat1 treat2
## 1  Ausbun, 1997  0.092 0.195    ind    grp
## 2  Crable, 1986 -0.675 0.350    ind    grp
## 3  Thiede, 2011 -0.107 0.198    ind    grp
## 4 Bonertz, 2015 -0.090 0.324    ind    grp
## 5     Joy, 2002 -0.135 0.453    ind    grp
## 6   Jones, 2013 -0.217 0.289    ind    grp
```

- The second column, `TE`, contains the effect size of all comparisons, and `seTE` the respective standard error. To use *{netmeta}*, all effect sizes in our data set must be pre-calculated already. In Chapter 3, we already covered how the most common effect sizes can be calculated, and additional tools can be found in Chapter 17.
- `treat1` and `treat2` represent the two conditions that are being compared. Our data set also contains two additional columns, which are not shown here: `treat1.long` and `treat2.long`. These columns simply contain the full name of the condition.
- The `studlab` column contains unique study labels, signifying from which study the specific treatment comparison was extracted. This column is helpful to check for multi-arm studies (i.e. studies with more than one comparison). We can do this using the `table` and `as.matrix` function:

```
as.matrix(table(TherapyFormats$author))
```

```
## [...]
## Bengston, 2004              1
## Blevins, 2003               1
## Bond, 1988                  1
## Bonertz, 2015               1
## Breiman, 2001               3
## [...]
```

Our `TherapyFormats` data set only contains one multi-arm study, the one by Breiman.

This study, as we see, contains three comparisons, while all other studies only contain one.

When we prepare network meta-analysis data, it is essential to always (1) include a study label column in the data set, (2) give each individual study a unique name in the column, and (3) to give studies which contribute two or more comparisons *exactly* the same name.

12.2.2.2 Model Fitting

We can now fit our first network meta-analysis model using the `netmeta` function. The most important arguments are:

- `TE`. The name of the column in our data set containing the effect sizes for each comparison.
- `seTE`. The name of the column which contains the standard errors of each comparison.
- `treat1`. The column in our data set which contains the name of the *first* treatment.
- `treat2`. The column in our data set which contains the name of the *second* treatment.
- `studlab`. The study from which a comparison was extracted. Although this argument is optional *per se*, we recommend to always specify it. It is the only way to let the function know if there are multi-arm trials in our network.
- `data`. The name of our data set.
- `sm`. The type of effect size we are using. Can be `"RD"` (risk difference), `"RR"` (risk ratio), `"OR"` (odds ratio), `"HR"` (hazard ratio), `"MD"` (mean difference), `"SMD"` (standardized mean difference), among others. Check the function documentation (`?netmeta`) for other available measures.
- `comb.fixed`. Should a fixed-effect network meta-analysis should be conducted? Must be `TRUE` or `FALSE`.
- `comb.random`. Should a random-effects model be used? Either `TRUE` or `FALSE`.
- `reference.group`. This lets us specify which treatment should be used as a reference treatment (e.g. `reference.group = "grp"`) for all other treatments.
- `tol.multiarm`. Effect sizes of comparisons from multi-arm studies are–by design–consistent. Sometimes, however, original papers may report slightly deviating results for each comparison, which may result in a violation of consistency. This argument lets us specify a *tolerance threshold* (a numeric value) for the inconsistency of effect sizes and their standard errors allowed in our model.
- `details.chkmultiarm`. Whether to print the estimates of multi-arm comparisons with inconsistent effect sizes (`TRUE` or `FALSE`).
- `sep.trts`. The character to be used as a separator in comparison labels (for example, `" vs. "`).

We save the results of our first network meta-analysis under the name m.netmeta. As reference group, we use the "care as usual" ("cau") condition. For now, let us assume that a fixed-effect model is appropriate. This gives the following code:

```
m.netmeta <- netmeta(TE = TE,
                     seTE = seTE,
                     treat1 = treat1,
                     treat2 = treat2,
                     studlab = author,
                     data = TherapyFormats,
                     sm = "SMD",
                     comb.fixed = TRUE,
                     comb.random = FALSE,
                     reference.group = "cau",
                     details.chkmultiarm = TRUE,
                     sep.trts = " vs ")
m.netmeta
```

```
## Original data (with adjusted standard errors for multi-arm studies):
##
##                      treat1 treat2    TE seTE seTE.adj narms multiarm
## [...]
## Burgan, 2012            ind    tel -0.31 0.13   0.1390     2
## Belk, 1986              ind    tel -0.17 0.08   0.0830     2
## Ledbetter, 1984         ind    tel -0.01 0.23   0.2310     2
## Narum, 1986             ind    tel  0.03 0.33   0.3380     2
## Breiman, 2001           ind    wlc -0.75 0.51   0.6267     3        *
## [...]
##
## Number of treatment arms (by study):
##                         narms
## Ausbun, 1997                2
## Crable, 1986                2
## Thiede, 2011                2
## Bonertz, 2015               2
## Joy, 2002                   2
## [...]
##
## Results (fixed effects model):
##
##                  treat1 treat2   SMD          95%-CI     Q leverage
## Ausbun, 1997        grp    ind  0.06 [ 0.00;  0.12]   0.64     0.03
## Crable, 1986        grp    ind  0.06 [ 0.00;  0.12]   3.05     0.01
## Thiede, 2011        grp    ind  0.06 [ 0.00;  0.12]   0.05     0.03
## Bonertz, 2015       grp    ind  0.06 [ 0.00;  0.12]   0.01     0.01
```

```
## Joy, 2002          grp    ind  0.06 [ 0.00;  0.12]   0.02    0.00
## [....]
##
## Number of studies: k = 182
## Number of treatments: n = 7
## Number of pairwise comparisons: m = 184
## Number of designs: d = 17
##
## Fixed effects model
##
## Treatment estimate (sm = 'SMD', comparison: other treatments vs
## 'cau'):
##         SMD            95%-CI
## cau       .                 .
## grp -0.57 [-0.63; -0.52]
## gsh -0.39 [-0.45; -0.32]
## ind -0.64 [-0.68; -0.59]
## tel -0.51 [-0.60; -0.41]
## ush -0.12 [-0.21; -0.04]
## wlc  0.25 [ 0.20;  0.31]
##
##
## Quantifying heterogeneity / inconsistency:
## tau^2 = 0.27; tau = 0.52; I^2 = 89.6% [88.3%; 90.7%]
##
## Tests of heterogeneity (within designs) and inconsistency
## (between designs):
##                            Q d.f.  p-value
## Total               1696.84  177 < 0.0001
## Within designs  1595.02  165 < 0.0001
## Between designs  101.83   12 < 0.0001
```

There is plenty to see in this output, so let us go through it step by step. The first thing we see are the calculated effect sizes for each comparison. The asterisk signifies our multi-arm study, for which the standard error has been corrected (to account for effect size dependency). Below that, we see an overview of the number of treatment arms in each included study.

The next table shows us the fitted values for each comparison in our (fixed-effect) network meta-analysis model. The Q column in this table is usually very interesting because it tells us which comparison contributes substantially to the overall inconsistency in our network. For example, we see that the Q value of `Crable, 1986` is rather high, with $Q = 3.05$.

Then, we get to the core of our network meta-analysis: the `Treatment estimate`. As specified, the effects of all treatments are displayed in comparison to the care as usual condition, which is why there is no effect shown for `cau`. Below that, we can see that the heterogeneity/inconsistency in our network model is very high, with $I^2 = 89.6\%$. This indicates that selecting a fixed-effect model was probably *not* appropriate (we will get back to this point later).

The last part of the output (`Tests of heterogeneity`) breaks down the total heterogeneity in our network. There are two components: *within-design* heterogeneity and inconsistency *between* designs. A "design" is defined as a selection of conditions included in one trial, for example, A − B, or A − B − C. When there are true effect size differences between studies which included exactly the same conditions, we can speak of within-design heterogeneity. Variation between designs, on the other hand, reflects the inconsistency in our network. Both the within-design heterogeneity and between-design inconsistency are highly significant ($ps < 0.001$).

This is yet another sign that the random-effects model may be indicated. To further corroborate this, we can calculate the total inconsistency based on the *full design-by-treatment interaction random-effects model* (Higgins et al., 2012). To do this, we only have to plug the `m.netmeta` object into the `decomp.design` function.

```
decomp.design(m.netmeta)
```

```
## Q statistics to assess homogeneity / consistency
##  [...]
## Design-specific decomposition of within-designs Q statistic
##
##      Design      Q df  p-value
##  cau vs grp   82.5 20 < 0.0001
##  cau vs gsh    0.7  7   0.9982
##  cau vs ind  100.0 29 < 0.0001
##  cau vs tel   11.4  5   0.0440
##  [...]
##
## Between-designs Q statistic after detaching of single designs
##
##    Detached design       Q df  p-value
##  [...]
##         ind vs wlc  77.09 11 < 0.0001
##         tel vs wlc  95.37 11 < 0.0001
##         ush vs wlc  95.67 11 < 0.0001
##  gsh vs ind vs wlc 101.71 10 < 0.0001
##
## Q statistic to assess consistency under the assumption of
## a full design-by-treatment interaction random effects model
##
##                    Q df p-value tau.within tau2.within
## Between designs 3.83 12  0.9864     0.5403      0.2919
```

In the output, we are first presented with Q values showing the individual contribution of each design to the within- and between-design heterogeneity/inconsistency in our model. The important part of the output is in the last section (`Q statistic to assess consistency under the assumption of a full design-by-treatment interaction random effects model`). We see that the value of Q decreases considerably when assuming a full design-by-treatment random-effects model ($Q = 101.83$ before, $Q = 3.83$ now), and that the between-design inconsistency is not significant anymore ($p = 0.986$). This also suggests that a random-effects model may be indicated to (at least partly) account for the inconsistency and heterogeneity in our network model.

12.2.2.3 Further Examination of the Network Model

12.2.2.3.1 *The Network Graph*

After a network meta-analysis model has been fitted using `netmeta`, it is possible to produce a *network graph*. This can be done using the `netgraph` function. The `netgraph` function has many arguments, which you can look up by running `?netgraph` in your console. Most of those arguments, however, have very sensible default values, so there is not too much to specify.

As a first step, we feed the function with our fitted model `m.netmeta`. Since we used the shortened labels in our model, we should replace them with the long version (stored in `treat1.long` and `treat2.long`) in the plot. This can be achieved using the `labels` argument, where we have to provide the full names of all treatments. The treatment labels should be in the same order as the ones stored in `m.netmeta$trts`.

```
# Show treatment order (shortened labels)
m.netmeta$trts
```

```
## [1] "cau" "grp" "gsh" "ind" "tel" "ush" "wlc"
```

```
# Replace with full name (see treat1.long and treat2.long)
long.labels <- c("Care As Usual", "Group",
                 "Guided Self-Help",
                 "Individual", "Telephone",
                 "Unguided Self-Help",
                 "Waitlist")

netgraph(m.netmeta,
         labels = long.labels)
```

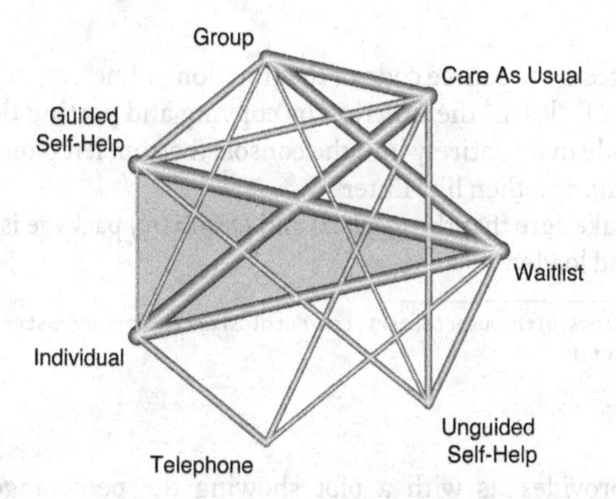

This network graph transports several kinds of information. First, we see the overall structure of comparisons in our network. This allows us to better understand which treatments were compared to each other in the original data.

Furthermore, we can see that the edges in the plot have a different *thickness*. The degree of thickness represents how often we find a specific comparison in our network. For example, we see that guided self-help formats have been compared to wait-lists in many trials. We also see the multi-arm trial in our network, which is represented by a shaded triangle. This is the study by Breiman, which compared guided self-help, individual therapy, and a wait-list.

The `netgraph` function also allows to plot a *3D graph*, which can be helpful to get a better grasp of complex network structures. The function requires the *{rgl}* package to be installed and loaded. To produce a 3D graph, we only have to set the `dim` argument to `"3d"`.

```
library(rgl)
netgraph(m.netmeta, dim = "3d")
```

12.2.2.3.2 *Visualizing Direct and Indirect Evidence*

In the next step, let us have a look at the proportion of *direct* and *indirect* evidence used to estimate each comparison. The `direct.evidence.plot` function in *{dmetar}* has been developed for this purpose.

The "direct.evidence.plot" Function

The `direct.evidence.plot` function is included in the *{dmetar}* package. Once *{dmetar}* is installed and loaded on your computer, the function is ready to be used. If you did <u>not</u> install *{dmetar}*, follow these instructions:

1. Access the source code of the function online[a].
2. Let R "learn" the function by copying and pasting the source code in its entirety into the console (bottom left pane of R Studio), and then hit "Enter".
3. Make sure that the *{ggplot2}* and *{gridExtra}* package is installed and loaded.

[a]`https://raw.githubusercontent.com/MathiasHarrer/dmetar/master/R/direct.evidence.plot.R`

The function provides us with a plot showing the percentage of direct and indirect evidence used for each estimated comparison. The only thing the `direct.evidence.plot` function requires as input is our fitted network meta-analysis model `m.netmeta`.

```
library(dmetar)

d.evidence <- direct.evidence.plot(m.netmeta)
plot(d.evidence)
```

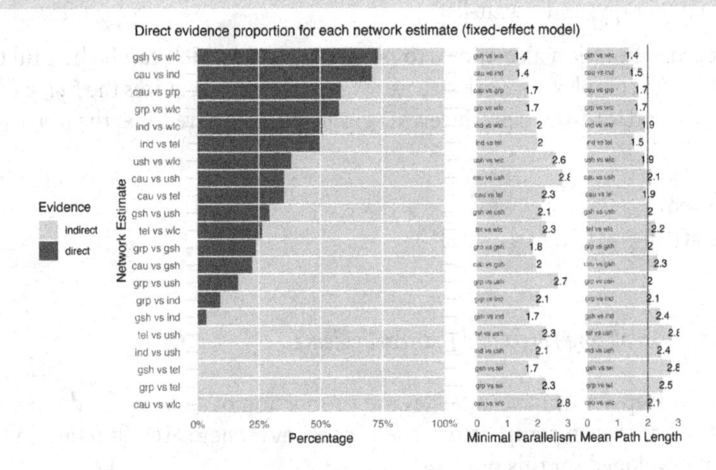

As we can see, there are several estimates in our network model which had to be inferred by indirect evidence alone. The plot also provides us with two additional metrics: the *minimal parallelism* and *mean path length* of each estimated comparison. According to König, Krahn, and Binder (2013), a mean path length > 2 means that a comparison estimate should be interpreted with particular caution.

12.2.2.3.3 *Effect Estimate Table*

Next, we can have a look at the estimates of our network for all possible treatment comparisons. To do this, we can use the matrix saved in `m.netmeta$TE.fixed` (if we use the fixed-effects model) or `m.netmeta$TE.random` (if we use the random-effects model). We need to make a few pre-processing steps to make the matrix easier to read. First, we extract the data from our `m.netmeta` object, and round the numbers in the matrix to two decimal places.

```
result.matrix <- m.netmeta$TE.fixed
result.matrix <- round(result.matrix, 2)
```

Given that one "triangle" in our matrix will hold redundant information, we replace the lower triangle with empty values using this code:

```
result.matrix[lower.tri(result.matrix, diag = FALSE)] <- NA
```

This gives the following result:

```
result.matrix
```

```
##     cau  grp   gsh  ind   tel   ush   wlc
## cau   0 0.58  0.39 0.64  0.51  0.13 -0.26
## grp  NA 0.00 -0.18 0.06 -0.06 -0.45 -0.84
## gsh  NA   NA  0.00 0.25  0.12 -0.27 -0.65
## ind  NA   NA    NA 0.00 -0.13 -0.51 -0.90
## tel  NA   NA    NA   NA  0.00 -0.38 -0.77
## ush  NA   NA    NA   NA    NA  0.00 -0.39
## wlc  NA   NA    NA   NA    NA    NA  0.00
```

If we want to report these results in our research paper, a good idea might be to also include the confidence intervals for each effect size estimate. These can be obtained the same way as before using the `lower.fixed` and `upper.fixed` (or `lower.random` and `upper.random`) matrices in `m.netmeta`.

An even more convenient way to export all estimated effect sizes is to use the `netleague` function. This function creates a table similar to the one we created above. Yet, in the matrix produced by `netleague`, the upper triangle will display only the pooled effect sizes of the *direct comparisons* available in our network, sort of like one would attain them if we had performed a conventional meta-analysis for each comparison. Because we do not have direct evidence for all comparisons, some fields in the upper triangle will remain empty. The lower triangle of the matrix produced by `netleague` contains the estimated effect sizes for *each* comparison (even the ones for which only indirect evidence was available).

The output of `netleague` can be easily exported into a .csv file. It can be used to report comprehensive results of our network meta-analysis in a single table. Another big plus of using this function is that effect size estimates *and* confidence intervals will be displayed together in each cell. Suppose that we want to produce such a treatment estimate table, and save it as a .csv file called "netleague.csv". This can be achieved using the following code:

```
# Produce effect table
netleague <- netleague(m.netmeta,
                       bracket = "(", # use round brackets
                       digits=2)      # round to two digits

# Save results (here: the ones of the fixed-effect model)
write.csv(netleague$fixed, "netleague.csv")
```

12.2.2.3.4 *Treatment Ranking*

The most interesting question we can answer in network meta-analysis is which treatment has the highest effects. The `netrank` function implemented in *{netmeta}* is helpful in this respect. It allows us to generate a *ranking* of treatments, indicating which treatment is more or less likely to produce the largest benefits.

The `netrank` function is, like the model used in `netmeta` itself, based on a frequentist approach. This frequentist method uses *P-scores* to rank treatments, which measure the certainty that one treatment is better than another treatment, averaged over all competing treatments. The P-score has been shown to be equivalent to the *SUCRA* score (Rücker and Schwarzer, 2015), which we will describe in the chapter on Bayesian network meta-analysis.

The `netrank` function requires our `m.netmeta` model as input. Additionally, we should also specify the `small.values` parameter, which defines if smaller (i.e. negative) effect sizes in a comparison indicate a beneficial ("good") or harmful ("bad") effect. Here, we use `small.values = "good"`, since negative effect sizes mean that a treatment was more effective in *reducing* depression.

```
netrank(m.netmeta, small.values = "good")
```

```
##     P-score
## ind  0.9958
## grp  0.8183
## tel  0.6837
## gsh  0.5022
## ush  0.3331
## cau  0.1669
## wlc  0.0000
```

We see that individual therapy (`ind`) has the highest P-score, indicating that this treatment format may be particularly helpful. Conversely, wait-lists (`wlc`) have a

P-score of zero, which seems to go along with our intuition that simply letting people wait for treatment is not the best option.

Nevertheless, one should never automatically conclude that one treatment is the "best", solely because it has the highest score in the ranking (Mbuagbaw et al., 2017). A way to better visualize the *uncertainty* in our network is to produce a forest plot, in which one condition is used as the comparison group.

In *{netmeta}*, this can be achieved using the `forest` function. The `forest` function in *{netmeta}* works very similar to the one of the *{meta}* package, which we already described in Chapter 6. The main difference is that we need to specify the reference group in the forest plot using the `reference.group` argument. We use care us usual (`"cau"`) again.

```
forest(m.netmeta,
       reference.group = "cau",
       sortvar = TE,
       xlim = c(-1.3, 0.5),
       smlab = paste("Therapy Formats vs. Care As Usual \n",
                     "(Depressive Symptoms)"),
       drop.reference.group = TRUE,
       label.left = "Favors Intervention",
       label.right = "Favors Care As Usual",
       labels = long.labels)
```

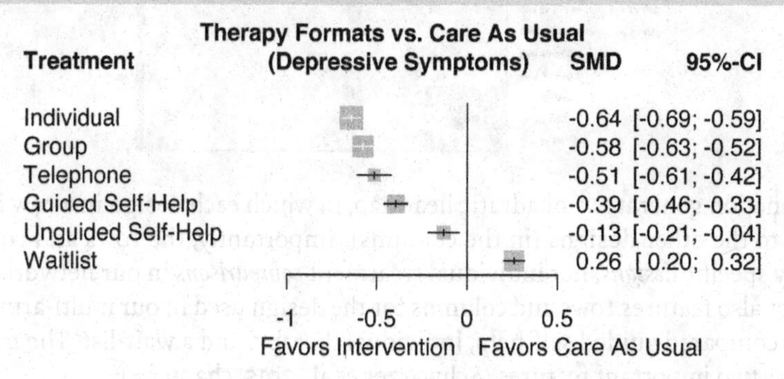

The forest plot shows that there are other high-performing treatments formats besides individual therapy. We also see that some of the confidence intervals are overlapping. This makes a clear-cut decision less easy. While individual treatments do seem to produce the best results, there are several therapy formats which also provide substantial benefits compared to care as usual.

12.2.2.4 Evaluating the Validity of the Results

12.2.2.4.1 *The Net Heat Plot*

The *{netmeta}* package has an in-built function, netheat, which allows us to produce a *net heat plot*. Net heat plots are very helpful to evaluate the inconsistency in our network model, and what designs contribute to it.

The netheat function only needs a fitted network meta-analysis object to produce the plot.

```
netheat(m.netmeta)
```

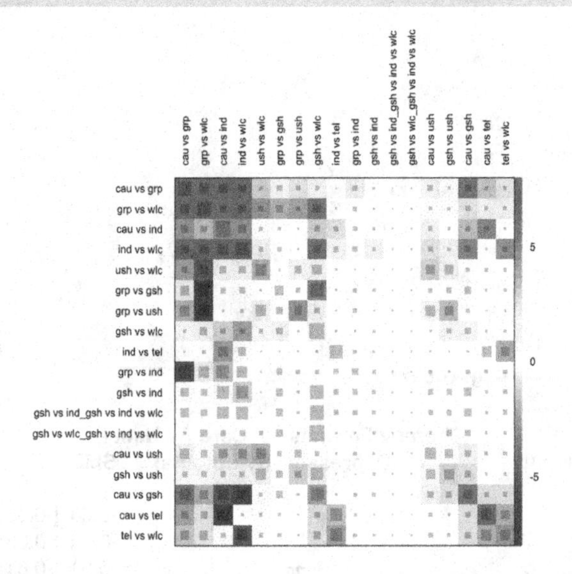

The function generates a quadratic heatmap, in which each design in a row is compared to the other designs (in the columns). Importantly, the rows and columns signify specific *designs*, not individual treatment *comparisons* in our network. Thus, the plot also features rows and columns for the design used in our multi-arm study, which compared guided self-help, individual therapy, and a wait-list. The net heat plot has two important features (Schwarzer et al., 2015, chapter 8):

- *Gray boxes*. The gray boxes signify how important a treatment comparison is for the estimation of another treatment comparison. The bigger the box, the more important the comparison. An easy way to analyze this is to go through the rows of the plot one after another and to check in each row which boxes are the largest. A common finding is that boxes are large in the diagonal of the heat map because this means that direct evidence was used. A particularly big box, for example, can be seen at the intersection of the "cau vs grp" row and the "cau vs grp" column.
- *Colored backgrounds*. The colored backgrounds signify the amount of *inconsistency* of the design in a *row* that can be attributed to the design in a *column*. Field colors

can range from a deep red (which indicates strong inconsistency) to blue (which indicates that evidence from this design supports evidence in the row). The `netheat` function uses an algorithm to sort rows and columns into clusters with higher versus lower inconsistency. In our plot, several inconsistent fields are displayed in the upper-left corner. For example, in the row "ind vs wlc", we see that the entry in column "cau vs grp" is displayed in red. This means that the evidence contributed by "cau vs grp" for the estimation of "ind vs wlc" is inconsistent. On the other hand, we see that the field in the "gsh vs wlc" column has a deep blue background, which indicates that evidence of this design *supports* the evidence of the row design "ind vs wlc".

We should remind ourselves that these results are based on the fixed-effect model, since we used it to fit our network meta-analysis model. Yet, from what we have learned so far, it has become increasingly clear that using the fixed-effect model was not appropriate–there is too much heterogeneity and design inconsistency.

Therefore, let us check how the net heat plot changes when we assume a random-effects model. We can do this by setting the `random` argument in `netheat` to `TRUE`.

```
netheat(m.netmeta, random = TRUE)
```

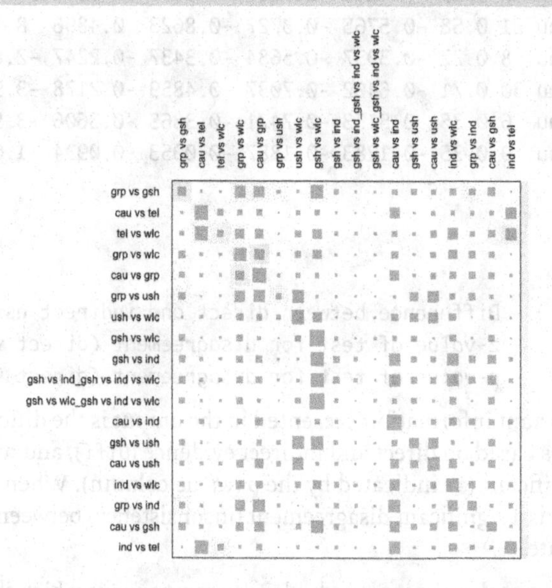

We see that this results in a substantial decrease of inconsistency in our network. There are no fields with a dark red background now, which indicates that the overall consistency of our model improves considerably once a random-effects model is used.

We can therefore conclude that the random-effects model is preferable for our data. In practice, this would mean that we re-run the model using `netmeta` while setting `comb.random` to `TRUE` (and `comb.fixed` to `FALSE`), and that we only report results of analyses based on the random-effects model. We omit this step here, since all the analyses we presented before can also be applied to random-effects network models, in exactly the same way.

12.2.2.4.2 Net Splitting

Another method to check for consistency in our network is *net splitting*. This method splits our network estimates into the contribution of direct and indirect evidence, which allows us to control for inconsistency in the estimates of individual comparisons in our network. To apply the net splitting technique, we only have to provide the `netsplit` function with our fitted model.

```
netsplit(m.netmeta)
```

```
## Separate indirect from direct evidence using back-calculation method
##
## Fixed effects model:
##
##   comparison  k prop     nma  direct  indir.    Diff       z  p-value
##   grp vs cau 21 0.58 -0.5765 -0.3727 -0.8623  0.4896  8.7074 < 0.0001
##   gsh vs cau  8 0.22 -0.3937 -0.5684 -0.3437 -0.2247 -2.8258   0.0047
##   ind vs cau 30 0.71 -0.6402 -0.7037 -0.4859 -0.2178 -3.9799 < 0.0001
##   tel vs cau  6 0.35 -0.5133 -0.7471 -0.3865 -0.3606 -3.5729   0.0004
##   ush vs cau  9 0.35 -0.1283 -0.1887 -0.0953 -0.0934 -1.0230   0.3063
##   [...]
##
## Legend:
##   [...]
##   Diff       - Difference between direct and indirect estimates
##   z          - z-value of test for disagreement (direct vs. indirect)
##   p-value    - p-value of test for disagreement (direct vs. indirect)
```

The most important information presented in the output is the difference between effect estimates based on direct and indirect evidence (`Diff`), and whether this difference is significant (as indicated by the `p-value` column). When a difference is $p < 0.05$, there is a significant disagreement (inconsistency) between the direct and indirect estimate.

We see in the output that there are indeed many comparisons which show significant inconsistency between direct and indirect evidence (when using the fixed-effects model). A good way to visualize the net split results is through a forest plot.

```
netsplit(m.netmeta) %>% forest()
```

Comparison	Number of Studies	Direct Evidence	SMD	95%-CI
grp vs cau				
Direct estimate	21	0.58	-0.37	[-0.44; -0.30]
Indirect estimate			-0.86	[-0.95; -0.78]
Network estimate			-0.58	[-0.63; -0.52]
gsh vs cau				
Direct estimate	8	0.22	-0.57	[-0.71; -0.43]
Indirect estimate			-0.34	[-0.42; -0.27]
Network estimate			-0.39	[-0.46; -0.33]
ind vs cau				
Direct estimate	30	0.71	-0.70	[-0.76; -0.65]
Indirect estimate			-0.49	[-0.58; -0.40]
Network estimate			-0.64	[-0.69; -0.59]
tel vs cau				
Direct estimate	6	0.35	-0.75	[-0.91; -0.59]
Indirect estimate			-0.39	[-0.50; -0.27]
Network estimate			-0.51	[-0.61; -0.42]
ush vs cau				
Direct estimate	9	0.35	-0.19	[-0.33; -0.04]
Indirect estimate			-0.10	[-0.20; 0.01]
Network estimate			-0.13	[-0.21; -0.04]
grp vs gsh				
Direct estimate	5	0.24	-0.23	[-0.36; -0.11]
Indirect estimate			-0.17	[-0.24; -0.10]
Network estimate			-0.18	[-0.24; -0.12]
grp vs ind				
Direct estimate	7	0.09	0.10	[-0.10; 0.30]
Indirect estimate			0.06	[0.00; 0.12]
Network estimate			0.06	[0.00; 0.13]
grp vs ush				
Direct estimate	1	0.17	-0.67	[-0.89; -0.46]
Indirect estimate			-0.40	[-0.50; -0.31]
Network estimate			-0.45	[-0.53; -0.36]
grp vs wlc				
Direct estimate	18	0.58	-0.98	[-1.05; -0.91]
Indirect estimate			-0.64	[-0.71; -0.56]
Network estimate			-0.84	[-0.89; -0.78]
gsh vs ind				
Direct estimate	4	0.03	0.11	[-0.26; 0.47]
Indirect estimate			0.25	[0.18; 0.32]
Network estimate			0.25	[0.18; 0.31]
gsh vs ush				
Direct estimate	5	0.29	-0.40	[-0.56; -0.24]
Indirect estimate			-0.21	[-0.31; -0.11]
Network estimate			-0.27	[-0.35; -0.18]
gsh vs wlc				
Direct estimate	36	0.73	-0.61	[-0.67; -0.56]
Indirect estimate			-0.76	[-0.85; -0.67]
Network estimate			-0.65	[-0.70; -0.60]
ind vs tel				
Direct estimate	4	0.50	-0.19	[-0.32; -0.05]
Indirect estimate			-0.07	[-0.20; 0.06]
Network estimate			-0.13	[-0.22; -0.03]
ind vs wlc				
Direct estimate	18	0.51	-0.76	[-0.84; -0.67]
Indirect estimate			-1.05	[-1.13; -0.97]
Network estimate			-0.90	[-0.96; -0.84]
tel vs wlc				
Direct estimate	1	0.26	-0.56	[-0.75; -0.37]
Indirect estimate			-0.85	[-0.96; -0.73]
Network estimate			-0.77	[-0.87; -0.67]
ush vs wlc				
Direct estimate	11	0.38	-0.52	[-0.65; -0.39]
Indirect estimate			-0.31	[-0.41; -0.20]
Network estimate			-0.39	[-0.47; -0.31]

Fixed effect model

-1 -0.5 0 0.5 1

12.2.2.4.3 Comparison-Adjusted Funnel Plots

Assessing publication bias in network meta-analysis models is difficult. Most of the techniques that we covered in Chapter 9 are not directly applicable once we make the step from conventional to network meta-analysis. *Comparison-adjusted funnel plots*, however, have been proposed to evaluate the risk of publication bias in network meta-analyses, and can be used in some contexts (Salanti et al., 2014). Such funnel plots are applicable when we have a *specific* hypothesis concerning how publication bias has affected our network model.

Publication bias may be created, for example, because studies with "novel" findings are more likely to get published–even if they have a small sample size. There is a natural incentive in science to produce "groundbreaking" results, for example, to show that a new type of treatment is superior to the current state of the art.

This would mean that something similar to small-study effects (see Chapter 9.2.1) exists in our data. We would expect that effects of comparisons in which a new treatment was compared to an older one are *asymmetrically* distributed in the funnel plot. This is because "disappointing" results (i.e. the new treatment is not better than the old one) end up in the file drawer. With decreasing sample size, the benefit of the new treatment must be increasingly large to become significant, and thus merit publication. In theory, this would create the characteristic asymmetrical funnel plot that we also find in standard meta-analyses.

Of course, such a pattern will only appear when the effect sizes in our plot are coded in a certain way. To test our "new versus old" hypothesis, for example, we have to make sure that each effect size used in the plot can has the same interpretation. We have to make sure that (for example) a positive effect size always indicates that the "new" treatment was superior, while a negative sign means the opposite. We can do this by defining a "ranking" of treatments from old to new, and by using this ranking to define the sign of each effect.

The `funnel` function in *{netmeta}* can be used to generate such comparison-adjusted funnel plots. Here are the most important arguments:

- `order`. This argument specifies the order of the hypothesized publication bias mechanism. We simply have to provide the names of all treatments in our network and sort them according to our hypothesis. For example, if we want to test if publication bias favored "new" treatments, we insert the names of all treatments, starting from the oldest treatment, and ending with the most novel type of intervention.

- `pch`. This lets us specify the symbol(s) to be used for the studies in the funnel plot. Setting this to 19 gives simple dots, for example.

- `col`. Using this argument, we can specify the colors used to distinguish different comparison. The number of colors we specify here must be the same as the number of *unique* comparisons in our funnel plot. In practice, this can mean that many different colors are needed. A complete list of colors that R can use for plotting can be found online[2].

- `linreg`. When set to `TRUE`, Egger's test for funnel plot asymmetry (Chapter 9.2.1.2) is conducted, and its p-value is displayed in the plot.

Arguments that are defined for the `funnel` function in *{meta}* can also be used additionally.

```
funnel(m.netmeta,
       order = c("wlc", "cau", "ind", "grp", # from old to new
                 "tel", "ush", "gsh"),
       pch = c(1:4, 5, 6, 8, 15:19, 21:24),
       col = c("blue", "red", "purple", "forestgreen", "grey",
               "green", "black", "brown", "orange", "pink",
               "khaki", "plum", "aquamarine", "sandybrown",
               "coral", "gold4"),
       linreg = TRUE)
```

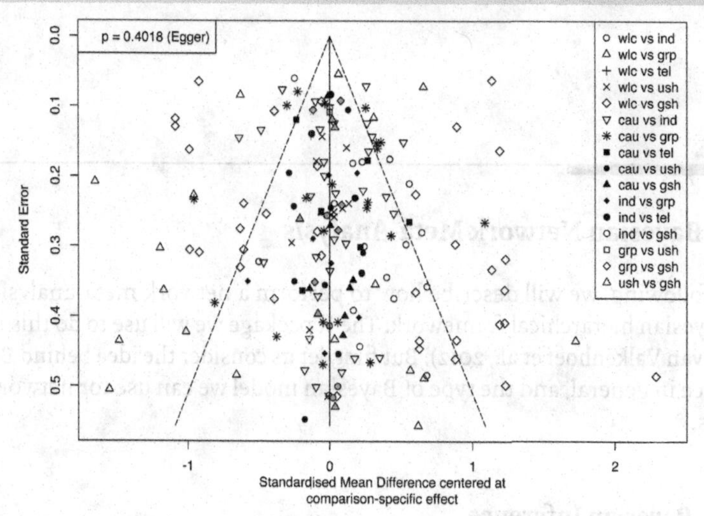

If our hypothesis is true, we can expect that studies with a small sample (and thus a higher standard error) are asymmetrically distributed around the zero line in the plot. This is because small studies comparing a novel treatment to an older one, yet finding that the new treatment is not better, are less likely to get published. Therefore, they are systematically missing on one side of the funnel.

The plot, however, looks quite symmetrical. This is corroborated by Egger's test, which is not significant ($p = 0.402$). Overall, this does not indicate that there are small-study effects in our network. At least not because "innovative" treatments with superior effects are more likely to be found in the published literature.

[2] `http://www.stat.columbia.edu/~tzheng/files/Rcolor.pdf`

Network Meta-Analysis using *{netmeta}*: Concluding Remarks

This has been a long chapter, and we have covered many new topics. We have shown the core ideas behind the statistical model used by *{netmeta}*, described how to fit a network meta-analysis model with this approach, how to visualize and interpret the results, and how to evaluate the validity of your findings. It cannot be stressed enough that (clinical) decision-making in network meta-analyses should not be based on one single test or metric. Instead, we have to explore our model and its results with open eyes, check the patterns we find for their consistency, and take into account the large uncertainty that is often associated with some of the estimates.

In the next chapter, we will try to (re-)think network meta-analysis from a Bayesian perspective. Although the philosophy behind this approach varies considerably from the one we described here, both techniques essentially try to achieve the same thing. In practice, the analysis "pipeline" is also surprisingly similar. Time to go Bayesian!

12.3 Bayesian Network Meta-Analysis

In the following, we will describe how to perform a network meta-analysis based on a Bayesian hierarchical framework. The R package we will use to do this is called *{gemtc}* (van Valkenhoef et al., 2012). But first, let us consider the idea behind Bayesian inference in general, and the type of Bayesian model we can use for network meta-analysis.

12.3.1 Bayesian Inference

Besides the frequentist approach, *Bayesian* inference is another important strand of inference statistics. Frequentist statistics is arguably used more often in most research fields. The Bayesian approach, however, is actually older; and while being increasingly picked up by researchers in recent years (Marsman et al., 2017), it has never really been "gone" (McGrayne, 2011).

The foundation of Bayesian statistics is *Bayes' Theorem*, first formulated by Reverend Thomas Bayes (1701-1761, Bellhouse et al., 2004). Bayesian statistics differs from frequentism because it also incorporates "subjective" *prior* knowledge to make inferences. Bayes' theorem allows us to estimate the probability of an event A, *given* that we already know that another event B has occurred. This results in a *conditional probability*, which can be denoted like this: $P(\text{A}|\text{B})$. The theorem is based on a formula that explains how this conditional probability can be calculated:

$$P(\text{A}|\text{B}) = \frac{P(\text{B}|\text{A}) \times P(\text{A})}{P(\text{B})} \tag{12.9}$$

In this formula, the two probabilities in the numerator of the fraction each have their own names. The $P(\text{B}|\text{A})$ part is known as the *likelihood*. It is the probability of event B, given that A is the case, or occurs (Etz, 2018). $P(\text{A})$ is the *prior* probability that A occurs. $P(\text{A}|\text{B})$, lastly, is the *posterior* probability: the probability of A given B. Since $P(\text{B})$ is a fixed constant, the formula above is often simplified:

$$P(\text{A}|\text{B}) \propto P(\text{B}|\text{A}) \times P(\text{A}) \tag{12.10}$$

Where the $\propto$ symbol means that, since we discarded the denominator of the fraction, the probability on the left remains at least *proportional* to the part on the right as values change.

It is easier to understand Bayes' theorem if we think of the formula above as a process, beginning on the right side of the equation. We simply combine the prior information we have on the probability of A, with the likelihood of B given that A occurs, to produce our posterior, or adapted, probability of A: $P(\text{A}|\text{B})$. The crucial point here is that we can produce a "better" (posterior) estimate of A's probability when we take previous knowledge into account. This knowledge is the assumed (prior) probability of A.

Bayes' Theorem is often explained in the way we just did, with A and B standing for specific events. However, we can also think of A and B as probability *distributions* of two variables. Imagine that A is a random variable following a normal distribution. This distribution can be characterized by a set of parameters, which we denote with $\boldsymbol{\theta}$. Since A is normally distributed, $\boldsymbol{\theta}$ contains two elements: the true mean μ and variance σ^2 of A. These parameters $\boldsymbol{\theta}$ are what we actually want to estimate.

Furthermore, imagine that for B, we have collected *actual data*, which we want to use to estimate $\boldsymbol{\theta}$. We store our observed data in a vector $\boldsymbol{Y}$. Our observed data also follows a normal distribution, represented by $P(\boldsymbol{Y})$. This leads to a formula that looks like this:

$$P(\boldsymbol{\theta}|\boldsymbol{Y}) \propto P(\boldsymbol{Y}|\boldsymbol{\theta}) \times P(\boldsymbol{\theta}) \tag{12.11}$$

The new equation contains $P(\boldsymbol{\theta})$, the assumed prior distribution of $\boldsymbol{\theta}$. This prior distribution can be defined by us *a priori*, either based on our previous knowledge, or even only an intuition concerning what $\boldsymbol{\theta}$ may look like. Together with the likelihood distribution $P(\boldsymbol{Y}|\boldsymbol{\theta})$, the probability of our collected data given the parameters $\boldsymbol{\theta}$, we can estimate the posterior distribution $P(\boldsymbol{\theta}|\boldsymbol{Y})$. This posterior distribution represents our estimate of $\boldsymbol{\theta}$ if we take both the observed data and our prior knowledge into account.

Importantly, the posterior is still a *distribution*, not one estimated "true" value. This means that even the results of Bayesian inference are still *probabilistic*. They are also *subjective*, in the sense that they represent our *beliefs* concerning the actual parameter values. Therefore, in Bayesian statistics, we do not calculate confidence intervals around our estimates, but *credible intervals* (CrI).

Here is a visualization of the three distributions we described before, and how they might look like in a concrete example:

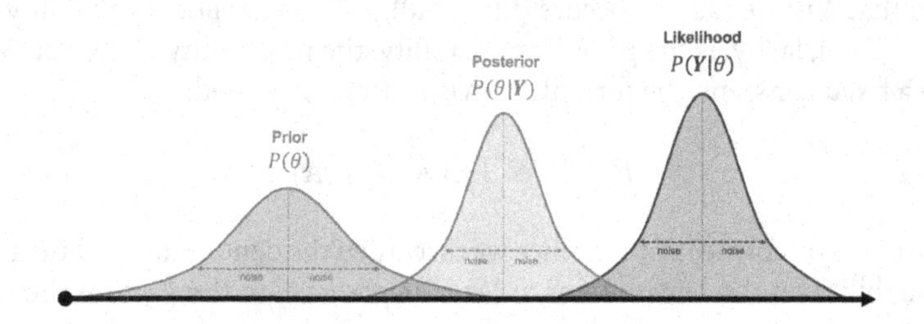

Another asset of Bayesian approaches is that the parameters do not have to follow a bell curve distribution, like the ones in our visualization. Other kinds of (more complex) distributions can also be modeled. A disadvantage of Bayesian inference, however, is that generating the (joint) distribution from our collected data can be very computationally expensive. Special *Markov Chain Monte Carlo* simulation procedures, such as the *Gibbs sampling algorithm*, have been developed to generate posterior distributions. Markov Chain Monte Carlo is also used in the *{gemtc}* package to run our Bayesian network meta-analysis model (van Valkenhoef et al., 2012).

12.3.2 The Bayesian Network Meta-Analysis Model

12.3.2.1 Pairwise Meta-Analysis

We will now formulate the Bayesian hierarchical model that *{gemtc}* uses for network meta-analysis. Let us start by defining the model for a conventional, pairwise meta-analysis first.

This definition is equivalent to the one provided in Chapter 4.1.2, where we discuss the "standard" random-effects model. What we describe in the following is simply the "Bayesian way" to conceptualize meta-analysis. On the other hand, this Bayesian definition of pairwise meta-analysis is already very informative, because it is directly applicable to network meta-analyses, without any further extension (Dias et al., 2013). We refer to this model as a Bayesian *hierarchical* model (Efthimiou et al., 2016, see Chapter 13.1 for a more detailed discussion). There is nothing mysterious about the word "hierarchical" here. Indeed, we already described in Chapter 10 that every meta-analysis model presupposes a hierarchical, or "multi-level" structure.

Suppose that we want to conduct a conventional meta-analysis. We have included K studies, and have calculated an observed effect size $\hat{\theta}_k$ for each one. We can then define the fixed-effect model like so:

$$\hat{\theta}_k \sim \mathcal{N}(\theta, \sigma_k^2) \tag{12.12}$$

This formula expresses the *likelihood* of our effect sizes–the $P(\boldsymbol{Y}|\boldsymbol{\theta})$ part in equation (12.11)–assuming that they follow a normal distribution. We assume that each effect size is a draw from the same distribution, the mean of which is the true effect size θ, and the variance of which is σ_k^2. In the fixed-effect model, we assume that the true effect size is identical across all studies, so θ stays the same for different studies k and their observed effect sizes $\hat{\theta}_k$.

An interesting aspect of the Bayesian model is that, while the true effect θ is unknown, we can still define a prior distribution for it. This prior distribution approximates how we think θ *may* look like. For example, we could assume a prior based on a normal distribution with a mean of zero, $\theta \sim \mathcal{N}(0, \sigma^2)$, where we specify σ^2. By default, the *{gemtc}* package uses so-called *uninformative priors*, which are prior distributions with a very large variance. This is done so that our prior "beliefs" do not have a big impact on the posterior results, and we primarily let the actually observed data "speak". We can easily extend the formula to a random-effects model:

$$\hat{\theta}_k \sim \mathcal{N}(\theta_k, \sigma_k^2) \tag{12.13}$$

This does not change much in the equation, except that now, we do not assume that each study is an estimator of the same true effect size θ. Instead, we assume that there are "study-specific" true effects θ_k estimated by each observed effect size $\hat{\theta}_k$. Furthermore, these study-specific true effects are part of an overarching distribution of true effect sizes. This true effect size distribution is defined by its mean μ and variance τ^2, our between-study heterogeneity.

$$\theta_k \sim \mathcal{N}(\mu, \tau^2) \tag{12.14}$$

In the Bayesian model, we also give an (uninformative) prior distribution to both μ and τ^2.

12.3.2.2 Extension to Network Meta-Analysis

Now that we have covered how a Bayesian meta-analysis model can be formulated for one pairwise comparison, we can start to extend it to network meta-analysis. The two formulas of the random-effects model from before can be re-used for this. We only have to conceptualize the model parameters a little differently. Since comparisons in network meta-analyses can consist of varying treatments, we denote an effect size found in some study k with $\hat{\theta}_{k,\text{A,B}}$. This signifies some effect size in study k in which treatment A was compared to treatment B. If we apply this new notation, we get these formulas:

$$\begin{aligned} \hat{\theta}_{k,\text{A,B}} &\sim \mathcal{N}(\theta_{k,\text{A,B}}, \sigma_k^2) \\ \theta_{k,\text{A,B}} &\sim \mathcal{N}(\theta_{\text{A,B}}, \tau^2) \end{aligned} \tag{12.15}$$

We see that the general idea expressed in the equations stays the same. We now assume that the (study-specific) true effect of the A − B comparison, $\theta_{k,\text{A,B}}$, is part of an overarching distribution of true effects with mean $\theta_{\text{A,B}}$. This mean true effect size $\theta_{\text{A,B}}$ is the result of subtracting $\theta_{1,\text{A}}$ from $\theta_{1,\text{B}}$, where $\theta_{1,\text{A}}$ is the effect of treatment A compared to some predefined reference treatment 1. Similarly, $\theta_{1,\text{B}}$ is defined as the effect of treatment B compared to the same reference treatment. In the Bayesian model, these effects compared to a reference group are also given a prior distribution.

As we have already mentioned in the previous chapter on frequentist network meta-analysis, inclusion of multi-arm studies into our network model is problematic, because the effect sizes will be correlated. In Bayesian network meta-analysis, this issue can be solved by assuming that effects of a multi-arm study stem from a *multivariate* (normal) distribution. Imagine that a multi-arm study k examined a total of $n = 5$ treatments: A, B, C, D, and E. When we choose E as the reference treatment, this leads to n - 1 = 4 treatment effects. Using a Bayesian hierarchical model, we assume that these observed treatment effects are draws from a multivariate normal distribution of the following form[3]:

$$\begin{bmatrix} \hat{\theta}_{k,\text{A,E}} \\ \hat{\theta}_{k,\text{B,E}} \\ \hat{\theta}_{k,\text{C,E}} \\ \hat{\theta}_{k,\text{D,E}} \end{bmatrix} = \mathcal{N}\left(\begin{bmatrix} \theta_{\text{A,E}} \\ \theta_{\text{B,E}} \\ \theta_{\text{C,E}} \\ \theta_{\text{D,E}} \end{bmatrix}, \begin{bmatrix} \tau^2 & \tau^2/2 & \tau^2/2 & \tau^2/2 \\ \tau^2/2 & \tau^2 & \tau^2/2 & \tau^2/2 \\ \tau^2/2 & \tau^2/2 & \tau^2 & \tau^2/2 \\ \tau^2/2 & \tau^2/2 & \tau^2/2 & \tau^2 \end{bmatrix} \right). \tag{12.16}$$

12.3.3 Bayesian Network Meta-Analysis in R

Now, let us use the *{gemtc}* package to perform our first Bayesian network meta-analysis. As always, we have to first install the package, and then load it from our library.

```
library(gemtc)
```

The *{gemtc}* package depends on *{rjags}* (Plummer, 2019), which is used for the Gibbs sampling procedure that we described before (Chapter 12.3.1). However, before we install and load this package, we first have to install another software called *JAGS* (short for "Just Another Gibbs Sampler"). The software is available for both Windows

[3] In practice, it is usually assumed that the between-study heterogeneity variances τ^2 in multi-arm trials are *homogeneous* (i.e. identical) across the comparisons. This allows us to define all covariances in the matrix as $\tau^2/2$.

and Mac, and you can download it for free from the Internet[4]. After this is completed, we can install and load the *{rjags}* package.

```
install.packages("rjags")
library(rjags)
```

12.3.3.1 Data Preparation

In our example, we will again use the `TherapyFormats` data set, which we already used to fit a frequentist network meta-analysis. However, it is necessary to tweak the structure of our data a little so that it can be used in *{gemtc}*.

The original `TherapyFormats` data set includes the columns `TE` and `seTE`, which contain the standardized mean difference and standard error, with each row representing one comparison. If we want to use such relative effect data in *{gemtc}*, we have to reshape our data frame so that each row represents a single *treatment arm*. Furthermore, we have to specify which treatment was used as the reference group in a comparison by filling in `NA` into the effect size column. We have saved this reshaped version of the data set under the name `TherapyFormatsGeMTC`[5].

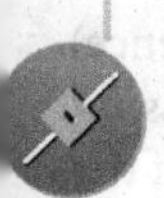

The "TherapyFormatsGeMTC" Data Set

The `TherapyFormatsGeMTC` data set is part of the *{dmetar}* package. If you have installed *{dmetar}*, and loaded it from your library, running `data(TherapyFormatsGeMTC)` automatically saves the data set in your R environment. The data set is then ready to be used. If you do not have *{dmetar}* installed, you can download the data set as an *.rda* file from the Internet[a], save it in your working directory, and then click on it in your R Studio window to import it.

[a] `https://www.protectlab.org/meta-analysis-in-r/data/TherapyFormatsGeMTC.rda`

The `TherapyFormatsGeMTC` data set is actually a list with two elements, one of which is called `data`. This element is the data frame we need to fit the model. Let us have a look at it.

[4] `https://sourceforge.net/projects/mcmc-jags/files/`

[5] We have also prepared an R vignette describing how one can transform network meta-analysis data in the "wider" *{netmeta}* format to the "longer" format required for relative effect size data in *{gemtc}*. The vignette can be found online: `https://www.protectlab.org/vignettes/reshape-gemtc/`

```
library(dmetar)
data(TherapyFormatsGeMTC)

head(TherapyFormatsGeMTC$data)
```

```
##           study   diff std.err treatment
## 1 Ausbun, 1997  0.092   0.195       ind
## 2 Ausbun, 1997     NA      NA       grp
## 3 Crable, 1986 -0.675   0.350       ind
## 4 Crable, 1986     NA      NA       grp
## 5 Thiede, 2011 -0.107   0.198       ind
## 6 Thiede, 2011     NA      NA       grp
```

The *{gemtc}* package also requires that the columns of our data frame are labeled correctly. If we are using effect sizes based on continuous outcomes (such as the mean difference or standardized mean difference), our data set has to contain these columns:

- `study`. This column contains a (unique) label for each study included in our network, equivalent to the `studlab` column used in *{netmeta}*.
- `treatment`. This column contains the label or shortened code for the treatment.
- `diff`. This column contains the effect size (e.g. the standardized mean difference) calculated for a comparison. Importantly, the `diff` column contains `NA`, a missing value, in the row of the reference treatment used in a comparison. The row of the treatment to which the reference treatment was compared then holds the actual effect size calculated for this comparison. Also keep in mind that the reference category is defined *study-wise*, not *comparison-wise*. This means that in multi-arm studies, we still have only one reference treatment to which all the other treatments are compared. For a three-arm study, for example, we need to include two effect sizes: one for the first treatment compared to the reference group, and a second one for the other treatment compared to the reference group.
- `std.err`. This column contains the standard error of the effect, sizes. It is also set to `NA` in the reference group and only defined in the row of the treatment that was compared to the reference group.

Please note that other data entry formats are also possible, for example, for binary outcome data. The way the data set needs to be structured for different types of effect size data is detailed in the *{gemtc}* documentation. You can access it by running `?mtc.model` in the console, and then scrolling to the "Details" section.

12.3.3.2 Network Graph

Now that we have our data ready, we feed it to the `mtc.network` function. This generates an object of class `mtc.network`, which we can use for later modeling steps.

Because we are using pre-calculated effect size data, we have to specify our data set using the `data.re` argument in `mtc.network`. For raw effect size data (e.g. mean, standard deviation and sample size), we would have used the `data.ab` argument.

The optional `treatments` argument can be used to provide *{gemtc}* with the actual names of all the treatments included in the network. This information should be prepared in a data frame with an `id` and `description` column. We have created such a data frame and saved it as `treat.codes` in `TherapyFormatsGeMTC`:

```
TherapyFormatsGeMTC$treat.codes
```

```
##    id        description
## 1 ind         Individual
## 2 grp              Group
## 3 gsh  Guided Self-Help
## 4 tel          Telephone
## 5 wlc           Waitlist
## 6 cau       Care As Usual
## 7 ush Unguided Self-Help
```

We use this data frame and our effect size data in `TherapyFormatsGeMTC` to build our `mtc.network` object. We save it under the name `network`.

```
network <- mtc.network(data.re = TherapyFormatsGeMTC$data,
                       treatments = TherapyFormatsGeMTC$treat.codes)
```

Plugging the resulting object into the `summary` function already provides us with some interesting information about our network.

```
summary(network)
```

```
## $Description
## [1] "MTC dataset: Network"
##
## $`Studies per treatment`
## ind grp gsh tel wlc cau ush
##  62  52  57  11  83  74  26
##
## $`Number of n-arm studies`
## 2-arm 3-arm
##   181     1
##
## $`Studies per treatment comparison`
##      t1  t2 nr
## 1  ind tel  4
## 2  ind wlc 18
```

```
## 3  grp ind  7
## [...]
```

We can also use the plot function to generate a network plot. Like the network generated by the *{netmeta}* package, the edge thickness corresponds with the number of studies we included for that comparison.

```
plot(network,
     use.description = TRUE) # Use full treatment names
```

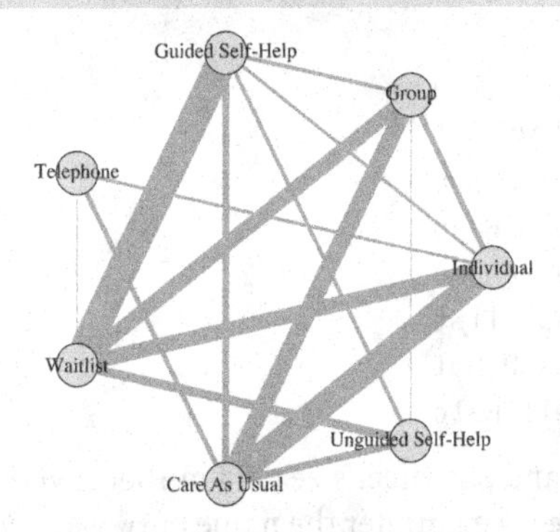

As an alternative, we can also check if we can create a better visualization of our network using the *Fruchterman-Reingold algorithm*. This algorithm comes with some inherent randomness, meaning that we have to set a seed to make our result reproducible. The network plots are created using the *{igraph}* package (Csardi and Nepusz, 2006). When this package is installed and loaded, we can also use other arguments to change the appearance of our plot. A detailed description of the different styling options can be found in the online *{igraph}* manual[6].

```
library(igraph)
set.seed(12345) # set seed for reproducibility

plot(network,
     use.description = TRUE,            # Use full treatment names
     vertex.color = "white",            # node color
     vertex.label.color = "gray10",     # treatment label color
     vertex.shape = "sphere",           # shape of the node
     vertex.label.family = "Helvetica", # label font
     vertex.size = 20,                  # size of the node
     vertex.label.dist = 2,             # distance label-node center
     vertex.label.cex = 1.5,            # node label size
```

[6] https://igraph.org/r/doc/plot.common.html

```
    edge.curved = 0.2,                  # edge curvature
    layout = layout.fruchterman.reingold)
```

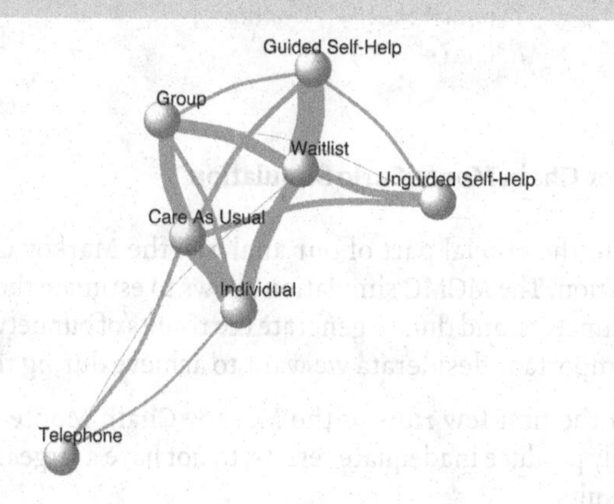

12.3.3.3 Model Compilation

Using our `mtc.network` object, we can now start to specify and compile our model. The great thing about the *{gemtc}* package is that it automates most parts of the Bayesian inference process, for example by choosing adequate prior distributions for all parameters in our model.

Thus, there are only a few arguments we have to specify when compiling our model using the `mtc.model` function. First, we have to specify the `mtc.network` object we have created before. Furthermore, we have to decide if we want to use a random- or fixed effects model using the `linearModel` argument. Given that our previous frequentist analysis indicated substantial heterogeneity and inconsistency (see Chapter 12.2.2.4.1), we will use `linearModel = "random"`. We also have to specify the number of *Markov chains* we want to use. A value between 3 and 4 is sensible here, and we take `n.chain = 4`.

There are two additional, optional arguments called `likelihood` and `link`. These two arguments vary depending on the type of effect size data we are using and are automatically inferred by *{gemtc}* unless explicitly specified. Since we are dealing with effect sizes based on continuous outcome data (viz. SMDs), we are assuming a `"normal"` likelihood along with an `"identity"` link. Had we been using binary outcome measures (e.g. log-odds ratios), the appropriate likelihood and link would have been `"binom"` (binomial) and `"logit"`, respectively. More details on this can be found in the documentation of `mtc.model`. However, when the data has been prepared correctly in the previous step, `mtc.model` usually selects the correct settings automatically.

```
# We give our compiled model the name `model`.
model <- mtc.model(network,
```

```
                     likelihood = "normal",
                     link = "identity",
                     linearModel = "random",
                     n.chain = 4)
```

12.3.3.4 Markov Chain Monte Carlo Simulation

Now we come to the crucial part of our analysis: the Markov Chain Monte Carlo (MCMC) simulation. The MCMC simulation allows to estimate the posterior distributions of our parameters, and thus to generate the results of our network meta-analysis. There are two important desiderata we want to achieve during this procedure:

- We want that the first few runs of the Markov Chain Monte Carlo simulations, which will likely produce inadequate results, to not have a large impact on the whole simulation results.
- The Markov Chain Monte Carlo process should run long enough for us to obtain accurate estimates of the model parameters (i.e. it should *converge*).

To address these points, we split the number of times the Markov Chain Monte Carlo algorithm iterates to infer the model results into *two phases*: first, we define a number of *burn-in* iterations (`n.adapt`), the results of which are discarded. For the following phase, we specify the number of actual simulation iterations (`n.iter`), which are actually used to estimate the model parameters.

Given that we typically simulate many, many iterations, we can also specify the `thin` argument, which allows us to only extract the values of every ith iteration. This can help to reduce the required memory.

The simulation can be performed using the `mtc.run` function. In our example, we will perform two separate runs with different settings to compare which one works better. We have to provide the function with our compiled `model`object, and specify the parameters we just described.

First, we conduct a simulation with only a few iterations, and then a second one in which the number of iterations is large. We save both objects as `mcmc1` and `mcmc2`, respectively. Note that, depending on the size of your network, the simulation may take some time to finish.

```
mcmc1 <- mtc.run(model, n.adapt = 50, n.iter = 1000, thin = 10)
mcmc2 <- mtc.run(model, n.adapt = 5000, n.iter = 1e5, thin = 10)
```

12.3.3.5 Assessing Model Convergence

To see if our simulations have resulted in the convergence of the algorithm, and to check which settings are preferable, we can evaluate some of the outputs of our `mcmc1` and `mcmc2` objects. A good start is to use the `plot` function. This provides us with a kind of "time series", commonly referred to as a *trace plot*, for each treatment comparison over all iterations. In this example, we only focus on the estimate of the individual therapy (`ind`) versus wait-list control (`wlc`) comparison.

```
plot(mcmc1)
plot(mcmc2)
```

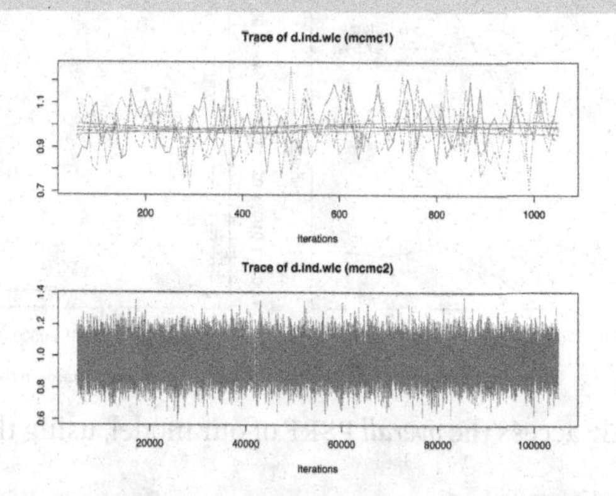

When comparing earlier to later iterations in `mcmc1`, we see that there is a slight discontinuity in the overall trend of the time series. The estimates of the four different chains (the four lines) slightly differ in their course when moving from the first half to the second half of the plot. In the plot for `mcmc2`, on the other hand, we see much more rapid up-and-down variation, but no real long-term trend. This delivers a first indication that the settings in `mcmc2` are more adequate.

We can continue with our convergence assessment by looking at the density plots of the posterior effect size estimate. We see that, while the distribution in `mcmc1` still diverges somewhat from a smooth normal distribution, the result of `mcmc2` comes closer to a classic bell curve.

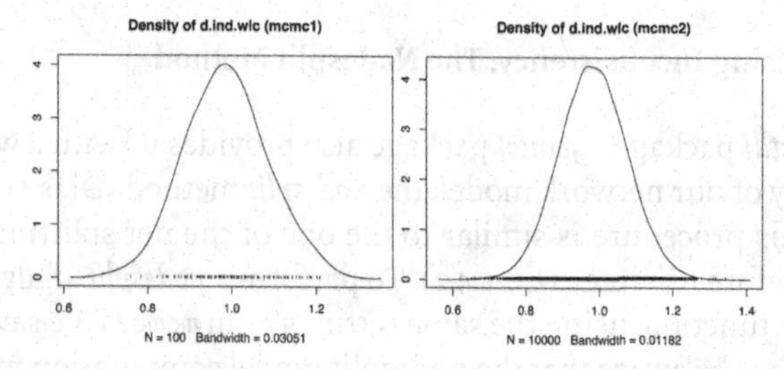

Another highly helpful method to assess convergence is the *Gelman-Rubin plot*. This plot shows the so-called *Potential Scale Reduction Factor* (PSRF), which compares the variation within each chain to the variation between chains, and how both develop over time. In case of convergence, the PRSF should gradually shrink down to zero with increasing numbers of iterations, and should at least be below 1.05 in the end.

To produce this plot, we simply have to plug in the `mtc.run` object into the `gelman.plot` function. Here is the result for both simulations (again only showing the `ind` versus `wlc` comparison).

```
gelman.plot(mcmc1)
gelman.plot(mcmc2)
```

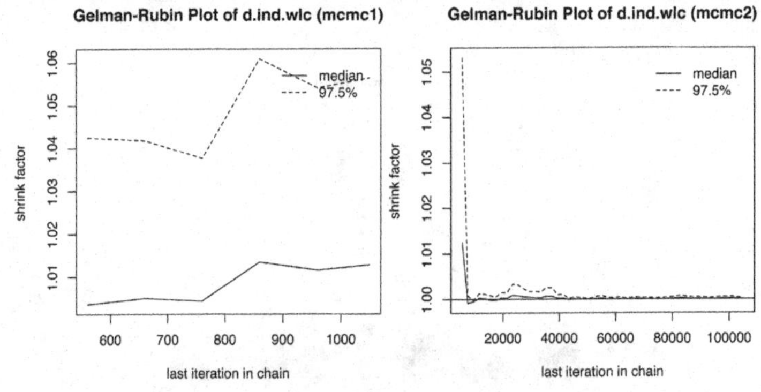

We can also directly access the *overall* PSRF of our model, using this code:

```
gelman.diag(mcmc1)$mpsrf
```

```
## [1] 1.034131
```

```
gelman.diag(mcmc2)$mpsrf
```

```
## [1] 1.000351
```

We see that, while the overall PRSF is below the threshold in both simulations, the value in `mcmc2` is much lower and very close to 1. This indicates that the second model should be used.

12.3.3.6 Assessing Inconsistency: The Nodesplit Method

Like the *{netmeta}* package, *{gemtc}* package also provides us with a way to evaluate the consistency of our network model: the *nodesplit* method (Dias et al., 2010). The idea behind this procedure is similar to the one of the net splitting method that we described before (Chapter 12.2.2.4.2). To perform a *nodesplit* analysis, we use the `mtc.nodesplit` function, using the same settings as in `mcmc2`. We save the result as `nodesplit`. Please be aware that the nodesplit model computation may take a long time, even up to several hours, depending on the complexity of your network.

```
nodesplit <- mtc.nodesplit(network,
                           linearModel = "random",
                           likelihood = "normal",
                           link = "identity",
                           n.adapt = 5000,
                           n.iter = 1e5,
                           thin = 10)
```

Using the summary function, we can print the results.

```
summary(nodesplit)
```

```
## Node-splitting analysis of inconsistency
## ========================================
##
##    comparison  p.value CrI
## 1  d.ind.tel   0.62785
## 2  -> direct           0.13 (-0.39, 0.64)
## 3  -> indirect         -0.037 (-0.46, 0.38)
## 4  -> network          0.034 (-0.30, 0.36)
## 5  d.ind.wlc   0.87530
## 6  -> direct           1.0 (0.74, 1.3)
## 7  -> indirect         0.97 (0.71, 1.2)
## 8  -> network          0.98 (0.80, 1.2)
## 9  d.ind.grp   0.61380
## 10 -> direct           0.14 (-0.29, 0.57)
## 11 -> indirect         0.26 (0.044, 0.48)
## 12 -> network          0.24 (0.041, 0.43)
## [...]
```

The function output shows us the results for the effects of different comparisons when using only direct, only indirect, and all available evidence. Different estimates using direct and indirect evidence suggest the presence of inconsistency. We can control for this by looking at the Bayesian `p.value` column. One or more comparisons with $p < 0.05$ are problematic, since this indicates inconsistency in our network. From the output, we see that this is not the case in our (random-effects model) example.

When the nodesplitting method *does* show inconsistencies in some of the estimates, it is important to again check *all* included evidence for potential differences between designs. It may be possible, for example, that studies comparing A and B included systematically different populations than other studies which also assessed A. Another approach is to check if the inconsistency persists when only a sensible subset of studies is included in the network. Lastly, it is also possible to assess reasons for inconsistency by running a network meta-regression, which we will cover later.

It is also possible to generate a forest plot for the nodesplit model, using the `plot` function. However, a forest is only generated when we plug the nodesplit object into `summary` first.

```
plot(summary(nodesplit))
```

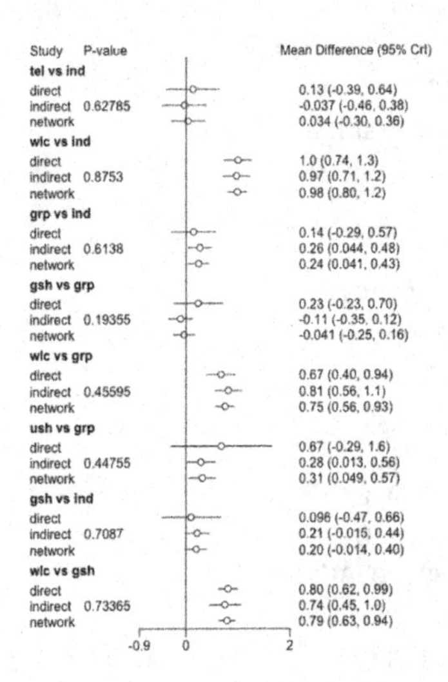

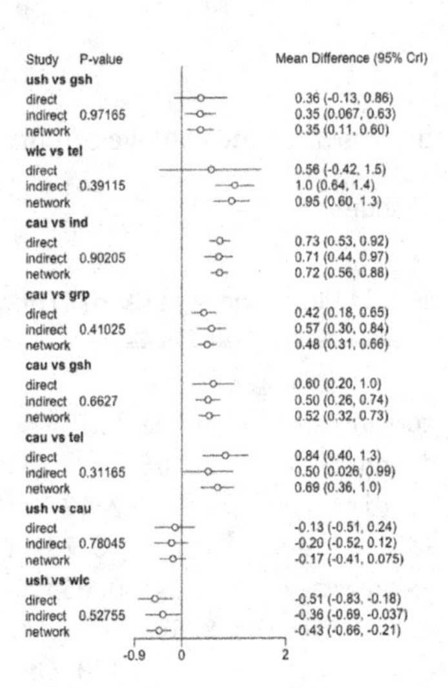

12.3.3.7 Generating the Network Meta-Analysis Results

Now that we fitted our network meta-analysis model, and have convinced ourselves that it is trustworthy, it is time to finally produce the results.

As mentioned before, the main question we may want to answer in network meta-analyses is which treatment performs the best. To answer this question, we can first run the `rank.probability` function. This function calculates the probability of a treatment being the best option, second best option, third best option, and so forth. The function needs our `mcmc2` object as input, and we additionally specify the `preferredDirection` argument. If smaller (i.e. negative) effect sizes indicate better outcomes, we set this argument to -1. Otherwise, we use 1. We save the results under the name `rank`, and then visualize them using a so-called *rankogram*.

```
rank <- rank.probability(mcmc2, preferredDirection = -1)

plot(rank, beside=TRUE)
```

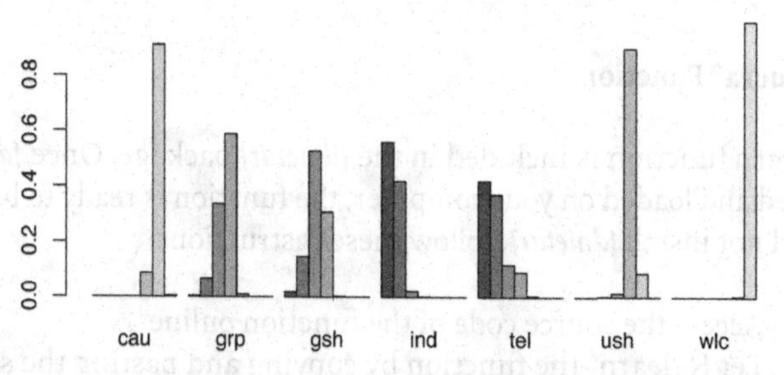

In this plot, we see that individual therapy (ind) is probably the best treatment option in our network, given that its first bar (signifying the first rank) is the largest. This finding is in agreement with the results of the frequentist analysis, where we found the same pattern.

Additionally, we can also produce a forest plot of our results using the `forest` function. To do this, we first have to put our results object into the `relative.effect` function and specify `t1`, the reference treatment. We use care as usual (`"cau"`) as the reference group again. Then, we call the `forest` function on the results to generate the plot.

```
forest(relative.effect(mcmc2, t1 = "cau"),
       use.description = TRUE, # Use long treatment names
       xlim = c(-1.5, 0.5))
```

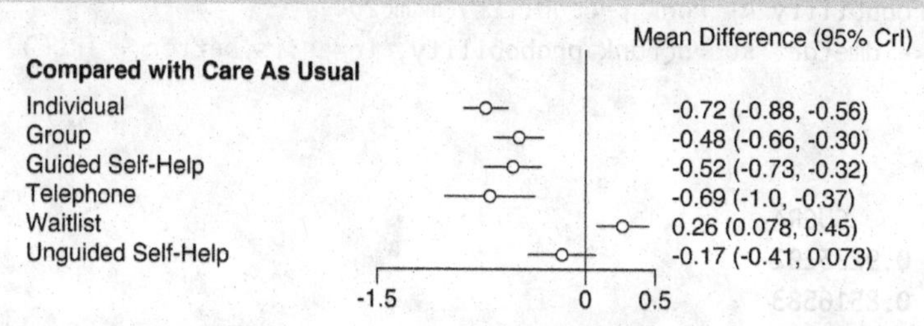

In the chapter on frequentist network meta-analysis, we already covered the P-score as a metric to evaluate which treatment in a network is likely to be the most efficacious. An equivalent to the P-score is the *Surface Under the Cumulative Ranking* (SUCRA) score, which can be calculated like this (Salanti et al., 2011):

$$\text{SUCRA}_j = \frac{\sum_{b=1}^{a-1} \text{cum}_{jb}}{a-1} \tag{12.17}$$

Where j is some treatment, a are all competing treatments, b are the $b = 1, 2, \dots, a-1$ best treatments, and cum represents the *cumulative probability* of a treatment being among the b best treatments. To calculate the SUCRA scores in R, we can use the `sucra` function.

The "sucra" Function

The `sucra` function is included in the *{dmetar}* package. Once *{dmetar}* is installed and loaded on your computer, the function is ready to be used. If you did not install *{dmetar}*, follow these instructions:

1. Access the source code of the function online[a].
2. Let R "learn" the function by copying and pasting the source code in its entirety into the console (bottom left pane of R Studio), and then hit "Enter".
3. Make sure that the *{ggplot2}* package is installed and loaded.

[a] `https://raw.githubusercontent.com/MathiasHarrer/dmetar/master/R/sucra.R`

The `sucra` function only needs a `rank.probability` object as input, and we need to specify if lower values indicate better outcomes. This can be done using the `lower.is.better` argument. Let us see what results we get.

```
library(dmetar)
rank.probability <- rank.probability(mcmc2)
sucra <- dmetar::sucra(rank.probability, lower.is.better = TRUE)

sucra
```

```
##         SUCRA
## ind 0.9225292
## tel 0.8516583
## gsh 0.6451292
## [...]
```

```
plot(sucra)
```

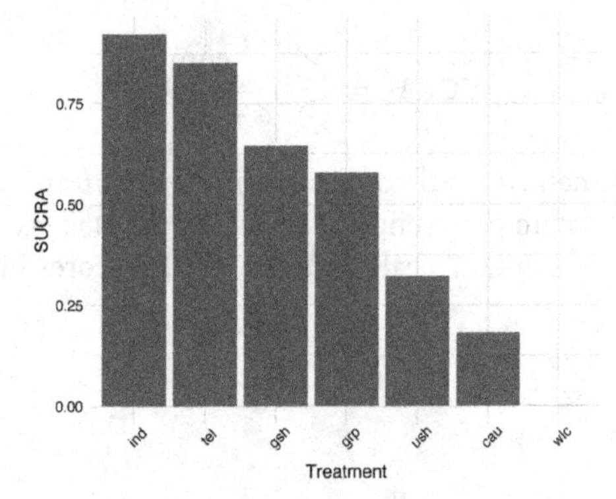

Looking at the SUCRA values of each treatment, we again see that individual treatment may be the best option, followed by telephone-based treatment and guided self-help.

Usually, we want to report the effect size estimate for each treatment comparison based on our model. A treatment effect table can be exported using the `relative.effect.table` function. We save the results of this function in an object called `result`, which we can then export as a .csv file. The `relative.effect.table` function automatically creates a treatment comparison matrix containing the estimated effect, as well as the credible intervals for each comparison.

```
results <- relative.effect.table(mcmc2)
save(results, file = "results.csv")
```

12.3.4 Network Meta-Regression

A big asset of the *{gemtc}* package is that it allows us to conduct *network meta-regression*. Much like conventional meta-regression (Chapter 8), we can use this functionality to determine if specific study characteristics influence the magnitude of effect sizes found in our network. It can also be a helpful tool to check for variables that may explain inconsistency.

Imagine we want to evaluate if the risk of bias of a study has an influence on the effects in our network meta-analysis. For example, it could be that studies with a high risk of bias generally report higher effects compared to the control group or alternative treatments. By including risk of bias as a predictor to our model, we can control for such an association, and assess its impact on our results.

To run a network meta-regression in *{gemtc}*, we have to follow similar steps as before, when we fitted a Bayesian network meta-analysis model without covariates. First, we need to set up our network using `mtc.network`. This time, however, we specify an additional argument called `studies`. This argument requires a data frame in which predictor information for each study is stored. The `TherapyFormatsGeMTC` data set includes an element `study.info`, which contains the risk of bias of each study. Let us have a quick look at the data.

```
TherapyFormatsGeMTC$study.info
```

```
##                  study rob
## 1       Campbell, 2000   1
## 2       Reynolds, 1989   1
## 3      Carpenter, 1994   0
## 4       Shrednik, 2000   1
## [...]
```

The data set contains two columns: `study`, the name of the study included in our network and `rob`, its risk of bias. Please note that the study labels must be completely identical to the ones used in the actual effect size data set. The `rob` variable is a dummy-coded predictor, where `0` indicates a low risk of bias, and `1` high risk of bias. Using the `study.info` data frame, we can now create a meta-regression network using `mtc.network`.

```
network.mr <- mtc.network(data.re = TherapyFormatsGeMTC$data,
                          studies = TherapyFormatsGeMTC$study.info,
                          treatments = TherapyFormatsGeMTC$treat.codes)
```

Now, we must define the *regressor* we want to include in our network meta-analysis model. This can be done by generating a list object with three elements:

- `coefficient`: We set this element to `"shared"` because we want to estimate one shared coefficient for the effect of (high) risk of bias across all treatments included in our network meta-analysis.
- `variable`: This specifies the name of the variable we want to use as the predictor (here: `"rob"`).
- `control`: We also have to specify the treatment which we want to use as the reference group. We use `"cau"` (care as usual) in our example.

```
regressor <- list(coefficient = "shared",
                  variable = "rob",
                  control = "cau")
```

Next, we compile our model. We provide the `mtc.model` function with the network we just generated, set the type of our model to `"regression"`, and provide the function with the `regressor` object we just generated. We save the output under the name `model.mr`.

```
model.mr <- mtc.model(network.mr,
                      likelihood = "normal",
                      link = "identity",
                      type = "regression",
                      regressor = regressor)
```

After this step, we can run the model using the `mtc.run` function. We use the same specifications as used for fitting the `mcmc2` model before. The results are saved as `mcmc3`.

```
mcmc3 <- mtc.run(model.mr,
                 n.adapt = 5000,
                 n.iter = 1e5,
                 thin = 10)
```

Now, we can analyze the results using the summary function.

```
summary(mcmc3)
```

```
## Results on the Mean Difference scale
## [...]
##
## 1. Empirical mean and standard deviation for each variable,
##    plus standard error of the mean:
##
##                Mean      SD  Naive SE Time-series SE
## d.ind.cau   0.6992 0.07970 0.0003985      0.0004201
## d.ind.grp   0.1933 0.10009 0.0005005      0.0005321
## [...]
## B          -0.3297 0.13047 0.0006523      0.0010379
##
## 2. Quantiles for each variable:
##
##                  2.5%      25%      50%     75%    97.5%
## d.ind.cau   0.542044  0.64602  0.69967  0.7529  0.85571
## d.ind.grp  -0.002622  0.12599  0.19353  0.2608  0.38962
## [...]
## B          -0.586266 -0.41790 -0.32957 -0.2417 -0.07455
##
## [...]
## -- Regression settings:
##
## Regression on "rob", shared coefficients, "cau" as control
## Input standardized: x' = (rob - 0.4340659) / 1
## Estimates at the centering value: rob = 0.4340659
```

The results for our predictor are reported next to `B`. Because our predictor is dummy-coded, the value of `B` represents the effect of a study having a *high* risk of bias. The estimate is $b =$ -0.33, and when looking at the second table (`Quantiles for each variable`), we see that the 95% credible interval of b ranges from -0.59 to -0.08. Since the credible interval does not include zero, we may conclude that risk of bias does indeed influence the results. When the risk of bias is high (`rob` = 1), we can predict higher overall effects (since negative effect sizes indicate "better" outcomes in our example).

We can explore the effect of the predictor further by generating two forest plots: one for the estimated treatment effects when the risk of bias is high, and another for when it is low. We can do this using the `relative.effect` function, where we specify the `covariate` value. A value of `covariate = 0` stands for studies with a low risk of bias, and `covariate = 1` for high risk of bias.

```
forest(relative.effect(mcmc3, t1 = "cau", covariate = 1),
       use.description = TRUE, xlim = c(-1.5, 1))
title("High Risk of Bias")

forest(relative.effect(mcmc3, t1 = "cau", covariate = 0),
       use.description = TRUE, xlim = c(-1.5, 1))
title("Low Risk of Bias")
```

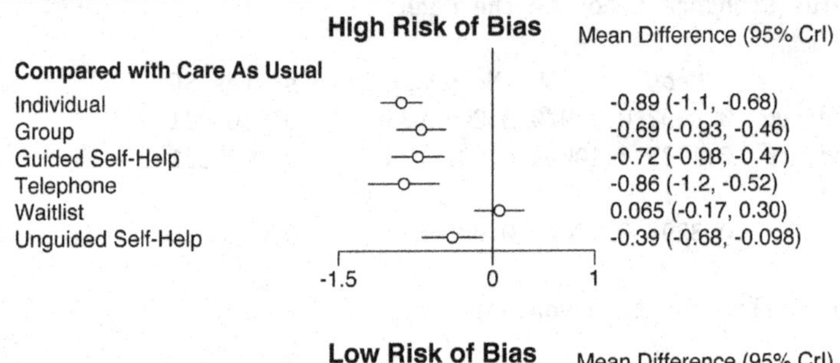

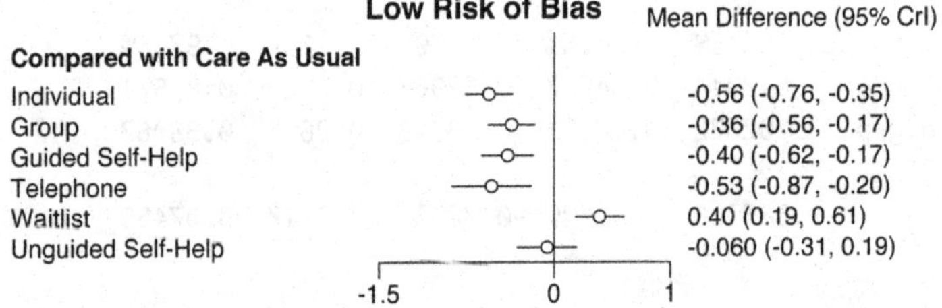

Comparing the forest plots, we can see that there is a pattern. The treatment effects based on high risk of bias studies are generally higher (i.e. more negative). This is in line with the estimate of our predictor b in the fitted model.

Lastly, we can also examine if the network meta-regression model we just generated fits the data better than the "normal" network meta-analysis model from before. To do this, we can compare the *deviance information criteria* (DICs), which are an equivalent to the AIC and BIC values in frequentist statistics. We can access the DIC of both `mcmc3` and `mcmc2` using this code:

```
summary(mcmc3)$DIC
```

```
##        Dbar          pD         DIC data points
##      185.82       75.37      261.19      183.00
```

```
summary(mcmc2)$DIC
```

```
##        Dbar          pD         DIC data points
##       185.6       138.0       323.6       183.0
```

We see in the output that the DIC value of our meta-regression model (261.19) is lower than the one of our previous model which did not control for risk of bias (DIC = 323.6). Lower DIC values indicate a better fit. Based on this finding, we can conclude that our network meta-regression model fits the data better than one without the covariate.

Further Reading

This is the end of our brief introduction to network meta-analysis using R. We have described the general idea behind network meta-analysis, the assumptions and some of the caveats associated with it, two different statistical approaches through which network meta-analysis can be conducted, and how they are implemented in R.

We would like to stress that what we covered here should only be seen as a rough overview. Although we have covered some of the main pitfalls, it is still possible that you may get stuck once you begin with your own network meta-analysis.

An excellent resource to learn more about network meta-analysis and how it can be applied in practice is *Network Meta-Analysis for Decision-Making*, written by Dias and colleagues (2018). The book also features several hands-on examples, and shows how to run network meta-analysis models using the open source software *WinBUGS*. A shorter (and rather technical) overview of the "state-of-art" in network meta-analysis can be found in an open-access paper by Efthimiou et al. (2016).

□

12.4 Questions & Answers

Test your knowledge!

1. When are network meta-analyses useful? What is their advantage compared to standard meta-analyses?
2. What is the difference between direct and indirect evidence in a treatment network? How can direct evidence be used to generate indirect evidence?
3. What is the main idea behind the assumption of transitivity in network meta-analyses?
4. What is the relationship between transitivity and consistency?
5. Name two modeling approaches that can be used to conduct network meta-analyses. Is one of them better than the other?
6. When we include several comparisons from one study (i.e. multi-arm studies), what problem does this cause?
7. What do we have to keep in mind when interpreting the P- or SUCRA score of different treatments?

Answers to these questions are listed in Appendix A at the end of this book.

12.5 Summary

- Network meta-analysis is a useful tool to jointly estimate the relative effectiveness of various treatments or interventions.
- To estimate the treatment effects, network meta-analysis combines both direct (i.e. observed) and indirect evidence. This, however, is based on the assumption of transitivity. Transitivity is fulfilled when we can combine direct evidence of two comparisons to derive valid indirect evidence about a third one.

- The statistical manifestation of transitivity is consistency, the opposite of which is inconsistency. Inconsistency arises when the true effect of some comparison based on direct evidence does not coalign with the one based on indirect evidence.
- Some methods, such as nodesplitting or net heat plots, can be used to identify inconsistencies in our network. When inconsistencies are found, this threatens the validity of our results as a whole. In such cases, the entire network should be checked for characteristics that may have caused systematic differences between studies/designs.
- Network meta-analysis is possible using either a frequentist or Bayesian approach. In practice, each of these methods has individual strengths, but the overall results are usually very similar.
- In network meta-analyses based on a Bayesian hierarchical model, we can also add study covariates that predict effect size differences. This results in a network meta-regression model.
- Indices such as the SUCRA or P-score can be used to examine which type of treatment may be the most effective in our network. However, it is also important to integrate uncertainty into our decision-making process. Confidence/credible intervals of different treatments often overlap, which makes it less clear if one format is truly superior to all the others.

13

Bayesian Meta-Analysis

In the last chapters, we have delved into somewhat more sophisticated extensions of meta-analysis, such as "multilevel" models (Chapter 10), meta-analytic structural equation modeling (Chapter 11), and network meta-analysis (Chapter 12). Now, we will take one step back and revisit "conventional" meta-analysis again–but this time from another angle. In this chapter, we deal with *Bayesian meta-analysis*.

We already covered a Bayesian model in the last chapter on network meta-analysis. There, we discussed the main ideas behind Bayesian statistics, including Bayes' theorem and the idea of prior distributions (see Chapter 12.3.1). In the present chapter, we build on this knowledge and try to get a more thorough understanding of the "Bayesian way" to do meta-analysis. When we set up our Bayesian network meta-analysis model, for example, the *{gemtc}* package specified the priors automatically for us. Here, we will do this ourselves.

While its background is slightly more involved, we will see that Bayesian meta-analysis essentially aims to do the same thing as any "conventional" meta-analysis: it pools observed effect sizes into one overall (true) effect. Using a Bayesian model, however, also comes with several practical advantages compared to frequentist approaches. This makes it worthwhile to learn how we can implement such models using R.

13.1 The Bayesian Hierarchical Model

To perform a Bayesian meta-analysis, we employ a so-called *Bayesian hierarchical model* (Röver, 2017; Higgins et al., 2009). We already briefly covered this type of model in the network meta-analysis chapter (Chapter 12.3.2).

In Chapter 10, we learned that every meta-analytic model comes with an inherent "multilevel", and thus *hierarchical,* structure. On the first level, we have the individual participants. Data on this level usually reaches us in the form of calculated effect sizes $\hat\theta_k$ of each study k. We assume that participants are nested within studies on the second level and that the true effect sizes θ_k of different studies in our meta-analysis follow their own distribution. This distribution of true effects has a mean

DOI: 10.1201/9781003107347-13

μ (the pooled "true" effect we want to estimate) and variance τ^2, representing the between-study heterogeneity.

Let us try to formalize this. On the first level, we assume that the observed effect size $\hat{\theta}_k$ reported in study k is an estimate of the "true" effect θ_k in this study. The observed effect $\hat{\theta}_k$ deviates from θ_k due to the sampling error ϵ_k. This is because we assume that $\hat{\theta}_k$ was drawn (sampled) from the population underlying k. This population can be seen as a distribution with mean θ_k, the "true" effect of the study, and a variance σ^2.

In the second step, we assume that the true effect sizes θ_k themselves are only samples of an overarching distribution of true effect sizes. The mean of this distribution μ is the pooled effect size we want to estimate. The study-specific true effects θ_k deviate from μ because the overarching distribution also has a variance τ^2, signifying the between-study heterogeneity. Taken together, this gives these two equations:

$$\begin{aligned} \hat{\theta}_k &\sim \mathcal{N}(\theta_k, \sigma_k^2) \\ \theta_k &\sim \mathcal{N}(\mu, \tau^2) \end{aligned} \tag{13.1}$$

Here, we use $\mathcal{N}$ to indicate that parameters to the left were sampled from a *normal* distribution. Some may argue that this is an unnecessarily strict assumption for the second equation (Higgins et al., 2009), but the formulation as shown here is the one that is used most of the time. As covered before, the fixed-effect model is simply a special case of this model in which we assume that $\tau^2 = 0$, meaning that there is no between-study heterogeneity, and that all studies share one single true effect size (i.e. that for all studies k: $\theta_k = \mu$). We can also simplify this formula by using the marginal form:

$$\hat{\theta}_k \sim \mathcal{N}(\mu, \sigma_k^2 + \tau^2) \tag{13.2}$$

You may have already detected that these formulas look a lot like the ones we defined when discussing the random-effects (Chapter 4.1.2) and three-level meta-analysis model. Indeed, there is nothing particularly "Bayesian" about this formulation. This changes, however, when we add the following equations (Williams et al., 2018):

$$\begin{aligned} (\mu, \tau^2) &\sim p(.) \\ \tau^2 &> 0 \end{aligned} \tag{13.3}$$

The first line is particularly important. It defines *prior distributions* for the parameters μ and τ^2. This allows us to specify *a priori* how we think the true pooled effect size μ and between-study heterogeneity τ^2 may look like, and how certain we are about this. The second equation adds the constraint that the between-study heterogeneity variance must be larger than zero. However, this formula does not specify the exact *type* of prior distribution used for μ and τ^2. It only tells us that *some* prior distribution is assumed. We will cover reasonable, specific priors for Bayesian meta-analysis models in more detail later.

In the chapter on network meta-analysis, we already covered the method through which Bayesian approaches estimate model parameters. To recap, this involves using *Markov Chain Monte Carlo*-based sampling procedures, for example, *Gibbs sampling*. In the *{brms}* package, which we will be using in this chapter, so-called *No-U-Turn* sampling (NUTS, Hoffman and Gelman, 2014) is used.

In the previous chapters, we primarily used the *{meta}* and *{metafor}* packages. These packages allow to conduct meta-analyses based on a non-Bayesian, or *frequentist* framework. Therefore, you might be wondering why one should start using Bayesian methods, given that we can already resort to such powerful tools using "conventional" approaches. The reason is that Bayesian meta-analysis comes with a few distinct advantages (Williams et al., 2018; McNeish, 2016; Chung et al., 2013):

- Bayesian methods allow to directly model the *uncertainty* in our estimate of τ^2. They can also be superior in estimating pooled effects, particularly when the number of included studies is small (which is very often the case in practice).
- Bayesian methods produce full *posterior distributions* for both μ and τ^2. This allows to calculate the exact *probability* that μ or τ^2 is smaller or larger than some specified value. This is in contrast to frequentist methods, where we only calculate confidence intervals. However, (95%) confidence intervals only state that, if data sampling were repeated many, many times, the true value of a population parameter (such as μ or τ^2) would fall into the range of the confidence interval in 95% of the samples. They do not tell us the *probability* that the true parameter lies between two specified values.
- Bayesian methods allow us to integrate *prior knowledge* and assumptions when calculating meta-analyses.

13.2 Setting Prior Distributions

Before, we formalized the hierarchical model we can use to pool effects in a Bayesian meta-analysis. However, to run such a model, we have to specify the prior distributions of μ and τ^2. Particularly when the number of studies is small, priors can have a considerable impact on the results, so we should choose them wisely.

Generally, a good approach is to use *weakly informative* priors (Williams et al., 2018). Weakly informative priors can be contrasted with *non-informative* priors. Non-informative priors are the simplest form of a prior distribution. They are usually based on *uniform* distributions, and are used to represent that all values are equally credible.

Weakly informative priors, on the other hand, are a little more sophisticated. They rely on distributions which represent that we have a *weak* belief that some values are more credible than others. However, they are still not making any specific statements concerning the value of the parameter to be estimated from our data.

Intuitively, this makes a lot of sense. In many meta-analyses, for example, it seems reasonable to assume that the true effect lies somewhere between SMD = -2.0 and 2.0, but will unlikely be SMD = 50. Based on this rationale, a good starting point for our μ prior may therefore be a normal distribution with mean 0 and variance 1. This means that we grant an approximate 95% prior probability that the true pooled effect size μ lies between –2.0 and 2.0:

$$\mu \sim \mathcal{N}(0,1) \tag{13.4}$$

The next prior we have to specify is the one for τ^2. This one is a little more difficult since we know that τ^2 should always be non-negative, but can be (close to) zero. A recommended distribution for this case, and one which is often used for variances such as τ^2, is the *Half-Cauchy* prior. The Half-Cauchy distribution is a special case of a Cauchy distribution, which is only defined for one "half" (i.e. the positive side) of the distribution.

The Half-Cauchy distribution is controlled by two parameters. The first one is the location parameter x_0, which specifies the peak of the distribution. The second one is s, the scaling parameter. It controls how *heavy-tailed* the distribution is (i.e. how much it "spreads out" to higher values). The Half-Cauchy distribution is denoted with $\mathcal{HC}(x_0, s)$.

The graph below visualizes the Half-Cauchy distribution for varying values of s, with the value of x_0 fixed at 0.

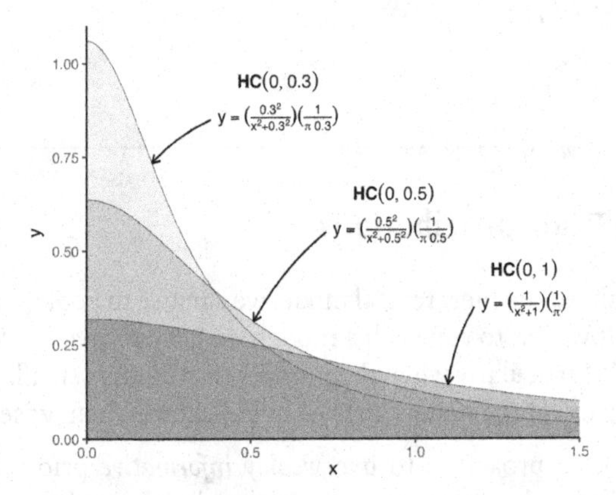

The Half-Cauchy distribution typically has rather heavy tails, which makes it particularly useful as a prior distribution for τ. The heavy tails ensure that we still give very high values of τ *some* probability, while at the same time assuming that lower values are more likely.

In many meta-analyses, τ (the square root of τ^2) lies somewhere around 0.3, or at least in the same ballpark. To specify the Half-Cauchy prior, we may therefore use $s = 0.3$. This ensures that a value of less than $\tau = 0.3$ has a 50% probability (Williams et al., 2018). We can confirm this using the Half-Cauchy distribution function implemented in the `phcauchy` function of the *{extraDistr}* package (Wołodźko, 2020).

```
library(extraDistr)
phcauchy(0.3, sigma = 0.3)
```

```
## [1] 0.5
```

However, this is already a quite specific assumption concerning the true value of τ. A more conservative approach, which we will follow in our hands-on example, is to set s to 0.5; this makes the distribution flatter. In general, it is advised to always conduct sensitivity analyses with different prior specifications to check if they affect the results substantially. Using $s = 0.5$ as our parameter of the Half-Cauchy distribution, we can write down our τ prior like this:

$$\tau \sim \mathcal{HC}(0, 0.5) \tag{13.5}$$

We can now put together the formulas of the hierarchical model, and our prior specifications. This leads to the complete model we can use for our Bayesian meta-analysis:

$$\begin{aligned} \hat{\theta}_k &\sim \mathcal{N}(\theta_k, \sigma_k^2) \\ \theta_k &\sim \mathcal{N}(\mu, \tau^2) \\ \mu &\sim \mathcal{N}(0, 1) \\ \tau &\sim \mathcal{HC}(0, 0.5) \end{aligned} \tag{13.6}$$

13.3 Bayesian Meta-Analysis in R

Now that we have defined the Bayesian model for our meta-analysis, it is time to implement it in R. Here, we use the *{brms}* package (Bürkner, 2017; Bürkner, 2017) to fit our model. The *{brms}* package is a very versatile and powerful tool to fit Bayesian regression models. It can be used for a wide range of applications, including multilevel (mixed-effects) models, generalized linear models, multivariate models, and generalized additive models, to name just a few. Most of these models require person-level data but *{brms}* can also be used for meta-analysis, where we deal with (weighted) study-level data[1].

[1]The *{brms}* package is based on *Stan*, a low-level programming language for Bayesian modeling. The Stan project has its own, actively maintained online forum (`https://discourse.mc-stan.org/`), where

Before we start fitting the model, we first have to install and load the *{brms}* package.

```
library(brms)
```

13.3.1 Fitting the Model

In our hands-on example, we again use the `ThirdWave` data set, which contains information from a meta-analysis investigating the effects of "third wave" psychotherapies in college students (Chapter 4.2.1). Before we fit the model, let us first specify the prior distribution of the overall effect size μ and the between-study heterogeneity τ. Previously, we defined that $\mu \sim \mathcal{N}(0,1)$ and $\tau \sim \mathcal{HC}(0,0.5)$.

We can use the `prior` function to specify the distributions. The function takes two arguments. In the first argument, we specify the distribution we want to assume for our prior, including the distribution parameters. In the second argument, we have to define the `class` of the prior. For μ, the appropriate class is `Intercept`, since it is a fixed population-level effect. For τ, the class is `sd`, because it is a variance (or, to be more precise, a *standard deviation*). We can define both priors using the `prior` function, then concatenate them, and save the resulting object as `priors`.

```
priors <- c(prior(normal(0,1), class = Intercept),
            prior(cauchy(0,0.5), class = sd))
```

Now, we can proceed and fit the model. To do this, we use the `brm` function in *{brms}*. The function has many arguments, but only a few are relevant for us.

In the `formula` argument, the formula for the model is specified. The *{brms}* package uses a regression formula notation, in which an outcome (in our case, an observed effect size) `y` is predicted by one or more predictors `x`. A tilde (`~`) is used to specify that there is a predictive relationship: `y ~ x`.

Meta-analyses are somewhat special, because we do not have a variable predicting the effect size (unless when we perform a meta-regression). This means that `x` has to be replaced with `1`, indicating an *intercept-only* model. Furthermore, we cannot simply use the effect size of each study in `y` *as is*. We also have to give studies with higher precision (i.e. sample size) a greater weight. This can be done by using `y|se(se_y)` instead of only `y`, where the `se(se_y)` part stands for the standard error of each effect size `y` in our data set.

If we want to use a random-effects model, the last step is to add a random-effects term `(1|study)` to the right side of the formula. This specifies that the effect sizes in `y` are assumed to be nested within studies, the true effects of which are themselves random draws from an overarching population of true effect sizes. If we want to use

issues pertaining to *{brms}* can also be discussed. The forum also has a "meta-analysis" tag, which allows to filter out potentially relevant threads.

a fixed-effect model, we can simply omit this term. The generic full formula for a random-effects model therefore looks like this: `y|se(se_y) ~ 1 + (1|random)`. To learn more about the formula setup in `brm` models, you can type `?brmsformula` in your console to open the documentation.

The other arguments are fairly straightforward. In `prior`, we specify the priors we want to define for our model. In our example, we can simply plug in the `priors` object we created previously. The `iter` argument specifies the number of iterations of the MCMC algorithm. The more complex your model, the higher this number should be. However, more iterations also mean that the function will take longer to finish. Lastly, we also have to specify `data`, where we simply provide the name of our data set.

We save our fitted Bayesian meta-analysis model under the name `m.brm`. The code looks like this:

```
m.brm <- brm(TE|se(seTE) ~ 1 + (1|Author),
             data = ThirdWave,
             prior = priors,
             iter = 4000)
```

Please be aware that Bayesian methods are much more computationally expensive compared to standard meta-analytic techniques we covered before. It may therefore take a few minutes until the sampling is completed.

13.3.2 Assessing Convergence

Before we start analyzing the results, we have to make sure that the model has *converged* (i.e. that the MCMC algorithm found the optimal solution). If this is not the case, the parameters are not trustworthy and should not be interpreted. Non-convergence happens frequently in Bayesian models and can often be resolved by re-running the model with more iterations (`iter`). Generally, we should always do two things. First, conduct *posterior predictive checks*, and secondly, check the $\hat{R}$ values of the parameter estimates.

In a posterior predictive check, data are simulated through random draws from the posterior predictive distribution and then compared to the observed data. If a model has converged and is valid, we can expect that the densities of the replications are roughly similar to the one of the observed data. This can easily be checked using the output of the `pp_check` function.

```
pp_check(m.brm)
```

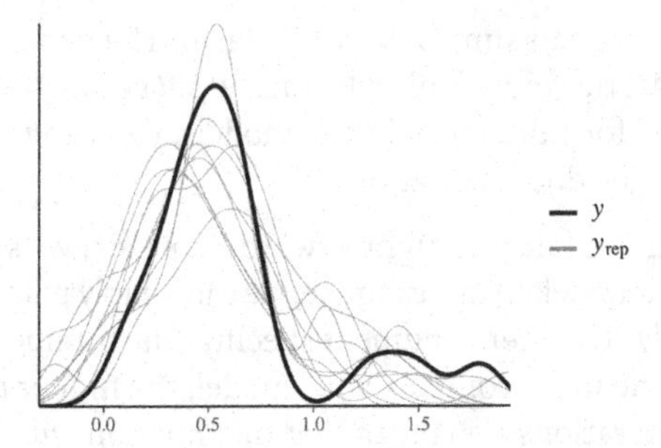

The $\widehat{R}$ value, on the other hand, represents the *Potential Scale Reduction Factor* (PSRF) we already covered when discussing Bayesian network meta-analysis (Chapter 12.3.3.5). The $\widehat{R}$ value of our estimates should be considerably smaller than 1.05. To check this, we can produce a summary of our `m.brm` object.

```
summary(m.brm)
```

```
## Family: gaussian
##   Links: mu = identity; sigma = identity
## Formula: TE | se(seTE) ~ 1 + (1 | Author)
##    Data: ThirdWave (Number of observations: 18)
## Samples: 4 chains, each with iter = 4000; warmup = 2000; thin = 1;
##          total post-warmup samples = 8000
##
## Group-Level Effects:
## ~Author (Number of levels: 18)
##               Estimate Est.Error l-95% CI u-95% CI Rhat Bulk_ESS
## sd(Intercept)     0.29      0.10     0.11     0.51 1.00     2086
##
## Population-Level Effects:
##           Estimate Est.Error l-95% CI u-95% CI Rhat Bulk_ESS
## Intercept     0.57      0.09     0.39     0.76 1.00     3660
##
##
## [...]
##
## Samples were drawn using sampling(NUTS). For each parameter,
## Bulk_ESS and Tail_ESS are effective sample size measures,
## and Rhat is the potential scale reduction factor on split
## chains (at convergence, Rhat = 1).
```

As we can see, the `Rhat` value for both parameters is 1, signifying convergence. This means that we can start interpreting the results.

13.3.3 Interpreting the Results

We can begin to interpret the results by looking at the `Group-Level Effects` in our summary output first. This section is reserved for the random effect we defined in our formula. Since we fitted a random-effects meta-analysis model, the variable `~Author`, signifying the individual studies, has been modeled with a random intercept. As we described before, this represents our assumption on level 2 that each study has its own "true" effect size, which has been sampled from an overarching distribution of true effect sizes. We also see that our group-level effect has 18 levels, corresponding with the $K = 18$ studies in our data.

The estimate of the between-study heterogeneity, `sd(Intercept)`, is $\tau = 0.29$, thus closely resembling our initial "best guess" when setting the priors. Using the `ranef` function, we can also extract the estimated deviation of each study's "true" effect size from the pooled effect:

```
ranef(m.brm)
```

```
## $Author
## , , Intercept
##
##                      Estimate Est.Error         Q2.5       Q97.5
## Call et al.        0.06836636 0.1991649 -0.327463365  0.47663987
## Cavanagh et al.   -0.14151644 0.1767123 -0.510165576  0.18799272
## DanitzOrsillo      0.48091338 0.2829719 -0.003425284  1.08636421
## de Vibe et al.    -0.31923470 0.1454819 -0.612269461 -0.03795683
## Frazier et al.    -0.11388029 0.1497128 -0.417029387  0.17085917
## [...]
```

The next part of the output we can interpret are the `Population-Level Effects`. This section represents the "fixed" population parameters we modeled. In our case, this is μ, the overall effect size of our meta-analysis.

In the output, we see that the estimate is a (bias-corrected) SMD of 0.57, with the 95% credible interval ranging from 95%CrI: 0.39–0.76. This indicates that the interventions studied in this meta-analysis have a moderate-sized overall effect.

Because this is a Bayesian model, we do not find any p-values here. But our example should underline that we can also make reasonable inferences without having to resort to classical significance testing. A great thing we can do in Bayesian, but not in frequentist meta-analysis, is to model the parameters we want to estimate *probabilistically*. The Bayesian model not only estimates the parameters of interest but a whole posterior distribution for τ^2 and μ, which we can access quite easily. We only have to use the `posterior_samples` function.

```
post.samples <- posterior_samples(m.brm, c("^b", "^sd"))
names(post.samples)
```

```
## [1] "b_Intercept"          "sd_Author__Intercept"
```

The resulting data frame contains two columns: `b_Intercept`, the posterior sample data for the pooled effect size, and `sd_Author_Intercept`, the one for the between-study heterogeneity τ. We rename the columns `smd` and `tau` to make the name more informative.

```
names(post.samples) <- c("smd", "tau")
```

Using the data in `post.samples`, we can now generate a *density plot* of the posterior distributions. We use the *{ggplot2}* package for plotting.

```
ggplot(aes(x = smd), data = post.samples) +
  geom_density(fill = "lightblue",                # set the color
               color = "lightblue", alpha = 0.7) +
  geom_point(y = 0,                               # add point at mean
             x = mean(post.samples$smd)) +
  labs(x = expression(italic(SMD)),
       y = element_blank()) +
  theme_minimal()

ggplot(aes(x = tau), data = post.samples) +
  geom_density(fill = "lightgreen",               # set the color
               color = "lightgreen", alpha = 0.7) +
  geom_point(y = 0,
             x = mean(post.samples$tau)) +        # add point at mean
    labs(x = expression(tau),
       y = element_blank()) +
  theme_minimal()
```

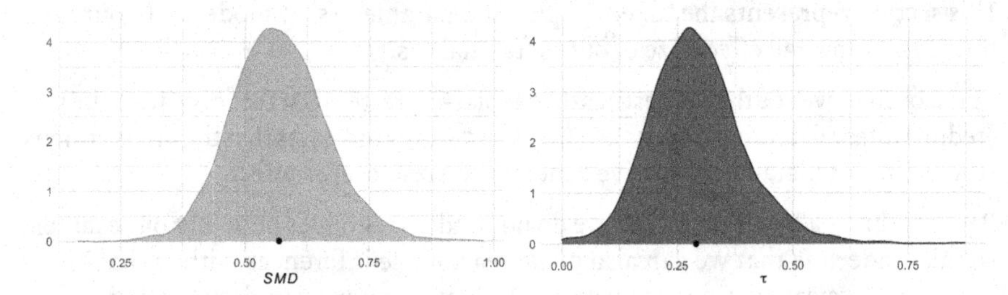

We see that the posterior distributions follow a unimodal, and roughly normal distribution, peaking around the estimated values for μ and τ.

The fact that Bayesian methods create an actual sampling distribution for our parameters of interest means that we can calculate *exact probabilities* that μ or τ is larger or smaller than some specific value. Imagine that found in previous literature that, if effects of an intervention are below SMD = 0.30, they are not meaningful anymore. We could therefore calculate the probability that the true overall effect in our meta-analysis is smaller than SMD = 0.30, based on our model.

This can be done by looking at the *empirical cumulative distribution function* (ECDF). The ECDF lets us select one specific value X, and returns the probability of some value x being smaller than X, based on provided data. The ECDF of μ's posterior distribution in our example can be seen below.

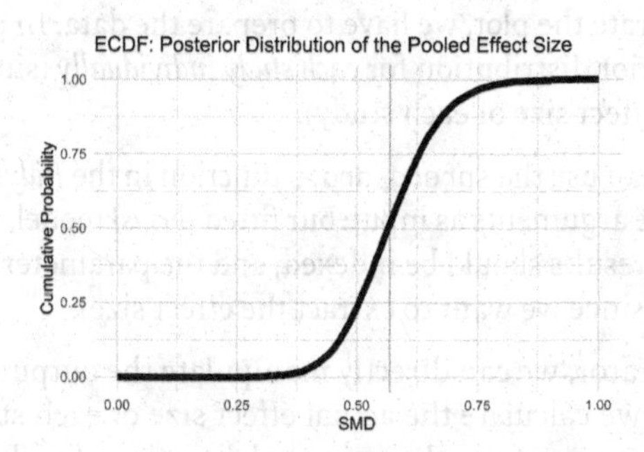

We can use the `ecdf` function to define an ECDF in R, and then check the probability of our pooled effect being smaller than 0.30. The code looks like this:

```
smd.ecdf <- ecdf(post.samples$smd)
smd.ecdf(0.3)
```

```
## [1] 0.002125
```

We see that with 0.21%, the probability of our pooled effect being smaller than 0.30 is very, very low. Assuming the cut-off is valid, this would mean that the overall effect of the intervention we find in this meta-analysis is very likely to be meaningful.

13.3.4 Generating a Forest Plot

As you have seen, Bayesian models allow us to extract their sampled posterior distribution. This can be extremely helpful to directly assess the probability of specific values given our model. We can also exploit this feature to create enhanced *forest plots* (Chapter 6), which are both very informative and pleasing to the eye.

Unfortunately, there is currently no maintained package to directly create forest plots from *{brms}* models. But it is possible to build them oneself using functions of the *{tidybayes}* package (Kay, 2020). So, let us first load this package along with a few other ones before we proceed.

```
library(tidybayes)
library(dplyr)
library(ggplot2)
library(ggridges)
library(glue)
```

```
library(stringr)
library(forcats)
```

Before we can generate the plot, we have to prepare the data. In particular, we need to extract the posterior distribution for *each study individually* (since forest plots also depict the specific effect size of each study).

To achieve this, we can use the `spread_draws` function in the *{tidybayes}* package. The function needs three arguments as input: our fitted *{brms}* model, the random-effects factor by which the results should be indexed, and the parameter we want to extract (here `b_Intercept`, since we want to extract the effect size).

Using the pipe operator, we can directly manipulate the output. Using the `mutate` function in *{dplyr}*, we calculate the actual effect size of each study by adding the pooled effect size `b_Intercept` to the estimated deviation of each study. We save the result as `study.draws`.

```
study.draws <- spread_draws(m.brm, r_Author[Author,], b_Intercept) %>%
  mutate(b_Intercept = r_Author + b_Intercept)
```

Next, we want to generate the distribution of the pooled effect in a similar way (since in forest plots, the summary effect is usually displayed in the last row). We therefore slightly adapt the code from before, dropping the second argument to only get the pooled effect.

The call to `mutate` only adds an extra column called `"Author"`. We save the result as `pooled.effect.draws`.

```
pooled.effect.draws <- spread_draws(m.brm, b_Intercept) %>%
  mutate(Author = "Pooled Effect")
```

Next, we bind `study.draws` and `pooled.effect.draws` together in one data frame. We then start a pipe again, calling `ungroup` first, and then use `mutate` to (1) clean the study labels (i.e. replace dots with spaces) and (2) reorder the study factor levels by effect size (high to low). The result is the data we need for plotting, which we save as `forest.data`.

```
forest.data <- bind_rows(study.draws,
                         pooled.effect.draws) %>%
   ungroup() %>%
   mutate(Author = str_replace_all(Author, "[.]", " ")) %>%
   mutate(Author = reorder(Author, b_Intercept))
```

Lastly, the forest plot should also display the effect size (SMD and credible interval) of each study. To do this, we use our newly generated `forest.data` data set, group it by Author, and then use the mean_qi function to calculate these values. We save the output as forest.data.summary.

```
forest.data.summary <- group_by(forest.data, Author) %>%
  mean_qi(b_Intercept)
```

We are now ready to generate the forest plot using the *{ggplot2}* package. The code to generate the plot looks like this:

```
ggplot(aes(b_Intercept,
           relevel(Author, "Pooled Effect",
                   after = Inf)),
       data = forest.data) +

  # Add vertical lines for pooled effect and CI
  geom_vline(xintercept = fixef(m.brm)[1, 1],
             color = "grey", size = 1) +
  geom_vline(xintercept = fixef(m.brm)[1, 3:4],
             color = "grey", linetype = 2) +
  geom_vline(xintercept = 0, color = "black",
             size = 1) +

  # Add densities
  geom_density_ridges(fill = "blue",
                      rel_min_height = 0.01,
                      col = NA, scale = 1,
                      alpha = 0.8) +
  geom_pointintervalh(data = forest.data.summary,
                      size = 1) +

  # Add text and labels
  geom_text(data = mutate_if(forest.data.summary,
                             is.numeric, round, 2),
    aes(label = glue("{b_Intercept} [{.lower}, {.upper}]"),
        x = Inf), hjust = "inward") +
  labs(x = "Standardized Mean Difference", # summary measure
       y = element_blank()) +
  theme_minimal()
```

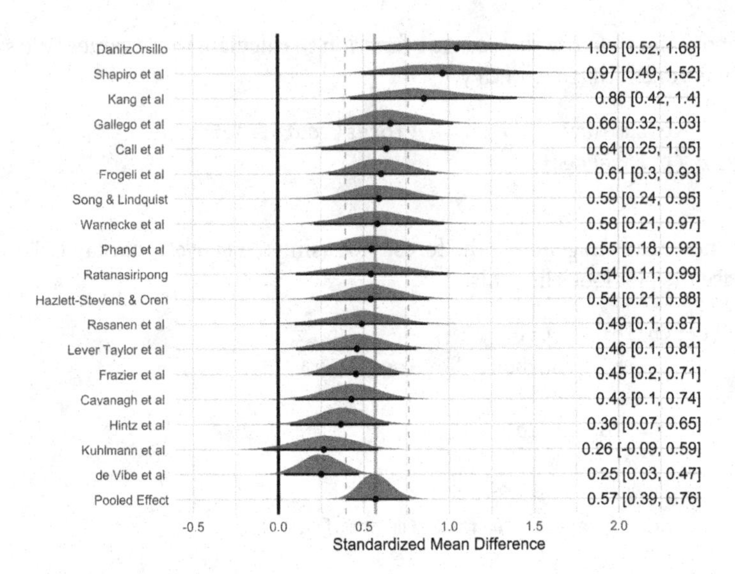

!

Observed Versus Model-Based Effect Sizes

One thing is very important to mention here. The effect sizes displayed in the forest plot do not represent the *observed* effect sizes of the original studies, but the estimate of the "true" effect size (θ_k) of a study based on the Bayesian model. The dots shown in the forest plot are equivalent to the study-wise estimates we saw when extracting the random effects using `ranef` (except that these values were centered around the pooled effect).

□

13.4 Questions & Answers

Test your knowledge!

1. What are the differences and similarities between the "conventional" random-effects model and a Bayesian hierarchical model?
2. Name three advantages of Bayesian meta-analyses compared to their frequentist counterpart.
3. Explain the difference between a weakly informative and uninformative prior.
4. What is a Half-Cauchy distribution, and why is it useful for Bayesian meta-analysis?
5. What is an ECDF, and how can it be used in Bayesian meta-analyses?

Answers to these questions are listed in Appendix A at the end of this book.

13.5 Summary

- While meta-analysis is usually conducted using frequentist statistics, it is also possible to conduct Bayesian meta-analyses.
- Bayesian meta-analysis is based on the Bayesian hierarchical model. The core tenets of this model are identical to the "conventional" random-effects model. The difference, however, is that (informative, weakly informative or uninformative) prior distributions are assumed for μ and τ^2.
- For Bayesian meta-analysis models, it is usually a good idea to assume *weakly informative* priors. Weakly informative priors are used to represent a *weak* belief that some values of more credible than others.
- To specify the prior distribution for the between-study heterogeneity variance τ^2, the a Half-Cauchy distribution can be used. Half-Cauchy distributions are particularly suited for this task because they are only defined for positive values

and possess heavier tails. This can be used to represent that very high values of τ^2 are less likely, but still very much possible.

- When fitting Bayesian meta-analysis models, it is important to (1) always check if the model included enough iterations to converge (for example, by checking the $\hat{R}$ values) and to (2) conduct sensitivity analyses with different prior specifications to evaluate the impact on the results.

Part IV

Helpful Tools

14

Power Analysis

One of the reasons why meta-analysis can be so helpful is because it allows us to combine several *imprecise* findings into a more *precise* one. In most cases, meta-analyses produce estimates with narrower confidence intervals than any of the included studies. This is particularly useful when the true effect is small. While primary studies may not be able to ascertain the significance of a small effect, meta-analytic estimates can often provide the statistical power needed to verify that such a small effect exists.

Lack of statistical power, however, may still play an important role–even in meta-analysis. The number of included studies in many meta-analyses is small, often below $K = 10$. The median number of studies in Cochrane systematic reviews, for example, is six (Borenstein et al., 2011). This becomes even more problematic if we factor in that meta-analyses often include subgroup analyses and meta-regression, for which even more power is required. Furthermore, many meta-analyses show high between-study heterogeneity. This also reduces the overall precision and thus the statistical power.

We already touched on the concept of statistical power in Chapter 9.2.2.2, where we learned about the p-curve method. The idea behind statistical power is derived from classical hypothesis testing. It is directly related to the two types of *errors* that can occur in a hypothesis test. The first error is to accept the alternative hypothesis (e.g. $\mu_1 \neq \mu_2$) while the null hypothesis ($\mu_1 = \mu_2$) is true. This leads to a *false positive*, also known as a *Type I* or α error. Conversely, it is also possible that we accept the null hypothesis, while the alternative hypothesis is true. This generates a *false negative*, known as a *Type II* or β error.

The power of a test directly depends on β. It is defined as Power = $1-\beta$. Suppose that our null hypothesis states that there is no difference between the means of two groups, while the alternative hypothesis postulates that a difference (i.e. an "effect") exists. The statistical power can be defined as the probability that the test will detect an effect (i.e. a mean difference), *if* it exists:

$$\text{Power} = P(\text{reject H}_0 \mid \mu_1 \neq \mu_2) = 1 - \beta \tag{14.1}$$

It is common practice to assume that a type I error is more grave than a type II error. Thus, the α level is conventionally set to 0.05 and the β level to 0.2. This leads to a threshold of $1-\beta$ = 1 - 0.2 = 80%, which is typically used to determine if the statistical power of a test is adequate or not. When researchers plan a new study, they usually select a sample size that guarantees a power of 80%. It is easier to obtain statistically significant results when the true effect is large. Therefore, when the power is fixed at

DOI: 10.1201/9781003107347-14

80%, the required sample size only depends on the size of the true effect. The smaller the assumed effect, the larger the sample size needed to ascertain 80% power.

Researchers who conduct a primary study can plan the size of their sample *a priori*, based on the effect size they expect to find. The situation is different in meta-analysis, where we can only work with the published material. However, we have some control over the number and type of studies we include in our meta-analysis (e.g. by defining more lenient or strict inclusion criteria). This way, we can also adjust the overall power. There are several factors that can influence the statistical power in meta-analyses:

- The total *number* of included or eligible studies, and their *sample size*. How many studies do we expect, and are they rather small or large?
- The effect size we want to find. This is particularly important, as we have to make assumptions about how big an effect size has to be to still be meaningful. For example, one study calculated that for depression interventions, even effects as small as SMD = 0.24 may still be meaningful to patients (Cuijpers et al., 2014). If we want to study negative effects of an intervention (e.g. death or symptom deterioration), even very small effect sizes are extremely important and should be detected.
- The expected between-study heterogeneity. Substantial heterogeneity affects the precision of our meta-analytic estimates, and thus our potential to find significant effects.

Besides that, it is also important to think about other analyses, such as subgroup analyses, that we want to conduct. How many studies are there for each subgroup? What effects do we want to find in each group?

Post-Hoc Power Tests: "The Abuse of Power"

Please note that power analyses should always be conducted *a priori*, meaning *before* you perform the meta-analysis.

Power analyses conducted *after* an analysis ("post-hoc") are based on a logic that is deeply flawed (Hoenig and Heisey, 2001). First, post hoc power analyses are *uniformative*–they tell us nothing that we do not already know. When we find that an effect is not significant based on our collected sample, the calculated post-hoc power will be, by definition, insufficient (i.e. 50% or lower). When we calculate the post hoc power of a test, we simply "play around" with a power function that is directly linked to the *p*-value of the result. There is nothing in the post hoc power estimate that the *p*-value would not already tell us. Namely that, based on the effect and sample size of our test, the power is insufficient to ascertain statistical significance.

When we interpret the post hoc power, this can also lead to the *power approach paradox* (PAP). This paradox arises because an analysis yielding no significant effect is thought to show *more* evidence that the null hypothesis is true when the p-value is *smaller*, since then, the power to detect a true effect would be *higher*.

14.1 Fixed-Effect Model

To determine the power of a meta-analysis under the fixed-effect model, we have to specify a distribution which represents that our alternative hypothesis is correct. To do this, however, it is not sufficient to simply say that $\theta \neq 0$ (i.e. that *some* effect exists). We have to assume a *specific* true effect that we want to be able to detect with sufficient (80%) power. For example SMD = 0.29.

We already covered previously (see Chapter 8.1.2) that dividing an effect size through its standard error creates a z score. These z scores follow a standard normal distribution, where a value of $|z| \geq 1.96$ means that the effect is significantly different from zero ($p < 0.05$). This is exactly what we want to achieve in our meta-analysis: no matter how large the exact effect size and standard error of our result, the value of $|z|$ should be at least 1.96, and thus statistically significant:

$$z = \frac{\theta}{\sigma_\theta} \quad \text{where} \quad |z| \geq 1.96. \tag{14.2}$$

The value of σ_θ, the standard error of the pooled effect size, can be calculated using this formula:

$$\sigma_\theta = \sqrt{\frac{\left(\frac{n_1+n_2}{n_1 n_2}\right) + \left(\frac{\theta^2}{2(n_1+n_2)}\right)}{K}} \tag{14.3}$$

Where n_1 and n_2 stand for the sample sizes in group 1 and group 2 of a study, where θ is the assumed effect size (expressed as a standardized mean difference), and K is the total number of studies in our meta-analysis. Importantly, as a simplification, this formula assumes that the sample sizes in both groups are identical across all included studies.

The formula is very similar to the one used to calculate the standard error of a standardized mean difference, with one exception. We now divide the the standard error by K. This means that the standard error of our pooled effect is reduced by factor K, representing the total number of studies in our meta-analysis. Put differently, when assuming a fixed-effect model, pooling the studies leads to a K-fold increase in the precision of our overall effect.

After we defined θ and calculated σ_θ, we end up with a value of z. This z score can be used to obtain the power of our meta-analysis, given a number of studies K with group sizes n_1 and n_2:

$$\begin{aligned} \text{Power} &= 1 - \beta \\ &= 1 - \Phi(c_\alpha - z) + \Phi(-c_\alpha - z) \\ &= 1 - \Phi(1.96 - z) + \Phi(-1.96 - z). \end{aligned} \tag{14.4}$$

Where c_α is the critical value of the standard normal distribution, given a specified α level. The Φ symbol represents the *cumulative distribution function* (CDF) of a standard normal distribution, $\Phi(z)$. In R, the CDF of the standard normal distribution is implemented in the `pnorm` function.

We can now use these formulas to calculate the power of a fixed-effect meta-analysis. Imagine that we expect $K=10$ studies, each with approximately 25 participants in both groups. We want to be able to detect an effect of SMD = 0.2. What power does such a meta-analysis have?

```
# Define assumptions
theta <- 0.2
K <- 10
n1 <- 25
n2 <- 25

# Calculate pooled effect standard error
sigma <- sqrt(((n1+n2)/(n1*n2)+(theta^2/(2*n1+n2)))/K)

# Calculate z
z = theta/sigma

# Calculate the power
1 - pnorm(1.96-z) + pnorm(-1.96-z)
```

```
## [1] 0.6059
```

We see that, with 60.6%, such a meta-analysis would be *underpowered*, even though 10 studies were included. A more convenient way to calculate the power of a (fixed-effect) meta-analysis is to use the `power.analysis` function.

The "power.analysis" Function

The `power.analysis` function is included in the *{dmetar}* package. Once *{dmetar}* is installed and loaded on your computer, the function is ready to be used. If you did not install *{dmetar}*, follow these instructions:

1. Access the source code of the function online[a].
2. Let R "learn" the function by copying and pasting the source code in its entirety into the console (bottom left pane of R Studio), and then hit "Enter".
3. Make sure that the *{ggplot2}* package is installed and loaded.

[a]`https://raw.githubusercontent.com/MathiasHarrer/dmetar/master/R/power.analysis.R`

The `power.analysis` function contains these arguments:

- `d`. The hypothesized, or plausible overall effect size, expressed as the standardized mean difference (SMD). Effect sizes must be positive numeric values.
- `OR`. The assumed effect of a treatment or intervention compared to control, expressed as an odds ratio (OR). If both `d` and `OR` are specified, results will only be computed for the value of `d`.
- `k`. The expected number of studies included in the meta-analysis.
- `n1`, `n2`. The expected average sample size in group 1 and group 2 of the included studies.
- `p`. The alpha level to be used. Default is α=0.05.
- `heterogeneity`. The level of between-study heterogeneity. Can either be `"fixed"` for no heterogeneity (fixed-effect model), `"low"` for low heterogeneity, `"moderate"` for moderate-sized heterogeneity, or `"high"` for high levels of heterogeneity. Default is `"fixed"`.

Let us try out this function, using the same input as in the example from before.

```
library(dmetar)
power.analysis(d = 0.2,
               k = 10,
               n1 = 25,
               n2 = 25,
               p = 0.05)
```

```
## Fixed-effect model used.
## Power: 60.66%
```

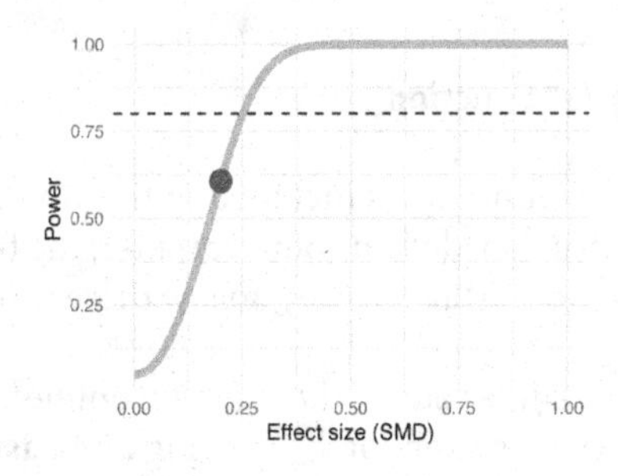

14.2 Random-Effects Model

For power analyses assuming a random-effects model, we have to take the between-study heterogeneity variance τ^2 into account. Therefore, we need to calculate an adapted version of the standard error, σ^*_θ:

$$\sigma^*_\theta = \sqrt{\frac{\left(\frac{n_1+n_2}{n_1 n_2}\right) + \left(\frac{\theta^2}{2(n_1+n_2)}\right) + \tau^2}{K}} \tag{14.5}$$

The problem is that the value of τ^2 is usually not known before seeing the data. Hedges and Pigott (2001), however, provide guidelines that may be used to model either low, moderate or large between-study heterogeneity:

Low heterogeneity:

$$\sigma^*_\theta = \sqrt{1.33 \times \frac{\sigma^2_\theta}{K}} \tag{14.6}$$

Moderate heterogeneity:

$$\sigma^*_\theta = \sqrt{1.67 \times \frac{\sigma^2_\theta}{K}} \tag{14.7}$$

Large heterogeneity:

$$\sigma^*_\theta = \sqrt{2 \times \frac{\sigma^2_\theta}{K}} \tag{14.8}$$

The `power.analysis` function can also be used for random-effects meta-analyses. The amount of assumed between-study heterogeneity can be controlled using the `heterogeneity` argument. Possible values are `"low"`, `"moderate"`, and `"high"`. Using the same values as in the previous example, let us now calculate the expected power when the between-study heterogeneity is moderate.

```
power.analysis(d = 0.2,
               k = 10,
               n1 = 25,
               n2 = 25,
               p = 0.05,
               heterogeneity = "moderate")
```

```
## Random-effects model used (moderate heterogeneity assumed).
## Power: 40.76%
```

We see that the estimated power is 40.76%. This is lower than the normative 80% threshold. It is also lower than the 60.66% we obtain we assuming a fixed-effect model. This is because between-study heterogeneity decreases the precision of our pooled effect estimate, resulting in a drop in statistical power.

Figure 14.1 visualizes the effect of the true effect size, number of studies, and amount of between-study heterogeneity on the power of a meta-analysis.[1]

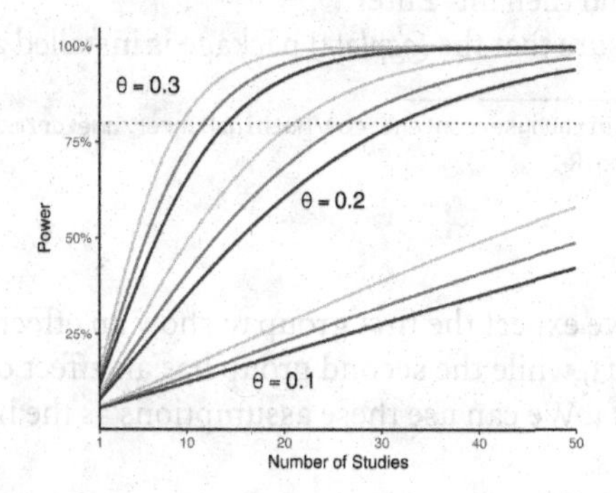

FIGURE 14.1: Power of random-effects meta-analyses (n=50 in each study). Darker colors indicate higher between-study heterogeneity.

[1] If you want to quickly check the power of a meta-analysis under varying assumptions, you can also use a *power calculator tool* developed for this purpose. The tool is based on the same R function that we cover in this chapter. It can be found online: `https://mathiasharrer.shinyapps.io/power_calculator_meta_analysis/`.

14.3 Subgroup Analyses

When planning subgroup analyses, it can be relevant to know how large the difference between two groups must be so that we can detect it, given the number of studies at our disposal. This is where a power analysis for subgroup differences can be applied. A subgroup power analysis can be conducted in R using the `power.analysis.subgroup` function, which implements an approach described by Hedges and Pigott (2004).

The "power.analysis.subgroup" Function

The `power.analysis.subgroup` function is included in the *{dmetar}* package. Once *{dmetar}* is installed and loaded on your computer, the function is ready to be used. If you did not install *{dmetar}*, follow these instructions:

1. Access the source code of the function online[a].
2. Let R "learn" the function by copying and pasting the source code in its entirety into the console (bottom left pane of R Studio), and then hit "Enter".
3. Make sure that the *{ggplot2}* package is installed and loaded.

[a] https://raw.githubusercontent.com/MathiasHarrer/dmetar/master/R/power.analysis.subgroup.R

Let us assume that we expect the first group to show an effect of SMD = 0.3 with a standard error of 0.13, while the second group has an effect of SMD = 0.66, and a standard error of 0.14. We can use these assumptions as the input to our function call:

```
power.analysis.subgroup(TE1 = 0.30, TE2 = 0.66,
                        seTE1 = 0.13, seTE2 = 0.14)
```

```
## Minimum effect size difference needed for sufficient
## power: 0.536 (input: 0.36)
## Power for subgroup difference test (two-tailed): 46.99%
```

In the output, we can see that the power of such a subgroup test (47%) would not be sufficient. The output also tells us that, all else being equal, the effect size difference needs to be at least 0.54 in order to reach sufficient power.

□

15

Risk of Bias Plots

When we conduct a meta-analysis, it is advisable to also examine the risk of bias of included studies (see Chapter 1.4.5). A frequently used method to do this are *domain-based* assessment tools, especially the *Cochrane risk of bias tool* for randomized controlled trials (Chapter 1.2).

Risk of bias results are typically reported in a *graphical format*, such as a summary figure displaying the proportion of low and high risk of bias studies on the assessed risk of bias domains. Another common type of visualization are *traffic light plots*, in which the risk of bias assessment is presented for each study individually.

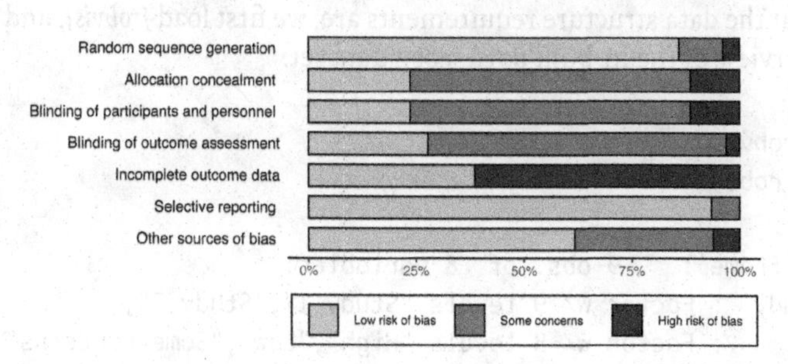

FIGURE 15.1: A risk of bias summary plot.

For a long time, the only program with an in-built functionality to create risk of bias plots that conform to the Cochrane guidelines was *RevMan*. This has changed with the release of the *{robvis}* package (McGuinness and Higgins, 2020; McGuinness, 2019), which allows to produce high-quality risk of bias plots directly in R.

In this chapter, we cover the core functionality of the *{robvis}* package, and how it can be used to produce plots for different types of domain-based risk of bias assessments.

DOI: 10.1201/9781003107347-15

15.1 Data Preparation

The structure of our risk of bias data set depends on the domain-based assessment tool we used. The *{robvis}* package includes templates for three different types of assessment methods[1]:

- *Revised Cochrane Risk of Bias Tool for Randomized Studies template* (RoB 2, Sterne et al., 2019). Using this tool, studies are evaluated on five domains: (D1) bias due to the randomization process, (D2) bias due to deviation from the intended intervention, (D3) bias due to missing data, (D4) bias in the measurement of the outcome, (D5) bias in the selection of the reported result. Based on these assessments, each study is encoded as showing either a low risk of bias, high risk of bias, or "some concerns". The manual for this assessment tool can be found online[2].
- *Risk of Bias In Non-randomized Studies–of Interventions tool template* (ROBINS-I, Sterne et al., 2016). This tool uses seven domains: (D1) confounding, (D2) participant selection, (D3) intervention classification, (D4) deviations from the intended intervention, (D5) missing data, (D6) outcome measurement bias, and (D7) selective reporting. Based on this information, studies can either have a low, moderate, serious or critical risk of bias. The ROBINS-I manual can also be found online[3].
- *"Generic" domain-based assessment template.* This template can be used for assessments based on the first version of the Cochrane risk of bias tool for randomized studies, or when different domains were specified.

To see what the data structure requirements are, we first load *{robvis}*, and then print out an overview of the in-built `data_rob2` data set.

```
library(robvis)
str(data_rob2)
```

```
## 'data.frame':	9 obs. of  8 variables:
##  $ Study  : Factor w/ 9 levels "Study 1","Study 2",...
##  $ D1     : Factor w/ 3 levels "High","Low","Some concerns",...
##  $ D2     : Factor w/ 3 levels "High","Low","Some concerns",...
##  $ D3     : Factor w/ 3 levels "High","Low","Some concerns",...
##  $ D4     : Factor w/ 2 levels "High","Low",...
##  $ D5     : Factor w/ 2 levels "Low","Some concerns",...
##  $ Overall: Factor w/ 3 levels "High","Low","Some concerns",...
##  $ Weight : num  33.333 33.333 0.143 9.091 12.5 ...
```

[1]There is also a fourth template for the *Quality Assessment of Diagnostic Test Accuracy Studies* (QUADAS-2, Whiting et al., 2011), which we do not cover here.

[2]https://www.riskofbias.info/welcome/rob-2-0-tool

[3]https://www.riskofbias.info/welcome/home

The data_rob2 data set contains risk of bias judgments based on the RoB 2 tool, and has 8 columns. The first column contains the study label, followed by five columns in which the risk of bias assessment on the five domains is stored. The name of these columns is not important–they just need to be in the same order as mentioned above.

To use the RoB 2 template, all risk assessment labels need to be identical to the one in this example: "High", "Low", or "Some concerns". The seventh column contains the overall risk of bias assessment for the study. The final column includes the fixed- or random-effects *weight* of each study (see Chapter 4.1.2). The Weight column is used to adjust the summary risk of bias plot based on the impact of a study on the pooled result. In *{meta}* meta-analysis objects, the study weights based on the fixed-effect and random-effects model are stored in w.fixed and w.random, respectively.

Now, let us have a look at the required data structure for the ROBINS-I tool, by printing the data_robins data set in *{robvis}*.

```
str(data_robins)
```

```
## 'data.frame':	12 obs. of  10 variables:
##  $ Study  : Factor w/ 12 levels "Study 1","Study 10",..
##  $ D1     : Factor w/ 4 levels "Critical","Low",..
##  $ D2     : Factor w/ 4 levels "Critical","Low",..
##  $ D3     : Factor w/ 4 levels "Critical","Low",..
##  $ D4     : Factor w/ 3 levels "Critical","Low",..
##  $ D5     : Factor w/ 4 levels "Critical","Low",..
##  $ D6     : Factor w/ 4 levels "Critical","Low",..
##  $ D7     : Factor w/ 4 levels "Critical","Low",..
##  $ Overall: Factor w/ 4 levels "Critical","Low",..
##  $ Weight : num  33.333 33.333 0.143 9.091 12.5 ...
```

The data structure looks very similar to the RoB 2 example, with a few subtle differences. Because the ROBINS-I tool contains seven domains, seven domain judgment columns are included (in the same order as mentioned above). Furthermore, the labels used in these columns differ: we have to use "Low", "Moderate", "Serious", or "Critical".

The data_rob1 data set is an example of how the data structure should look like if we want to use the generic risk of bias template.

```
str(data_rob1)
```

```
## 'data.frame':	9 obs. of  10 variables:
## $ Study                  : Factor w/ 9 levels "Study 1","Study 2",...
## $ Random.sequence.ge...: Factor w/ 3 levels "High","Low","Unclear",...
## $ Allocation.conceal...: Factor w/ 3 levels "High","Low","Unclear",...
## $ Blinding.of.partic...: Factor w/ 3 levels "High","Low","Unclear",...
## $ Blinding.of.outco....: Factor w/ 3 levels "High","Low","Unclear",...
```

```
## $ Incomplete.outcome...: Factor w/ 2 levels "High","Low",...
## $ Selective.reporti....: Factor w/ 2 levels "Low","Unclear",...
## $ Other.sources.of.b...: Factor w/ 3 levels "High","Low","Unclear"
## $ Overall              : Factor w/ 3 levels "High","Low","Unclear"
## $ Weight               : num  33.333 33.333 0.143 9.091 12.5 ...
```

For this template, the exact number and order of the domain-specific risk of bias assessments is less important. However, we now need to specify the name of the risk of bias criteria that we have chosen for our meta-analysis. The generic template can only be used when the risk of bias labels are either "Low", "Unclear", or "High".

15.2 Summary Plots

As soon as the risk of bias data has been prepared and imported, generating a plot becomes fairly straightforward. To produce a summary risk of bias plot, we can use the rob_summary function in *{robvis}*. Summary plots are a good way to visualize the overall risk of bias on certain important domains. A disadvantage is that they do not show the risk of bias of individual studies in our meta-analysis.

In the rob_summary function, we only need to specify two arguments. First, we have to provide our prepared data set. In the second argument, we specify the template we want to use. Available options are "ROB2" (for the RoB 2 tool), "ROBINS-I" (for the ROBINS-I tool), or "ROB1" (for the generic template). As an example, let us produce a summary plot for data_rob1 and data_rob2.

```
# ROB1 (generic template with user-defined domains)
rob_summary(data_rob1, "ROB1")

# ROB2 (revised risk of bias tool)
rob_summary(data_rob2, "ROB2")
```

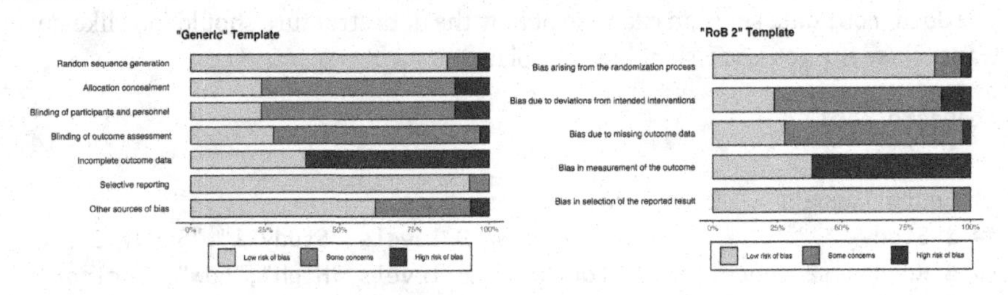

15.3 Traffic Light Plots

The *{robvis}* package also allows us to produce a traffic light plot. This type of plot is essentially a table displaying the risk of bias judgments separately for each individual study. The `rob_traffic_light` function can be used in the same way as `rob_summary`, with the first argument reserved for the data set, and the second one specifying the template. Let us again produce a plot for `data_rob1` and `data_rob2`.

```
# ROB1 (generic template with user-defined domains)
rob_traffic_light(data_rob1, "ROB1")

# ROB2 (revised risk of bias tool)
rob_traffic_light(data_rob2, "ROB2")
```

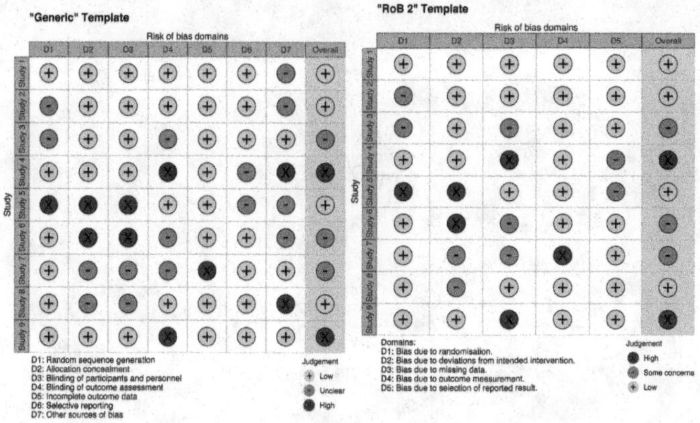

The *{robvis}* Web App

The functionality of *{robvis}* has also been made available as a web application[a]. The App comes with a graphical interface, which can be used to create, style and download risk of bias plots based on uploaded data. It also provides more information concerning the allowed data structures for different templates.

[a] `https://mcguinlu.shinyapps.io/robvis`

16

Reporting & Reproducibility

In the previous chapters, we discussed various techniques, approaches, and strategies we can use to conduct a meta-analysis in R. However, running the statistical analyses only makes up a small proportion of the entire meta-analysis "process" in practice. "In the wild", it is common that:

- We find an error in our R code, and therefore have to redo parts of the analysis with a few changes.
- Collaborators or reviewers suggest using a different approach or model, or performing an additional sensitivity analysis.
- We need to delegate some parts of the analysis to one of our collaborators and have to send her the current status of our work.
- We had to stop working on our project for some time, which means that we have forgotten many things by the time we resume working on it.
- We want to share results of our analysis with project collaborators but they do not know R and do not have R Studio installed.

These are just a few scenarios, but they illustrate that a *reproducible workflow* when conducting meta-analyses in R is beneficial to you and the people you work with. Aiming for reproducibility is also a cornerstone of *open science* practices. Fully reproducible meta-analyses make it as transparent as possible to others how we ended up with our results.

R Studio is an optimal tool to create a reproducible workflow and to facilitate cooperation. In this chapter, we introduce three tools to reproduce, report and disseminate our analyses: R Projects, *R Markdown*, and the *Open Science Framework*.

16.1 Using R Projects

A good way to start with your analysis is to first set up an R *project* in R Studio. R projects create a new environment in a folder on our computer. In this folder, all the data and R code you need for your analyses is saved. Conducting analyses in an R project means that all objects we create are temporarily saved in the project

DOI: 10.1201/9781003107347-16

environment and will be accessible the next time we reopen it. To create a new R project, we can click on the *R project* field in the upper right corner of the R Studio window, and then on *New Project...* in the drop-down menu.

Then we create a *New Directory*, a new folder on our computer, which will become the working directory of the project.

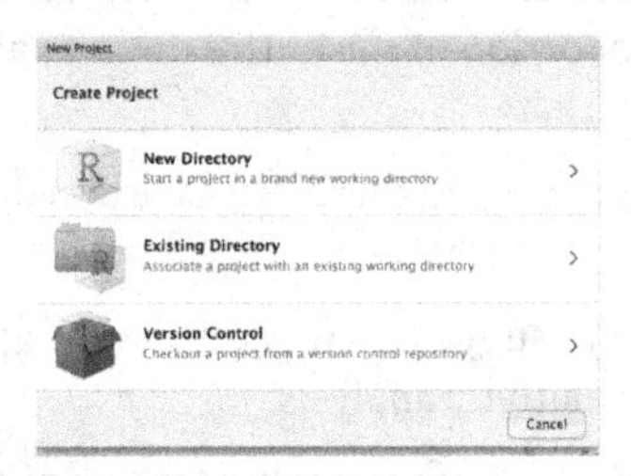

Then, we click on *New Project.*

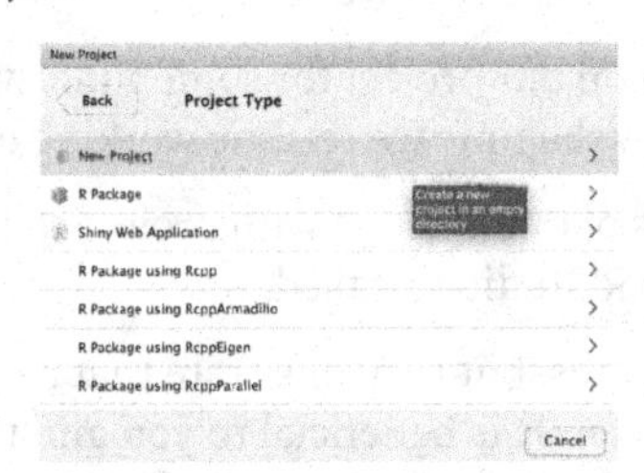

We give our new project the name "Meta-Analysis Project". The project folder will be stored in *~Documents/R.*

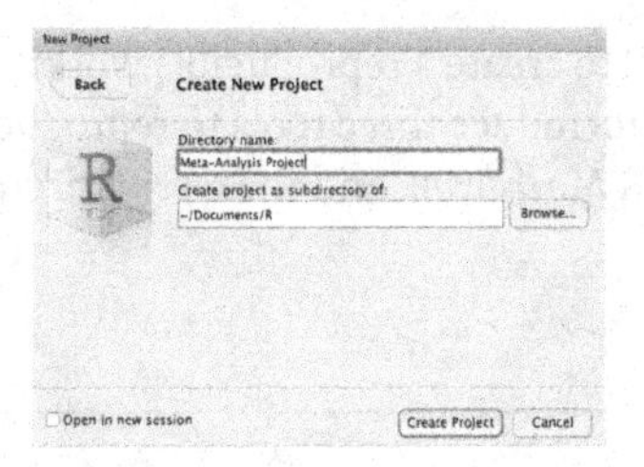

After clicking on *Create Project,* the R project is set. A great feature of R projects is that we do not have to use *absolute paths* to the files we want to reference. We only use the file name, or, if the file is in a (sub-)folder, the folder and file name. Suppose that we stored our data set *data.xlsx* in the sub-folder "data". Using the *{openxlsx}* package (Chapter 2.4), we can import the data set with a relative path.

```
read_excel("data/data.xlsx")
```

16.2 Writing Reproducible Reports with R Markdown

Markdown is a simple markup language for text formatting. *R Markdown* (Xie et al., 2018) is an extension of Markdown and makes it easy to combine plain text, R code, and R output in one document. This makes R Markdown an extremely helpful reporting tool. Using R Markdown, we can create HTML or PDF files containing all code used in our analyses, the output produced by the code, and can add detailed information on what we did in each analysis step.

It is very easy to build R Markdown files in R Studio. We only have to click on the white symbol with the green "plus" sign in the top left corner of the R Studio window. Then, in the drop-down menu, we click on *R Markdown...*.

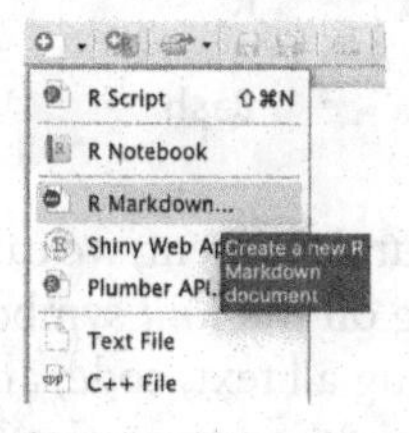

After defining the name of the new R Markdown document, it should pop up in the upper-left corner of the R Studio window.

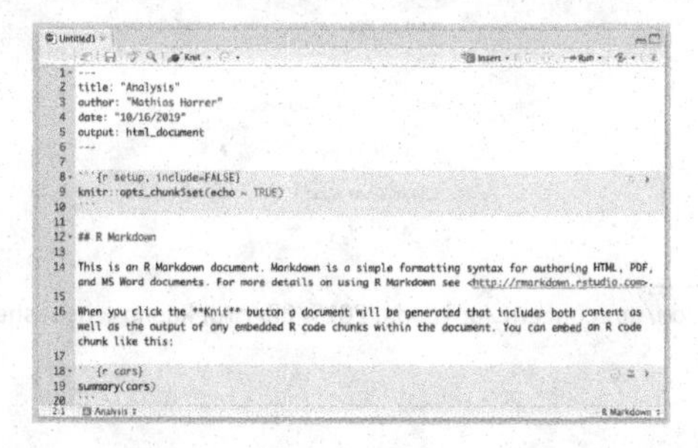

The file already contains some exemplary content, which we can delete, except for the first six lines:

```
---
title: "Analysis"
author: "Author Name"
```

```
date: "10/16/2019"
output: html_document
---
```

This part is the so-called *YAML* header. It controls the title, author, date, and export format of the document. The output format we chose for our document is `html_document`, meaning that the document will be exported as an HTML page once it is rendered.

All R Markdown documents consist of two parts: plain Markdown text, and so-called *R chunks*, shown in grey. We will not go into detail how the text parts in the R Markdown document are formatted, but there is an online cheat sheet[1], which is a great resource to start learning Markdown syntax (this should only take about twenty minutes). The R code chunks, on the other hand, simply contain all the code we would usually type into the console. By clicking on the *Insert* field in the upper right corner of the document, we can add new code chunks. The code can be run by clicking on the little green triangle above each chunk.

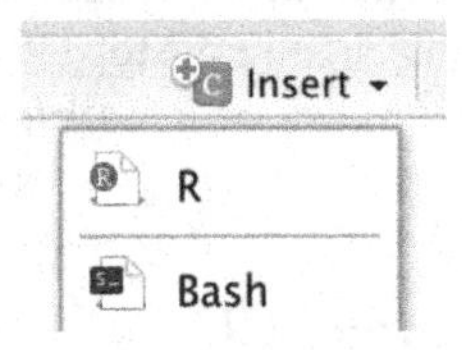

Once we are finished writing our document, we can export it as an HTML, PDF, or MS *Word* document by clicking on the *Knit* symbol in the upper left corner. This renders the document, including all text, code, and output, and exports it in the defined format. The final document is automatically saved in our project folder.

[1] `https://rstudio.com/wp-content/uploads/2015/02/rmarkdown-cheatsheet.pdf`

16.3 OSF Repositories

The *Open Science Framework* (OSF[2]) is an open-source online platform to facilitate collaboration and reproducibility in research. The OSF includes an online *repository*, where researchers deposit their study material to collaborate and make all steps of the research process (more) transparent. The OSF is a spearhead of the open science movement, which has gathered much momentum in the last decade.

It is encouraged that all meta-analysts make their research and analysis process transparent to the public, by providing open access to the collected data and R code used for their analyses. The OSF is a great tool to do this–all repositories you created for yourself are private by default, and it is up to you to decide if, when, and what you want to make public. In the following, we will show you how to set up an OSF repository in R, upload and download files, and how to add collaborators.

16.3.1 Access Token

To start using the OSF, we first have to create a personal account on the website[3]. After the account has been created, we also have to generate an *access token* so that we can manipulate our repository directly using R. To get the access token, we have to navigate to *Profile > Settings > Personal access tokens*. There, we click on *Create token*.

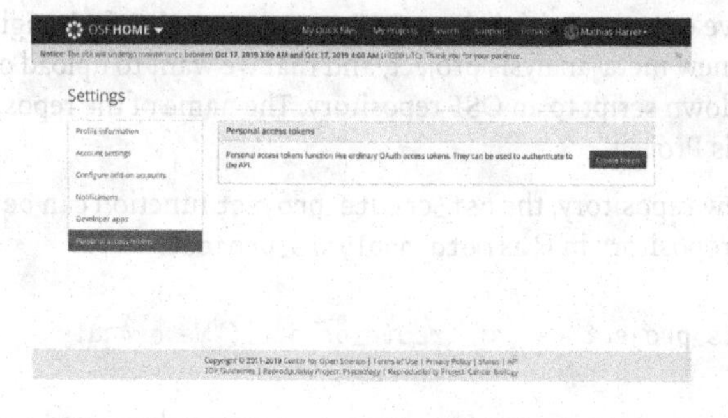

Then, under *Scopes*, we check all boxes, and click on *Create token* again. After that, our personal access token should appear. We copy the token and save it for later.

[2]https://www.osf.io
[3]https://osf.io/register

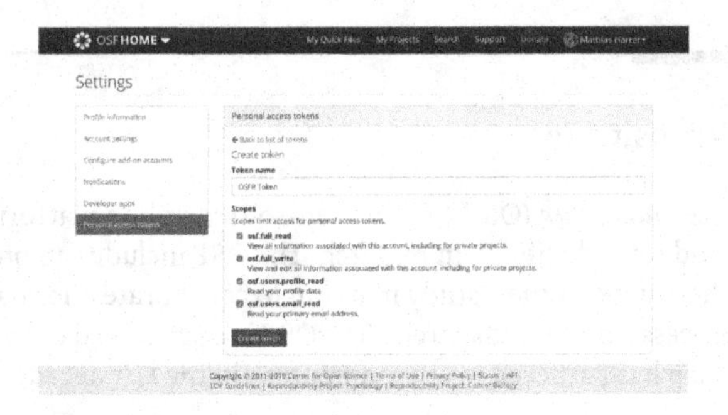

16.3.2 The *{osfr}* Package & Authentication

To access our OSF repository directly via R, we can use the *{osfr}* package (Wolen et al., 2020). Before we can use the functionality of this package, we first have to use our access token to authenticate. To do this, we use the `osf_auth` function, providing it with the access token we just received (the token displayed below is made up):

```
library(osfr)
osf_auth("AtmuMZ3pSuS7tceSMz2NNSAmVDNTzpm2Ud87")
```

16.3.3 Repository Setup

Using *{osfr}*, we can now initialize an OSF repository using R. Imagine that we are working on a new meta-analysis project, and that we want to upload our data as well as an R Markdown script to an OSF repository. The name of the repository should be "Meta-Analysis Project".

To create a new repository, the `osf_create_project` function can be used. We save the new OSF repository in R as `meta_analysis_project`.

```
meta_analysis_project <- osf_create_project("Meta-Analysis Project")
```

Using the `osf_open` function, we can then access the newly created repository online:

```
osf_open(meta_analysis_project)
```

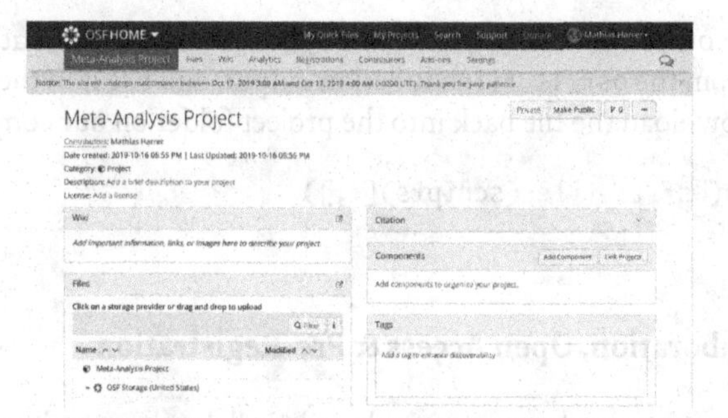

Now that the repository has been created, we can proceed by adding *components* to it. In OSF, components work like folders on a computer. Suppose we want to create two components: one for our data sets, and one for our R Markdown scripts. To do this, we can use the `osf_create_component` function. We have to provide the function with the R repository object (`meta_analysis_project`), and then set the title of the new component.

```
scripts <- osf_create_component(meta_analysis_project,
                                title = "Analysis Scripts")
datasets <- osf_create_component(meta_analysis_project,
                                 title = "Datasets")
```

When we go to the online page of the repository now, we see that the two components have been added.

16.3.4 Upload & Download

To upload data to the OSF repository, we can use the `osf_upload` function. The function requires us to specify the component to which we want to add the file, and the path to the file that should be uploaded. Suppose that we want to upload an R Markdown script called "Analysis.rmd", which is currently saved in our R project sub-folder "scripts". To upload, we can use the following code:

```
osf_upload(scripts, "scripts/Analysis.rmd")
```

To see if the file has been uploaded successfully, we can access contents of the component using the `osf_ls_files` function.

```
osf_ls_files(scripts)
```

```
## # A tibble: 2 x 3
##   name           id                       meta
##   <chr>          <chr>                    <list>
## 1 Analysis.rmd   1db74s7bfcf91f00125675721 <named list [3]>
```

We see in the output that the upload was successful. To download a file, we can select a row from the `osf_ls_files` function output, and use it in the `osf_download` function to download the file back into the project folder on our computer.

```
osf_download(osf_ls_files(scripts)[1,])
```

16.3.5 Collaboration, Open Access & Pre-Registration

On the OSF repository website, we can also add collaborators under the *Contributors* field.

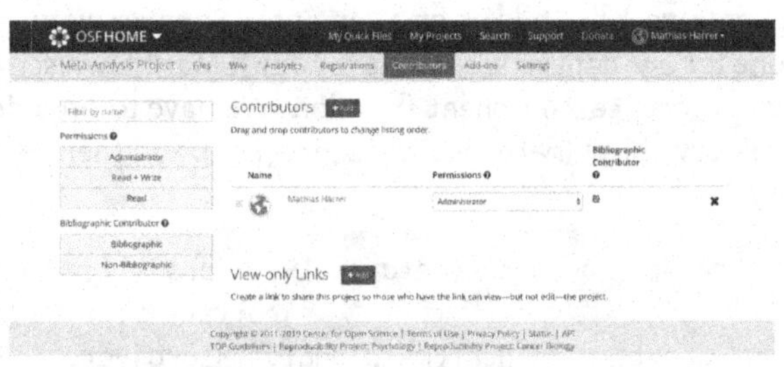

At any time, it is possible to make the repository *public* by clicking on the *Make Public* button in the upper right corner of the website.

In Chapter 1.4.2, we discussed that analysis plans and pre-registration are essential parts of a high-quality meta-analysis. The OSF makes it very convenient to also create an openly accessible pre-registration for our project. We simply have to click on the *Registrations* button on top, and then create a *New registration*. This leads us to the *OSF Registries* website, where we can provide detailed information on our planned study, including our analysis plan.

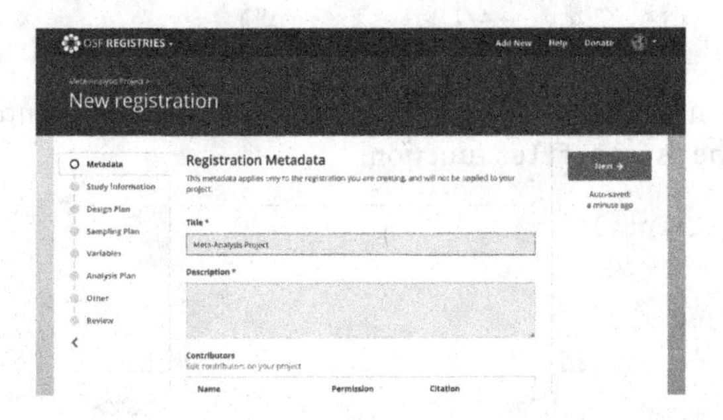

After specifying all the required details, the study can be registered. This creates a register entry that can be accessed through a unique ID (e.g. *osf.io/q2jp7*). After the registration is completed, it is not possible to change the stated search plan, hypotheses and/or analysis strategy anymore.

□

17

Effect Size Calculation & Conversion

A problem meta-analysts frequently face is that suitable "raw" effect size data cannot be extracted from all included studies. Most functions in the *{meta}* package, such as `metacont` (Chapter 4.2.2) or `metabin` (Chapter 4.2.3.1), can only be used when complete raw effect size data is available.

In practice, this often leads to difficulties. Some published articles, particularly older ones, do not report results in a way that allows to extract the needed (raw) effect size data. It is not uncommon to find that a study reports the results of a t-test, one-way ANOVA, or χ^2-test, but not the group-wise mean and standard deviation, or the number of events in the study conditions, that we need for our meta-analysis.

The good news is that we can sometimes *convert* reported information into the desired effect size format. This makes it possible to include affected studies in a meta-analysis with *pre-calculated* data (Chapter 4.2.1) using `metagen`. For example, we can convert the results of a two-sample t-test to a standardized mean difference and its standard error, and then use `metagen` to perform a meta-analysis of pre-calculated SMDs. The *{esc}* package (Lüdecke, 2019) provides several helpful functions which allow us to perform such conversions directly in R.

17.1 Mean & Standard Error

When calculating SMDs or Hedges' g from the mean and *standard error*, we can make use of the fact that the standard deviation of a mean is defined as its standard error, with the square root of the sample size "factored out" (Thalheimer and Cook, 2002):

$$\text{SD} = \text{SE}\sqrt{n} \tag{17.1}$$

We can calculate the SMD or Hedges' g using the `esc_mean_se` function. Here is an example:

```
library(esc)

esc_mean_se(grp1m = 8.5,   # mean of group 1
```

DOI: 10.1201/9781003107347-17

```
           grp1se = 1.5,  # standard error of group 1
           grp1n = 50,    # sample in group 1
           grp2m = 11,    # mean of group 2
           grp2se = 1.8,  # standard error of group 2
           grp2n = 60,    # sample in group 2
           es.type = "d") # convert to SMD; use "g" for Hedges' g
```

```
##
## Effect Size Calculation for Meta Analysis
##
##      Conversion: mean and se to effect size d
##     Effect Size:  -0.2012
##  Standard Error:   0.1920
##        Variance:   0.0369
##        Lower CI:  -0.5774
##        Upper CI:   0.1751
##          Weight:  27.1366
```

17.2 Regression Coefficients

It is possible to calculate SMDs, Hedges' g or a correlation r from standardized or unstandardized regression coefficients (Lipsey and Wilson, 2001). For unstandardized coefficients, we can use the esc_B function in *{esc}*. Here is an example:

```
library(esc)

esc_B(b = 3.3,         # unstandardized regression coefficient
      sdy = 5,         # standard deviation of predicted variable y
      grp1n = 100,     # sample size of the first group
      grp2n = 150,     # sample size of the second group
      es.type = "d") # convert to SMD; use "g" for Hedges' g
```

```
##
## Effect Size Calculation for Meta Analysis
##
##      Conversion: unstandardized regression coefficient to effect size
##     Effect Size:   0.6962
##  Standard Error:   0.1328
##        Variance:   0.0176
##        Lower CI:   0.4359
```

```
##          Upper CI:   0.9565
##            Weight:  56.7018
```

```
esc_B(b = 2.9,        # unstandardized regression coefficient
      sdy = 4,        # standard deviation of the predicted variable y
      grp1n = 50,     # sample size of the first group
      grp2n = 50,     # sample size of the second group
      es.type = "r") # convert to correlation
```

```
## Effect Size Calculation for Meta Analysis
##
##      Conversion: unstandardized regression coefficient
##                  to effect size correlation
##     Effect Size:   0.3611
##  Standard Error:   0.1031
##        Variance:   0.0106
##        Lower CI:   0.1743
##        Upper CI:   0.5229
##          Weight:  94.0238
##      Fisher's z:   0.3782
##       Lower CIz:   0.1761
##       Upper CIz:   0.5803
```

Standardized regression coefficients can be transformed using esc_beta.

```
esc_beta(beta = 0.32,   # standardized regression coefficient
         sdy = 5,       # standard deviation of the predicted variable y
         grp1n = 100,   # sample size of the first group
         grp2n = 150,   # sample size of the second group
         es.type = "d") # convert to SMD; use "g" for Hedges' g
```

```
##
## Effect Size Calculation for Meta Analysis
##
##      Conversion: standardized regression coefficient to effect size d
##     Effect Size:   0.6867
##  Standard Error:   0.1327
##        Variance:   0.0176
##        Lower CI:   0.4266
##        Upper CI:   0.9468
##          Weight:  56.7867
```

```
esc_beta(beta = 0.37,   # standardized regression coefficient
         sdy = 4,       # standard deviation of predicted variable y
```

```
         grp1n = 50,    # sample size of the first group
         grp2n = 50,    # sample size of the second group
         es.type = "r") # convert to correlation
```

```
## Effect Size Calculation for Meta Analysis
##
##      Conversion: standardized regression coefficient
##                  to effect size correlation
##     Effect Size:   0.3668
##  Standard Error:   0.1033
##        Variance:   0.0107
##        Lower CI:   0.1803
##        Upper CI:   0.5278
##          Weight:  93.7884
##      Fisher's z:   0.3847
##       Lower CIz:   0.1823
##       Upper CIz:   0.5871
```

17.3 Correlations

For *equally* sized groups ($n_1 = n_2$), we can use the following formula to derive the SMD from the *point-biserial* correlation (Lipsey and Wilson, 2001, chapter 3).

$$r_{pb} = \frac{\text{SMD}}{\sqrt{\text{SMD}^2 + 4}} \tag{17.2}$$

A different formula has to be used for *unequally* sized groups (Aaron et al., 1998):

$$r_{pb} = \frac{\text{SMD}}{\sqrt{\text{SMD}^2 + \dfrac{(N^2 - 2N)}{n_1 n_2}}} \tag{17.3}$$

To convert r_{pb} to an SMD or Hedges' g, we can use the esc_rpb function.

```
library(esc)

esc_rpb(r = 0.25,       # point-biserial correlation
        grp1n = 99,     # sample size of group 1
        grp2n = 120,    # sample size of group 2
        es.type = "d")  # convert to SMD; use "g" for Hedges' g
```

```
##
## Effect Size Calculation for Meta Analysis
##
##      Conversion: point-biserial r to effect size d
##     Effect Size:   0.5188
##  Standard Error:   0.1380
##        Variance:   0.0190
##        Lower CI:   0.2483
##        Upper CI:   0.7893
##          Weight:  52.4967
```

17.4 One-Way ANOVAs

We can also derive the SMD from the F-value of a one-way ANOVA with *two* groups. Such ANOVAs can be identified by looking at the *degrees of freedom*. In a one-way ANOVA with two groups, the degrees of freedom should always start with 1 (e.g. $F_{1,147}$=5.31). The formula used for the transformation looks like this (based on Rosnow and Rosenthal, 1996; Rosnow et al., 2000; see Thalheimer and Cook, 2002):

$$\text{SMD} = \sqrt{F\left(\frac{n_1+n_2}{n_1 n_2}\right)\left(\frac{n_1+n_2}{n_1+n_2-2}\right)} \tag{17.4}$$

To calculate the SMD or Hedges' g from F-values, we can use the esc_f function. Here is an example:

```
esc_f(f = 5.04,        # F value of the one-way anova
      grp1n = 519,     # sample size of group 1
      grp2n = 528,     # sample size of group 2
      es.type = "g") # convert to Hedges' g; use "d" for SMD
```

```
##
## Effect Size Calculation for Meta Analysis
##
##      Conversion: F-value (one-way-Anova) to effect size Hedges' g
##     Effect Size:   0.1387
##  Standard Error:   0.0619
##        Variance:   0.0038
##        Lower CI:   0.0174
##        Upper CI:   0.2600
##          Weight: 261.1022
```

17.5 Two-Sample t-Tests

An effect size expressed as a standardized mean difference can also be derived from an *independent* two-sample t-test value, using the following formula (Rosnow et al., 2000; Thalheimer and Cook, 2002):

$$\text{SMD} = \frac{t(n_1+n_2)}{\sqrt{(n_1+n_2-2)(n_1n_2)}} \tag{17.5}$$

In R, we can calculate the SMD or Hedges' g from a t-value using the `esc_t` function. Here is an example:

```
esc_t(t = 3.3,      # t-value
      grp1n = 100, # sample size of group1
      grp2n = 150, # sample size of group 2
      es.type="d") # convert to SMD; use "g" for Hedges' g
```

```
##
## Effect Size Calculation for Meta Analysis
##
##      Conversion: t-value to effect size d
##     Effect Size:   0.4260
##  Standard Error:   0.1305
##        Variance:   0.0170
##        Lower CI:   0.1703
##        Upper CI:   0.6818
##          Weight:  58.7211
```

17.6 p-Values

At times, studies only report the effect size (e.g. a value of Cohen's d), the p-value of that effect, and nothing more. Yet, to pool results in a meta-analysis, we need a measure of the *precision* of the effect size, preferably the standard error.

In such cases, we must estimate the standard error from the p-value of the effect size. This is possible for effect sizes based on *differences* (i.e. SMDs), or *ratios* (i.e. risk or odds ratios), using the formulas by Altman and Bland (2011). These formulas are implemented in the `se.from.p` function in R.

The "se.from.p" Function

The `se.from.p` function is included in the *{dmetar}* package. Once *{dmetar}* is installed and loaded on your computer, the function is ready to be used. If you did not install *{dmetar}*, follow these instructions:

1. Access the source code of the function online[a].
2. Let R "learn" the function by copying and pasting the source code in its entirety into the console (bottom left pane of R Studio), and then hit "Enter".

[a] https://raw.githubusercontent.com/MathiasHarrer/dmetar/master/R/SE_from_p.R

Assuming a study with $N = 71$ participants, reporting an effect size of $d = 0.71$ for which $p = 0.013$, we can calculate the standard error like this:

```
library(dmetar)

se.from.p(0.71,
          p = 0.013,
          N = 71,
          effect.size.type = "difference")
```

```
##   EffectSize StandardError StandardDeviation  LLCI  ULCI
## 1       0.71         0.286             2.410 0.149 1.270
```

For a study with $N = 200$ participants reporting an effect size of OR = 0.91 with $p = 0.38$, the standard error is calculated this way:

```
library(magrittr) # for pipe

se.from.p(0.91, p = 0.38, N = 200,
          effect.size.type = "ratio") %>% t()
```

```
##                         [,1]
## logEffectSize         -0.094
## logStandardError       0.105
## logStandardDeviation   1.498
## logLLCI               -0.302
## logULCI                0.113
## EffectSize             0.910
## LLCI                   0.739
## ULCI                   1.120
```

When `effect.size.type = "ratio"`, the function automatically also calculates the *log-transformed* effect size and standard error, which are needed to use the `metagen` function (Chapter 4.2.1).

17.7 χ^2 Tests

To convert a χ^2 statistic to an odds ratio, the `esc_chisq` function can be used (assuming that d.f. = 1; e.g. $\chi^2_1 = 8.7$). Here is an example:

```
esc_chisq(chisq = 7.9,          # chi-squared value
          totaln = 100,         # total sample size
          es.type = "cox.or") # convert to odds ratio
```

```
##
## Effect Size Calculation for Meta Analysis
##
##      Conversion: chi-squared-value to effect size Cox odds ratios
##     Effect Size:   2.6287
##  Standard Error:   0.3440
##        Variance:   0.1183
##        Lower CI:   1.3394
##        Upper CI:   5.1589
##          Weight:   8.4502
```

17.8 Number Needed to Treat

Effect sizes such as Cohen's d or Hedges' g are often difficult to interpret from a practical standpoint. Imagine that we found an intervention effect of $g = 0.35$ in our meta-analysis. How can we communicate what such an effect *means* to patients, public officials, medical professionals, or other stakeholders?

To make it easier for others to understand the results, meta-analyses also often report the *number needed to treat* (NNT). This measure is most commonly used in medical research. It signifies how many additional patients must receive the treatment under study to *prevent* one additional *negative event* (e.g. relapse) or *achieve* one additional *positive* event (e.g. symptom remission, response). If NNT = 3, for example, we can say that three individuals must receive the treatment to avoid one additional relapse case; or that three patients must be treated to achieve one additional case of reliable symptom remission, depending on the research question.

When we are dealing with binary effect size data, calculation of NNTs is relatively easy. The formula looks like this:

$$\text{NNT} = (p_{e_{\text{treat}}} - p_{e_{\text{control}}})^{-1} \tag{17.6}$$

In this formula, $p_{e_{\text{treat}}}$ and $p_{e_{\text{control}}}$ are the proportions of participants who experienced the event in the treatment and control group, respectively. These proportions are identical to the "risks" used to calculate the risk ratio (Chapter 3.3.2.1), and also known as the *experimental group event rate* (EER) and *control group event rate* (CER). Given its formula, the NTT can also be described as the inverse of the (absolute) risk difference.

Converting standardized mean differences or Hedges' g to a NNT is more complicated. There are two commonly used methods:

- The method by *Kraemer and Kupfer* (2006), which calculates the NNT from an *area under the curve* (AUC), defined as the probability that a patient in the treatment group has an outcome preferable to the one in the control group. This method allows to calculate the NNT directly from an SMD or g without any extra information.
- The method by *Furukawa and Leucht* calculates NNT values from SMDs using the CER, or a reasonable estimate thereof. Furukawa's method has been shown to be superior in estimating the true NNT value compared to the Kraemer & Kupfer method (Furukawa and Leucht, 2011). If we can make reasonable estimates of the CER, Furukawa's method should therefore always be preferred.

When we use risk or odds ratios as effect size measures, NNTs can be calculated directly from *{meta}* objects using the `nnt` function. After running our meta-analysis using `metabin` (Chapter 4.2.3.1), we only have to plug the results into the `nnt` function. Here is an example:

```
library(meta)
data(Olkin1995)

# Run meta-analysis with binary effect size data
m.b <- metabin(ev.exp, n.exp, ev.cont, n.cont,
               data = Olkin1995,
               sm = "RR")
nnt(m.b)
```

```
## Fixed effect model:
##
##      p.c     NNT lower.NNT upper.NNT
##   0.0000     Inf       Inf       Inf
##   0.1440 30.5677   26.1222   37.2386
##   0.3750 11.7383   10.0312   14.3001
```

```
##
## Random effects model:
##
##      p.c     NNT lower.NNT upper.NNT
##  0.0000     Inf       Inf       Inf
##  0.1440 29.7622   23.5092   41.8838
##  0.3750 11.4290    9.0278   16.0839
```

The nnt function provides the number needed to treat for different assumed CERs. The three lines show the result for the minimum, mean, and maximum CER in our data set. The mean CER estimate is the "typical" NNT that is usually reported.

It is also possible to use nnt with metagen models, as long as the summary measure sm is either "RR" or "OR". For such models, we also need to specify the assumed CER in the p.c argument in nnt. Here is an example using the m.gen_bin meta-analysis object we created in Chapter 4.2.3.1:

```
# Also show fixed-effect model results
m.gen_bin <- update.meta(m.gen_bin,
                         comb.fixed = TRUE)

nnt(m.gen_bin,
    p.c = 0.1) # Use a CER of 0.1
```

```
## Fixed effect model:
##
##      p.c     NNT lower.NNT upper.NNT
##  0.1000 -9.6906  -11.6058   -8.2116
##
## Random effects model:
##
##      p.c     NNT lower.NNT upper.NNT
##  0.1000 -9.7870  -16.4843   -6.4761
```

Standardized mean differences or Hedges' g can be converted to the NNT using the NNT function in *{dmetar}*.

The "NNT" Function

If you did not install *{dmetar}*, follow these instructions:

1. Access the source code of the NNT function online[a].
2. Let R "learn" the function by copying and pasting the source code in its entirety into the console (bottom left pane of R Studio), and then hit "Enter".

[a] https://raw.githubusercontent.com/MathiasHarrer/dmetar/master/R/NNT.R

To use the Kraemer & Kupfer method, we only have to provide the NNT function with an effect size (SMD or g). Furukawa's method is automatically used as soon as a CER value is supplied.

```
NNT(d = 0.245)
```

```
## Kraemer & Kupfer's method used.
## [1] 7.271
```

```
NNT(d = 0.245, CER = 0.35)
```

```
## Furukawa's method used.
## [1] 10.62
```

A Number to be Treated with Care: Criticism of the NNT

While common, usage of NNTs to communicate the results of clinical trials is not uncontroversial. Criticisms include that lay people often misunderstand it (despite purportedly being an "intuitive" alternative to other effect size measures, Christensen and Kristiansen, 2006); and that researchers often calculate NNTs incorrectly (Mendes et al., 2017).

Furthermore, it is not possible to calculate reliable standard errors (and confidence intervals) of NNTs, which means that they can not be used in meta-analyses (Hutton, 2010). It is only possible to convert results to the NNT after pooling has been conducted using another effect size measure.

17.9 Multi-Arm Studies

To avoid unit-of-analysis errors (Chapter 3.5.2), it is sometimes necessary to pool the mean and standard deviation of two or more trial arms before calculating a (standardized) mean difference. To pool continuous effect size data of two groups, we can use these equations:

$$
\begin{aligned}
n_{\text{pooled}} &= n_1 + n_2 \\
m_{\text{pooled}} &= \frac{n_1 m_1 + n_2 m_2}{n_1 + n_2} \\
SD_{\text{pooled}} &= \sqrt{\frac{(n_1-1)SD_1^2 + (n_2-1)SD_2^2 + \frac{n_1 n_2}{n_1+n_2}(m_1^2 + m_2^2 - 2m_1 m_2)}{n_1+n_2-1}}
\end{aligned}
\tag{17.7}
$$

We can apply this formula in R using the `pool.groups` function.

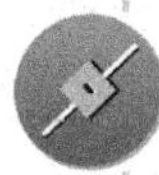

The "pool.groups" Function

The `pool.groups` function is included in the *{dmetar}* package. Once *{dmetar}* is installed and loaded on your computer, the function is ready to be used. If you did not install *{dmetar}*, follow these instructions:

1. Access the source code of the function online[a].
2. Let R "learn" the function by copying and pasting the source code in its entirety into the console (bottom left pane of R Studio), and then hit "Enter".

[a] https://raw.githubusercontent.com/MathiasHarrer/dmetar/master/R/pool.groups.R

Here is an example:

```
library(dmetar)

pool.groups(n1 = 50,   # sample size group 1
            n2 = 50,   # sample size group 2
            m1 = 3.5,  # mean group 1
            m2 = 4,    # mean group 2
            sd1 = 3,   # sd group 1
            sd2 = 3.8) # sd group2
```

```
##    Mpooled SDpooled Npooled
## 1     3.75    3.415     100
```

□

A

Questions & Answers

Chapter 1: Introduction

1. How can meta-analysis be defined? What differentiates a meta-analysis from other types of literature reviews?

Meta-analysis can be defined as an *analysis of analyses* (definition by Glass). In contrast to other types of (systematic) reviews, meta-analysis aims to synthesize evidence in a quantitative way. Usually, the goal is to derive a numerical estimate that describes a clearly circumscribed research field *as a whole.*

2. Can you name one of the founding mothers and fathers of meta-analysis? What achievement can be attributed to her or him?

Karl Pearson: combination of typhoid inoculation data across the British empire. Ronald Fisher: approaches to synthesize data of agricultural research studies. Mary Smith and Gene Glass: coined the term "meta-analysis", first meta-analysis of psychotherapy trials; John Hunter and Frank Schmidt: meta-analysis with correction of measurement artifacts (psychometric meta-analysis); Rebecca DerSimonian and Nan Laird: method to calculate random-effects model meta-analyses; Peter Elwood and Archie Cochrane: pioneer meta-analysis in medicine.

3. Name three common problems of meta-analyses and describe them in one or two sentences.

"Apples and Oranges": studies are too different to be synthesized; "Garbage In, Garbage Out": invalid evidence is only reproduced by meta-analyses; "File Drawer": negative results are not published, leading to biased findings in meta-analyses; "Researcher Agenda": researchers can tweak meta-analyses to prove what they want to prove.

4. Name qualities that define a good research question for a meta-analysis.

FINER: feasible, interesting, novel, ethical, relevant. PICO: clearly defined population, intervention/exposure, control group/comparison, and analyzed outcome.

5. Have a look again at the eligibility criteria of the meta-analysis on sleep interventions in college students (end of Chapter 1.4.1). Can you extract the PICO from the inclusion and exclusion criteria of this study?

Population: tertiary education students. Intervention: sleep-focused psychological interventions. Comparison: passive control condition. Outcome: sleep disturbance, as measured by standardized symptom measures.

6. Name a few important sources that can be used to search studies.

Review articles, references in studies, "forward search" (searching for studies that have cited a relevant article), searching relevant journals, bibliographic database search.

7. Describe the difference between "study quality" and "risk of bias" in one or two sentences.

DOI: 10.1201/9781003107347-A

A study can fulfill all study quality criteria that are considered important in a research field and still have a high risk of bias (e.g. because bias is difficult to avoid for this type of study or research topic).

Chapter 2: Discovering R

1. Show the variable 'Author'.

```
data$Author
```

2. Convert 'subgroup' to a factor.

```
data$subgroup <- as.factor(data$subgroup)
```

3. Select all the data of the "Jones" and "Martin" study.

```
library(tidyverse)
data %>%
  filter(Author %in% c("Jones", "Martin"))
```

4. Change the name of the study "Rose" to "Bloom".

```
data[5,1] <- "Bloom"
```

5. Create a new variable 'TE_seTE_diff' by subtracting 'seTE' from 'TE'. Save the results in 'data'.

```
data$TE_seTE_diff <- data$TE - data$seTE
```

6. Use a pipe to (1) filter all studies in subgroup "one" or "two", (2) select the variable 'TE_seTE_diff', (3) take the mean of the variable, and then apply the 'exp' function to it.

```
data %>%
  dplyr::filter(subgroup %in%
           c("one","two")) %>%
  pull(TE_seTE_diff) %>%
  mean() %>% exp()
```

Chapter 3: Effect Sizes

1. Is there a clear definition of the term "effect size"? What do people refer to when they speak of effect sizes?

No, there is no universally accepted definition. Some reserve the term "effect size" for differences between intervention and control groups. Others use a more liberal definition, and only exclude "one-variable" measures (e.g. means and proportions).

2. Name a primary reason why observed effect sizes deviate from the true effect size of the population. How can it be quantified?

Observed effect sizes are asssumed to deviate from the true effect size because of sampling error. The expected size of a study's sampling error can be expressed by its standard error.

3. Why are large studies better estimators of the true effect than small ones?

Because they are assumed to have a smaller sampling error, which leads to more precise effect estimates.

4. What criteria does an effect size metric have to fulfill to be usable for meta-analyses?

It needs to be comparable, computable, reliable, and interpretable.

5. What does a standardized mean difference of 1 represent?

It represents that the means of the two groups differ by one pooled standard deviation.

6. What kind of transformation is necessary to pool effect sizes based on ratios (e.g. an odds ratio)?

The effect size needs to be log-transformed (in order to use the inverse-variance pooling method).

7. Name three types of effect size corrections.

Small sample bias correction of standardized mean differences (Hedges' g); correction for unreliability; correction for range restriction.

8. When does the unit-of-analysis problem occur? How can it be avoided?

When effect sizes in our data set are correlated (for example because they are part of the same study). The unit-of-analysis problem can be (partly or fully) avoided by (1) splitting the sample size of the shared group, (2) removing comparisons, (3) combining groups, or (4) using models that account for the effect size dependencies (e.g. three-level models).

Chapter 4: Pooling Effect Sizes

1. What is the difference between a fixed-effect model and a random-effects model?

The fixed-effect model assumes that all studies are estimators of the same true effect size. The random-effects model assumes that the true effect sizes of studies vary because of between-study heterogeneity (captured by the variance τ^2), which needs to be estimated.

2. Can you think of a case in which the results of the fixed- and random-effects model are identical?

When the between-study heterogeneity variance τ^2 is zero.

3. What is τ^2? How can it be estimated?

The between-study heterogeneity variance. It can be estimated using different methods, for example, restricted maximum likelihood (REML), the Paule-Mandel estimator, or the DerSimonian-Laird estimator.

4. Which distribution is the Knapp-Hartung adjustment based on? What effect does it have?

It is based on a t-distribution. The Knapp-Hartung adjustment usually leads to more conservative (i.e. wider) confidence intervals.

5. What does "inverse-variance" pooling mean? When is this method not the best solution?

The method is called inverse-variance pooling because it uses the inverse of a study's variance as the pooling weight. The generic inverse-variance method is not the preferred option for meta-analyses of binary outcome data (e.g. risk or odds ratios).

6. You want to meta-analyze binary outcome data. The number of observations in the study arms is roughly similar, the observed event is very rare, and you do no expect the treatment effect to be large. Which pooling method would you use?

This is a scenario in which the Peto method may perform well.

7. For which outcome measures can GLMMs be used?

Proportions. It is also possible to use them for other binary outcome measures, but not generally recommended.

Chapter 5: Between-Study Heterogeneity

1. Why is it important to examine the between-study heterogeneity of a meta-analysis?

When the between-study heterogeneity is large, the true effect sizes can be assumed to vary considerably. In this case, a point estimate of the average true effect may not represent the data well in their totality. Between-study heterogeneity can also lead to effect estimates that are not robust, for example because a few outlying studies distort the overall result.

2. Can you name the two types of heterogeneity? Which one is relevant in the context of calculating a meta-analysis?

Baseline/design-related heterogeneity and statistical heterogeneity. Only statistical heterogeneity is assessed quantitatively in meta-analyses.

3. Why is the significance of Cochran's Q not a sufficient measure of between-study heterogeneity?

Because the significance of the Q test heavily depends on the number of studies included in our meta-analysis, and their size.

4. What are the advantages of using prediction intervals to express the amount of heterogeneity in a meta-analysis?

Prediction intervals allow to express the impact of between-study heterogeneity on future studies on the same scale as the summary measure.

5. What is the difference between statistical outliers and influential studies?

Statistical outliers are studies with *extreme* effect sizes. Studies are influential when their impact on the overall result is large. It is possible that a study can be defined as a statistical outlier without being very influential, and vice versa. For example, a large study may have a big impact on the pooled results, even though its effect size is not particularly small or large.

6. For what can GOSH plots be used?

GOSH plots can be used to explore patterns of heterogeneity in our data, and which studies contribute to them.

Chapter 6: Forest Plots

1. What are the key components of a forest plot?

Graphical representation of each study's observed effect size, with confidence intervals; the weight of each study, represented by the size of squares around the observed effect sizes; the numeric value of each study's observed effect and weight; the pooled effect, represented by a diamond; a reference line, usually representing no effect.

2. What are the advantages of presenting a forest plot of our meta-analysis?

They allow to quickly examine the number, effect size, and precision of all included studies, and how the observed effects "add up" to the pooled effect.

3. What are the limitations of forest plots, and how do drapery plots overcome this limitation?

Forest plots can only show the confidence intervals of effects assuming a fixed significance threshold (usually $p < 0.05$). Drapery plots can be used to show the confidence intervals (and thus the significance) of effect sizes for varying p-values.

Chapter 7: Subgroup Analyses

1. In the best case, what can a subgroup analysis tell us that influence and outlier analyses cannot?

Subgroup analyses can potentially explain *why* certain heterogeneity patterns exist in our data, versus only telling us *that* they exist.

2. Why is the model behind subgroup analyses called the fixed-effects (plural) model?

Because it assumes that, while studies within subgroups follow a random-effects model, the subgroup levels themselves are fixed. There are several fixed subgroup effects.

3. As part of your meta-analysis, you want to examine if the effect of an educational training program differs depending on the school district in which it was delivered. Is a subgroup analysis using the fixed-effects (plural) model appropriate to answer this question?

Probably not. It makes more sense to assume that the school districts represent draws from a larger population of districts, not all school districts there are.

4. A friend of yours conducted a meta-analysis containing a total of nine studies. Five of these studies fall into one subgroup, four into the other. She asks you if it makes sense to perform a subgroup analysis. What would you recommend?

It is probably not a good idea to conduct a subgroup analysis, since the total number of studies is smaller than ten.

5. You found a meta-analysis in which the authors claim that the analyzed treatment is more effective in women than men. This finding is based on a subgroup analysis, in which studies were divided into subgroups based on the share of females included in the study population. Is this finding credible, and why (not)?

The finding is based a subgroup variable that has been created using aggregated study data. This may introduce ecological bias, and the results are therefore questionable.

Chapter 8: Meta-Regression

1. What is the difference between a conventional regression analysis used in primary studies, and meta-regression?

The unit of analysis is studies (instead of persons), the effect sizes of which are more or less precise. In meta-regression, we have to build regression models that account for the fact that some studies should have a greater weight than others.

2. Subgroup analyses and meta-regression are closely related. How can the meta-regression formula be adapted to subgroup data?

By using dummy/categorical predictors.

3. Which method is used in meta-regression to give individual studies a differing weight?

Meta-regression uses *weighted least squares* to give studies with higher precision a greater weight.

4. What characteristics mark a meta-regression model that fits our data well? Which index can be used to examine this?

A "good" meta-regression model should lead to a large reduction in the amount of unexplained between-study heterogeneity variance. An index which covers this increase in explained variance is the R^2 analog.

5. When we calculate a subgroup analysis using meta-regression techniques, do we assume a separate or common value of τ^2 in the subgroups?

A common estimate of τ^2 is assumed in the subgroups.

6. What are the limitations and pitfalls of (multiple) meta-regression?

Overfitting meta-regression can lead to false positive results; multicollinearity can lead to parameter estimates that are not robust.

7. Name two methods that can be used to improve the robustness of (multiple) meta-regression models, and why they are helpful.

We can conduct a permutation test or use multi-model inference.

Chapter 9: Publication Bias

1. How can the term "publication bias" be defined? Why is it problematic in meta-analyses?

Publication bias exists when the probability of a study to get published depends on its results. This is problematic because it can lead to biased results in meta-analyses. Because not all evidence is considered, meta-analyses may result in findings that would not have materialized when all existing information had been considered.

2. What other reporting biases are there? Name and explain at least three.

Citation bias: studies with negative findings are less likely to be cited; time-lag bias: studies with negative findings are published later; multiple publication bias: studies with positive findings are more likely to be reported in several articles; language bias: evidence may be omitted because it is not published in English; outcome reporting bias: positive outcomes of a study are more likely to be reported than negative outcomes.

3. Name two questionable research practices (QRPs), and explain how they can threaten the validity of our meta-analysis.

P-hacking, HARKing. Both lead to an inflation of positive findings, even when there is no true effect.

4. *Explain the core assumptions behind small-study effect methods.*

Large studies (i.e. studies with a small standard error) are very likely to get published, no matter what their findings are. Smaller studies have a smaller precision, which means that very high effect sizes are needed to attain statistical significance. Therefore, only small studies with very high effects are published, while the rest ends up in the "file drawer".

5. *When we find out that our data displays small-study effects, does this automatically mean that there is publication bias?*

No. There are several other explanations why we find small-study effects, including between-study heterogeneity, effects of covariates (e.g. treatment fidelity is higher in smaller studies), or chance.

6. *What does p-curve estimate: the true effect of all studies included in our meta-analysis, or just the true effect of all significant effect sizes?*

P-curve only estimates the true effect of all significant effect sizes. This is one of the reasons why it does not perform well when there is between-study heterogeneity.

7. *Which publication bias method has the best performance?*

No publication bias method consistently outperforms all the others. Therefore, it is helpful to apply several methods, and see if their results coalign.

Chapter 10: "Multilevel" Meta-Analysis

1. *Why is it more accurate to speak of "three-level" instead of "multilevel" models?*

Because the "conventional" random-effects model is already a multilevel model. It assumes that participants are nested within studies, and that the studies themselves are drawn from a population of true effect sizes.

2. *When are three-level meta-analysis models useful?*

When we are dealing with correlated or nested data. Three-level models are particularly useful when studies contribute multiple effect sizes, or when there is good reason to believe that studies themselves fall into larger clusters.

3. *Name two common causes of effect size dependency.*

Dependence caused by the researchers involved in the primary studies; dependency created by the meta-analyst herself.

4. *How can the multilevel I^2 statistic be interpreted?*

It tells us the amount of variance not attributable to sampling error, and differentiates between heterogeneity variance *within* clusters, and heterogeneity variance *between* clusters.

5. *How can a three-level model be expanded to incorporate the effect of moderator variables?*

By integrating a fixed-effect term to the model formula.

Chapter 11: Structural Equation Modeling Meta-Analysis

1. *What is structural equation modeling, and what is used for?*

Structural equation modeling is a statistical method that can be used to test assumed relationships between manifest and latent variables.

2. *What are the two ways through which SEM can be represented?*

SEM can be represented graphically or through matrices.

3. Describe a random-effects meta-analysis from a SEM perspective.

From a SEM perspective, the true overall effect size in a random-effects meta-analysis can be seen as a latent variable. It is "influenced" by two arms: the sampling error on level 1 and the true effect size heterogeneity variance on level 2.

4. What is a multivariate meta-analysis, and when is it useful?

Multivariate meta-analysis allows to simultaneously pool two (or more) outcomes of studies. An asset of jointly estimating the two outcome variables is that the correlation between outcomes can be taken into account.

5. When we find that our proposed meta-analytic SEM fits the data well, does this automatically mean that this model is the "correct" one?

No. Frequently, there is more than one model that fits the data well.

Chapter 12: Network Meta-Analysis

1. When are network meta-analyses useful? What is their advantage compared to standard meta-analyses?

Network meta-analyses are useful when there are several competing treatments for some problem area, and we want to estimate which one has the largest benefits. In contrast to conventional meta-analyses, network meta-analysis models can integrate both direct and indirect evidence.

2. What is the difference between direct and indirect evidence in a treatment network? How can direct evidence be used to generate indirect evidence?

Direct evidence is information provided by comparisons that have actually been investigated in the included studies. Indirect evidence is derived from direct evidence by subtracting the effect of one (directly observed) comparison from the one of a related comparison (e.g. a comparison that used the same control group).

3. What is the main idea behind the assumption of transitivity in network meta-analyses?

The assumption of transitivity stipulates that direct evidence can be used to infer unobserved, indirect evidence, and that direct and indirect evidence is consistent.

4. What is the relationship between transitivity and consistency?

Transitivity is a pre-requisite to conduct network meta-analyses and cannot be tested directly. The statistical manifestation of transitivity is consistency and is fulfilled when effect size estimates based on direct evidence are identical/similar to estimates based on indirect evidence.

5. Name two modeling approaches that can be used to conduct network meta-analyses. Is one of them better than the other?

Network meta-analysis can be conducted using a frequentist or Bayesian model. Both models are equivalent and produce converging results with increasing sample size.

6. When we include several comparisons from one study (i.e. multi-arm studies), what problem does this cause?

This means that the effect estimates are correlated, causing a unit-of-analysis error.

7. What do we have to keep in mind when interpreting the P- or SUCRA score of different treatments?

That the effect estimates of different treatments often overlap. This means that P-/SUCRA scores should always be interpreted with some caution.

Chapter 12: Bayesian Meta-Analysis

1. What are differences and similarities between the "conventional" random-effects model and a Bayesian hierarchical model?

The random-effects model underlying frequentist meta-analysis is conceptually identical to the Bayesian hierarchical model. The main difference is that the Bayesian hierarchical model includes (weakly informative) prior distributions for the overall true effect size μ and between-study heterogeneity τ.

2. Name three advantages of Bayesian meta-analyses compared to their frequentist counterpart.

Uncertainty of the τ^2 estimate is directly modeled; a posterior distribution for μ is produced, which can be used to calculate the probability of μ lying below a certain value; prior knowledge or beliefs can be integrated into the model.

3. Explain the difference between a weakly informative and non-informative prior.

Non-informative priors assume that all or a range of possible values are equally likely. Weakly informative priors represent a *weak* belief that some values are more probable than others.

4. What is a Half-Cauchy distribution, and why is it useful for Bayesian meta-analysis?

The Half-Cauchy distribution is a Cauchy distribution that is only defined for positive values. It is controlled by a location and scaling parameter, the latter determining how heavy the tails of the distribution are. Half-Cauchy distributions can be used as priors for τ.

5. What is an ECDF, and how can it be used in Bayesian meta-analyses?

ECDF stands for *empirical cumulative distribution function*. ECDFs based on the posterior distribution of μ (or τ) can be used to determine the (cumulative) probability that the estimated parameter is below or above some specified threshold.

□

B

Effect Size Formulas

	Effect Size ($\hat{\theta}$)	Standard Error (SE)	Function
Arithmetic Mean	$\bar{x} = \frac{\sum_{i=1}^{n} x_i}{n}$	$\text{SE}_{\bar{x}} = \frac{s}{\sqrt{n}}$	mean
Proportion	$p = \frac{k}{n}$ $p_{\text{logit}} = \log_e \left(\frac{p}{1-p} \right)$	$\text{SE}_p = \sqrt{\frac{p(1-p)}{n}}$ $\text{SE}_{p_{\text{logit}}} = \sqrt{\frac{1}{np} + \frac{1}{n(1-p)}}$	
Correlation			
Product-Moment Correlation	$r_{xy} = \frac{\sigma^2_{xy}}{\sigma_x \sigma_y}$ $z = 0.5 \log_e \left(\frac{1+r}{1-r} \right)$	$SE_{r_{xy}} = \frac{1-r^2_{xy}}{\sqrt{n-2}}$ $\text{SE}_z = \frac{1}{\sqrt{n-3}}$	cor
Point-Biserial Correlation[1]	$r_{\text{pb}} = \frac{(\bar{y_1} - \bar{y_2})\sqrt{\frac{n_1}{N}\left(1 - \frac{n_1}{N}\right)}}{s_y}$		cor
(Standardized) Mean Difference			
Between-Group Mean Difference	$\text{MD} = \bar{x}_1 - \bar{x}_2$	$\text{SE}_{\text{MD}} = s_{\text{pooled}}{}^{*} \sqrt{\frac{1}{n_1} + \frac{1}{n_2}}$	
Between-Group Standardized Mean Difference	$\text{SMD} = \frac{\bar{x}_1 - \bar{x}_2}{s_{\text{pooled}}{}^{*}}$	$\text{SE}_{\text{SMD}} = \sqrt{\frac{n_1 + n_2}{n_1 n_2} + \frac{\text{SMD}^2_{\text{between}}}{2(n_1 + n_2)}}$	esc_mean_sd
Within-Group Mean Difference	$\text{MD} = \bar{x}_{t_2} - \bar{x}_{t_1}$	$SE_{\text{MD}} = \sqrt{\frac{s^2_{t_2} + s^2_{t_1} - (2 r_{t_1 t_2} s_{t_1} s_{t_2})}{n}}$	

DOI: 10.1201/9781003107347-B

(continued)

	Effect Size ($\hat{\theta}$)	Standard Error (SE)	Function
Within-Group Standardized Mean Difference	$\text{SMD} = \frac{\bar{x}_{t_2} - \bar{x}_{t_1}}{\left(\frac{\sqrt{(s^2_{t_1} + s^2_{t2})/2}}{\sqrt{2(1 - r_{t_1 t_2})}}\right)}$	$\text{SE}_{\text{SMD}} = \sqrt{\frac{2(1 - r_{t_1 t_2})}{n} + \frac{\text{SMD}^2_{\text{within}}}{2n}}$	
Binary Outcome Effect Size			
Risk Ratio	$p_{e_{\text{treat}}} = \frac{a}{n_{\text{treat}}}$ $p_{e_{\text{control}}} = \frac{c}{n_{\text{control}}}$ $\text{RR} = \frac{p_{e_{\text{treat}}}}{p_{e_{\text{control}}}}$ $\log \text{RR} = \log_e(\text{RR})$	$\text{SE}_{\log \text{RR}} = \sqrt{\frac{1}{a} + \frac{1}{c} - \frac{1}{a+b} - \frac{1}{c+d}}$	
Odds Ratio	$\text{Odds}_{\text{treat}} = \frac{a}{b}$ $\text{Odds}_{\text{control}} = \frac{c}{d}$ $\text{OR} = \frac{a/b}{c/d}$ $\log \text{OR} = \log_e(\text{OR})$	$\text{SE}_{\log \text{OR}} = \sqrt{\frac{1}{a} + \frac{1}{b} + \frac{1}{c} + \frac{1}{d}}$	esc_2x2
Incidence Rate Ratio	$\text{IRR} = \frac{E_{\text{treat}}/T_{\text{treat}}}{E_{\text{control}}/T_{\text{control}}}$ $\log \text{IRR} = \log_e(\text{IRR})$	$\text{SE}_{\log \text{IRR}} = \sqrt{\frac{1}{E_{\text{treat}}} + \frac{1}{E_{\text{control}}}}$	
Effect Size Correction			
Small Sample Bias	$g = \text{SMD} \times \left(1 - \frac{3}{4n-9}\right)$		hedges_g
Unreliability	$r_{xy_c} = \frac{r_{xy}}{\sqrt{r_{xx}}}$ $r_{xy_c} = \frac{r_{xy}}{\sqrt{r_{xx}}\sqrt{r_{yy}}}$ $\text{SMD}_c = \frac{\text{SMD}}{\sqrt{r_{xx}}}$	$\text{SE}_c = \frac{\text{SE}}{\sqrt{r_{xx}}}$ $\text{SE}_c = \frac{\text{SE}}{\sqrt{r_{xx}}\sqrt{r_{yy}}}$	

(continued)

	Effect Size ($\hat{\theta}$)	Standard Error (SE)	Function
Range Restriction	$U = \frac{s_{\text{unrestricted}}}{s_{\text{restricted}}}$ $r_{xy_c} = \frac{U \times r_{xy}}{\sqrt{(U^2-1)r_{xy}^2+1}}$	$SE_{r_{xy_c}} = \frac{r_{xy_c}}{r_{xy}} SE_{r_{xy}}$	
	$SMD_c = \frac{U \times SMD}{\sqrt{(U^2-1)SMD^2+1}}$	$SE_{SMD_c} = \frac{SMD_c}{SMD} SE_{SMD}$	

[1] Point-biserial correlations may be converted to SMDs for meta-analysis (see Chapter 3.2.3.2).

* The pooled standard deviation is defined as $s_{\text{pooled}} = \sqrt{\frac{(n_1-1)s_1^2+(n_2-1)s_2^2}{(n_1-1)+(n_2-1)}}$.

C

R & Package Information

This book was compiled using R version 4.0.3 ("Bunny-Wunnies Freak Out", 2020-10-10) running under macOS Catalina 10.15.4 (Apple Darwin 17.0 64-bit x86-64). The following package versions are used in the book:

```
brms 2.13.0                      dmetar 0.0.9000
dplyr 1.0.1                      esc 0.5.1
extraDistr 1.9.1                 forcats 0.5.0
gemtc 0.8-6                      ggplot2 3.3.2
ggridges 0.5.2                   glue 1.4.1
igraph 1.2.5                     meta 4.15-1
metafor 2.5-62                   metaSEM 1.2.5
metasens 0.5-0                   netmeta 1.2-1
openxlsx 4.1.5                   osfr 0.2.8
PerformanceAnalytics 2.0.4       rjags 4-10
robvis 0.3.0                     semPlot 1.1.2
stringr 1.4.0                    tidybayes 2.1.1
tidyverse 1.3.0
```

Attached base packages:

```
base 4.0.3         datasets 4.0.3    graphics 4.0.3
grDevices 4.0.3    methods 4.0.3     stats 4.0.3
utils 4.0.3
```

Locale: en_US.UTF-8

Attributions

Figure 1.2: Sirswindon at English Wikipedia[1], CC BY-SA 3.0[2], via Wikimedia Commons. Desaturated from original.

[1] https://commons.wikimedia.org/wiki/File:Hans.Eysenck.jpg
[2] https://creativecommons.org/licenses/by-sa/3.0

DOI: 10.1201/9781003107347-C

Bibliography

Aaron, B., Kromrey, J. D., and Ferron, J. (1998). Equating r-based and d-based effect size indices: Problems with a commonly recommended formula. `https://files.eric.ed.gov/fulltext/ED433353.pdf`.

Adler, R. A. (2014). Osteoporosis in men: A review. *Bone Research*, 2:14001.

Alexander, R. A., Scozzaro, M. J., and Borodkin, L. J. (1989). Statistical and empirical examination of the chi-square test for homogeneity of correlations in meta-analysis. *Psychological Bulletin*, 106(2):329.

Altman, D. G. and Bland, J. M. (2011). How to obtain the confidence interval from a P value. *BMJ*, 343:d2090.

Appelbaum, M., Cooper, H., Kline, R. B., Mayo-Wilson, E., Nezu, A. M., and Rao, S. M. (2018). Journal article reporting standards for quantitative research in psychology: The APA publications and communications board task force report. *American Psychologist*, 73(1):3.

Aronow, P. M. and Miller, B. T. (2019). *Foundations of agnostic statistics*. Cambridge University Press.

Assink, M., Wibbelink, C. J., et al. (2016). Fitting three-level meta-analytic models in R: A step-by-step tutorial. *The Quantitative Methods for Psychology*, 12(3):154–174.

Bakbergenuly, I., Hoaglin, D. C., and Kulinskaya, E. (2020). Methods for estimating between-study variance and overall effect in meta-analysis of odds ratios. *Research Synthesis Methods*, 11(3):426–442.

Bakbergenuly, I. and Kulinskaya, E. (2018). Meta-analysis of binary outcomes via generalized linear mixed models: A simulation study. *BMC Medical Research Methodology*, 18(1):70.

Balduzzi, S., Rücker, G., and Schwarzer, G. (2019). How to perform a meta-analysis with R: A practical tutorial. *Evidence-Based Mental Health*, 22(4):153–160.

Bauer, D. J. (2003). Estimating multilevel linear models as structural equation models. *Journal of Educational and Behavioral Statistics*, 28(2):135–167.

Baujat, B., Mahé, C., Pignon, J.-P., and Hill, C. (2002). A graphical method for exploring heterogeneity in meta-analyses: Application to a meta-analysis of 65 trials. *Statistics in Medicine*, 21(18):2641–2652.

Beck, A. T., Steer, R. A., and Brown, G. (1996). Beck Depression Inventory–II. *Psychological Assessment.*

Becker, B. J. (1988). Synthesizing standardized mean-change measures. *British Journal of Mathematical and Statistical Psychology*, 41(2):257–278.

Bellhouse, D. R. et al. (2004). The Reverend Thomas Bayes, FRS: A biography to celebrate the tercentenary of his birth. *Statistical Science*, 19(1):3–43.

Berlin, J. A. and Antman, E. M. (1994). Advantages and limitations of metaanalytic regressions of clinical trials data. *The Online Journal of Current Clinical Trials.*

Björk, B.-C., Roos, A., and Lauri, M. (2008). Global annual volume of peer reviewed scholarly articles and the share available via different open access options. In *Proceedings ELPUB 2008 Conference on Electronic Publishing*, pages 178–186.

Bonett, D. G. (2020). Point-biserial correlation: Interval estimation, hypothesis testing, meta-analysis, and sample size determination. *British Journal of Mathematical and Statistical Psychology*, 73:113–144.

Borenstein, M., Hedges, L. V., Higgins, J. P., and Rothstein, H. R. (2011). *Introduction to meta-analysis*. John Wiley & Sons.

Borenstein, M. and Higgins, J. P. (2013). Meta-analysis and subgroups. *Prevention Science*, 14(2):134–143.

Borenstein, M., Higgins, J. P., Hedges, L. V., and Rothstein, H. R. (2017). Basics of meta-analysis: I^2 is not an absolute measure of heterogeneity. *Research Synthesis Methods*, 8(1):5–18.

Bradburn, M. J., Deeks, J. J., Berlin, J. A., and Russell Localio, A. (2007). Much ado about nothing: A comparison of the performance of meta-analytical methods with rare events. *Statistics in Medicine*, 26(1):53–77.

Bramer, W. M., de Jonge, G. B., Rethlefsen, M. L., Mast, F., and Kleijnen, J. (2018). A systematic approach to searching: An efficient and complete method to develop literature searches. *Journal of the Medical Library Association: JMLA*, 106(4):531.

Breiman, L. (2001). Statistical modeling: The two cultures (with comments and a rejoinder by the author). *Statistical Science*, 16(3):199–231.

Browne, M. and Cudeck, R. (1993). Alternative ways of assessing model fit. In Bollen, K. and Long, J., editors, *Testing structural equation models*. Sage Publications.

Bürkner, P.-C. (2017). Advanced Bayesian multilevel modeling with the R package brms. *ArXiv Preprint 1705.11123*.

Büscher, R., Torok, M., and Sander, L. (2019). The effectiveness of internet-based self-help interventions to reduce suicidal ideation: Protocol for a systematic review and meta-analysis. *JMIR Research Protocols*, 8(7):e14174.

Bürkner, P.-C. (2017). brms: An R package for Bayesian multilevel models using Stan. *Journal of Statistical Software, Articles*, 80(1):1–28.

Campbell Collaboration (2016). Methodological expectations of Campbell Collaboration intervention reviews (MECCIR): Conduct standards. `https://onlinelibrary.wiley.com/page/journal/18911803/homepage/author-guidelines`.

Carter, E. C., Schönbrodt, F. D., Gervais, W. M., and Hilgard, J. (2019). Correcting for bias in psychology: A comparison of meta-analytic methods. *Advances in Methods and Practices in Psychological Science*, 2(2):115–144.

Chan, A.-W., Song, F., Vickers, A., Jefferson, T., Dickersin, K., Gøtzsche, P. C., Krumholz, H. M., Ghersi, D., and Van Der Worp, H. B. (2014). Increasing value and reducing waste: Addressing inaccessible research. *The Lancet*, 383(9913):257–266.

Chatfield, C. (1995). Model uncertainty, data mining and statistical inference. *Journal of the Royal Statistical Society: Series A (Statistics in Society)*, 158(3):419–444.

Cheung, M. W. and Chan, W. (2009). A two-stage approach to synthesizing covariance matrices in meta-analytic structural equation modeling. *Structural Equation Modeling: A Multidisciplinary Journal*, 16(1):28–53.

Cheung, M. W.-L. (2008). A model for integrating fixed-, random-, and mixed-effects meta-analyses into structural equation modeling. *Psychological Methods*, 13(3):182.

Cheung, M. W.-L. (2014). Modeling dependent effect sizes with three-level meta-analyses: A structural equation modeling approach. *Psychological Methods*, 19(2):211.

Cheung, M. W.-L. (2015a). *Meta-analysis: A structural equation modeling approach*. John Wiley & Sons.

Cheung, M. W.-L. (2015b). metasem: An R package for meta-analysis using structural equation modeling. *Frontiers in Psychology*, 5:1521.

Christensen, P. M. and Kristiansen, I. S. (2006). Number-Needed-to-Treat (NNT)–needs treatment with care. *Basic & Clinical Pharmacology & Toxicology*, 99(1):12–16.

Chung, Y., Rabe-Hesketh, S., Dorie, V., Gelman, A., and Liu, J. (2013). A nondegenerate penalized likelihood estimator for variance parameters in multilevel models. *Psychometrika*, 78(4):685–709.

Cipriani, A., Higgins, J. P., Geddes, J. R., and Salanti, G. (2013). Conceptual and technical challenges in network meta-analysis. *Annals of Internal Medicine*, 159(2):130–137.

Cochran, W. G. (1954). Some methods for strengthening the common χ^2 tests. *Biometrics*, 10(4):417–451.

Cohen, J. (1988). *Statistical power analysis for the behavioral sciences*. Erlbaum Press.

Csardi, G. and Nepusz, T. (2006). The igraph software package for complex network research. *InterJournal*, Complex Systems:1695.

Cuijpers, P. (2016). Meta-analyses in mental health research. A practical guide. *Amsterdam, the Netherlands: Pim Cuijpers Uitgeverij*.

Cuijpers, P., Karyotaki, E., Reijnders, M., and Ebert, D. (2019a). Was Eysenck right after all? A reassessment of the effects of psychotherapy for adult depression. *Epidemiology and Psychiatric Sciences*, 28(1):21–30.

Cuijpers, P., Noma, H., Karyotaki, E., Cipriani, A., and Furukawa, T. A. (2019b). Effectiveness and acceptability of cognitive behavior therapy delivery formats in adults with depression: A network meta-analysis. *JAMA Psychiatry*, 76(7):700–707.

Cuijpers, P., Reijnders, M., and Huibers, M. J. (2019c). The role of common factors in psychotherapy outcomes. *Annual Review of Clinical Psychology*, 15:207–231.

Cuijpers, P. and Smit, F. (2002). Excess mortality in depression: A meta-analysis of community studies. *Journal of Affective Disorders*, 72(3):227–236.

Cuijpers, P., Turner, E. H., Koole, S. L., Van Dijke, A., and Smit, F. (2014). What is the threshold for a clinically relevant effect? The case of major depressive disorders. *Depression and Anxiety*, 31(5):374–378.

Cuijpers, P., Weitz, E., Cristea, I., and Twisk, J. (2017). Pre-post effect sizes should be avoided in meta-analyses. *Epidemiology and Psychiatric Sciences*, 26(4):364–368.

Cummings, S. R., Browner, W. S., and Hulley, S. B. (2013). Conceiving the research question and developing the study plan. *Designing Clinical Research*, 4:14–22.

Dahlke, J. A. and Wiernik, B. M. (2019). psychmeta: An R package for psychometric meta-analysis. *Applied Psychological Measurement*, 43(5):415–416.

Dechartres, A., Atal, I., Riveros, C., Meerpohl, J., and Ravaud, P. (2018). Association between publication characteristics and treatment effect estimates: A meta-epidemiologic study. *Annals of Internal Medicine*, 169(6):385–393.

DerSimonian, R. and Laird, N. (1986). Meta-analysis in clinical trials. *Controlled Clinical Trials*, 7(3):177–188.

Dias, S., Ades, A. E., Welton, N. J., Jansen, J. P., and Sutton, A. J. (2018). *Network meta-analysis for decision-making*. John Wiley & Sons.

Dias, S., Sutton, A. J., Ades, A., and Welton, N. J. (2013). Evidence synthesis for decision making 2: A generalized linear modeling framework for pairwise and network meta-analysis of randomized controlled trials. *Medical Decision Making*, 33(5):607–617.

Dias, S., Welton, N. J., Caldwell, D., and Ades, A. E. (2010). Checking consistency in mixed treatment comparison meta-analysis. *Statistics in Medicine*, 29(7-8):932–944.

DiCiccio, T. J. and Efron, B. (1996). Bootstrap confidence intervals. *Statistical Science*, pages 189–212.

Duval, S. and Tweedie, R. (2000). Trim and fill: A simple funnel-plot–based method of testing and adjusting for publication bias in meta-analysis. *Biometrics*, 56(2):455–463.

Ebrahim, S., Bance, S., Athale, A., Malachowski, C., and Ioannidis, J. P. (2016). Meta-analyses with industry involvement are massively published and report no caveats for antidepressants. *Journal of Clinical Epidemiology*, 70:155–163.

Edwards, S., Clarke, M., Wordsworth, S., and Borrill, J. (2009). Indirect comparisons of treatments based on systematic reviews of randomised controlled trials. *International Journal of Clinical Practice*, 63(6):841–854.

Efthimiou, O. (2018). Practical guide to the meta-analysis of rare events. *Evidence-Based Mental Health*, 21(2):72–76.

Efthimiou, O., Debray, T. P., van Valkenhoef, G., Trelle, S., Panayidou, K., Moons, K. G., Reitsma, J. B., Shang, A., Salanti, G., and Group, G. M. R. (2016). GetReal in network meta-analysis: A review of the methodology. *Research Synthesis Methods*, 7(3):236–263.

Egger, M., Smith, G. D., Schneider, M., and Minder, C. (1997). Bias in meta-analysis detected by a simple, graphical test. *BMJ*, 315(7109):629–634.

Elwood, P. (2006). The first randomized trial of aspirin for heart attack and the advent of systematic overviews of trials. *Journal of the Royal Society of Medicine*, 99(11):586–588.

Epskamp, S. (2019). semplot: Path diagrams and visual analysis of various SEM packages' output. R package version 1.1.2. `https://CRAN.R-project.org/package=semPlot`.

Etz, A. (2018). Introduction to the concept of likelihood and its applications. *Advances in Methods and Practices in Psychological Science*, 1(1):60–69.

Eysenck, H. J. (1978). An exercise in mega-silliness. *American Psychologist*, 33(5).

Fanelli, D. (2012). Negative results are disappearing from most disciplines and countries. *Scientometrics*, 90(3):891–904.

Fisher, R. A. (1935). *The Design of Experiments*. Oliver & Boyd, Edinburgh, UK.

Follmann, D. A. and Proschan, M. A. (1999). Valid inference in random effects meta-analysis. *Biometrics*, 55(3):732–737.

Fraley, C. and Raftery, A. E. (2002). Model-based clustering, discriminant analysis, and density estimation. *Journal of the American Statistical Association*, 97(458):611–631.

Friese, M., Loschelder, D. D., Gieseler, K., Frankenbach, J., and Inzlicht, M. (2019). Is ego depletion real? An analysis of arguments. *Personality and Social Psychology Review*, 23(2):107–131.

Furukawa, T. A. and Leucht, S. (2011). How to obtain NNT from Cohen's *d*: comparison of two methods. *PLOS ONE*, 6(4):e19070.

Furukawa, T. A., McGuire, H., and Barbui, C. (2003). Low dosage tricyclic antidepressants for depression. *Cochrane Database of Systematic Reviews*, (3).

Furukawa, T. A., Reijnders, M., Kishimoto, S., Sakata, M., DeRubeis, R. J., Dimidjian, S., Dozois, D. J., Hegerl, U., Hollon, S. D., Jarrett, R. B., Lespérance, F., Segal, Z. V., Mohr, D. C., Simons, A. D., Quilty, L. C., Reynolds, C. F., Gentili, C., Leucht, S., Engel, R., and Cuijpers, P. (2020). Translating the BDI and BDI-II into the HAMD and vice versa with equipercentile linking. *Epidemiology and Psychiatric Sciences*, 29.

Gart, J. J. and Zweifel, J. R. (1967). On the bias of various estimators of the logit and its variance with application to quantal bioassay. *Biometrika*, 181–187.

Gasparrini, A., Armstrong, B., and Kenward, M. G. (2012). Multivariate meta-analysis for non-linear and other multi-parameter associations. *Statistics in Medicine*, 31(29):3821–3839.

Gigerenzer, G. (2004). Mindless statistics. *The Journal of Socio-Economics*, 33(5):587–606.

Glass, G. V. (1976). Primary, secondary, and meta-analysis of research. *Educational Researcher*, 5(10):3–8.

Good, P. (2013). *Permutation tests: A practical guide to resampling methods for testing hypotheses*. Springer Science & Business.

Greco, T., Zangrillo, A., Biondi-Zoccai, G., and Landoni, G. (2013). Meta-analysis: Pitfalls and hints. *Heart, Lung and Vessels*, 5(4):219.

Grolemund, G. (2014). *Hands-on programming with R: Write your own functions and simulations*. O'Reilly.

Hamberg, K. (2008). Gender bias in medicine. *Women's Health*, 4(3):237–243.

Harrer, M., Adam, S. H., Messner, E.-M., Baumeister, H., Cuijpers, P., Bruffaerts, R., Auerbach, R. P., Kessler, R. C., Jacobi, C., Taylor, C. B., and Ebert, D. D. (2020). Prevention of eating disorders at universities: A systematic review and meta-analysis. *International Journal of Eating Disorders*, 53(3):823–833.

Hartigan, J. A. and Wong, M. A. (1979). Algorithm AS 136: A k-means clustering algorithm. *Journal of the Royal Statistical Society. Series C (Applied Statistics)*, 28(1):100–108.

Hartung, J. (1999). An alternative method for meta-analysis. *Biometrical Journal: Journal of Mathematical Methods in Biosciences*, 41(8):901–916.

Hartung, J. and Knapp, G. (2001a). On tests of the overall treatment effect in meta-analysis with normally distributed responses. *Statistics in Medicine*, 20(12):1771–1782.

Hartung, J. and Knapp, G. (2001b). A refined method for the meta-analysis of controlled clinical trials with binary outcome. *Statistics in Medicine*, 20(24):3875–3889.

Hedges, L. and Olkin, I. (2014). *Statistical methods for meta-analysis*. Academic Press.

Hedges, L. V. (1981). Distribution theory for Glass's estimator of effect size and related estimators. *Journal of Educational Statistics*, 6(2):107–128.

Hedges, L. V. (1984). Estimation of effect size under nonrandom sampling: The effects of censoring studies yielding statistically insignificant mean differences. *Journal of Educational Statistics*, 9(1):61–85.

Hedges, L. V. (1992). Modeling publication selection effects in meta-analysis. *Statistical Science*, 7(2):246–255.

Hedges, L. V. and Pigott, T. D. (2001). The power of statistical tests in meta-analysis. *Psychological Methods*, 6(3):203.

Hedges, L. V. and Pigott, T. D. (2004). The power of statistical tests for moderators in meta-analysis. *Psychological Methods*, 9(4):426.

Hedges, L. V. and Vevea, J. L. (1996). Estimating effect size under publication bias: Small sample properties and robustness of a random effects selection model. *Journal of Educational and Behavioral Statistics*, 21(4):299–332.

Hedges, L. V. and Vevea, J. L. (1998). Fixed-and random-effects models in meta-analysis. *Psychological Methods*, 3(4):486.

Henrich, J., Heine, S. J., and Norenzayan, A. (2010). Most people are not WEIRD. *Nature*, 466(7302):29.

Higgins, J., Jackson, D., Barrett, J., Lu, G., Ades, A., and White, I. (2012). Consistency and inconsistency in network meta-analysis: Concepts and models for multi-arm studies. *Research Synthesis Methods*, 3(2):98–110.

Higgins, J., Thompson, S., Deeks, J., and Altman, D. (2002). Statistical heterogeneity in systematic reviews of clinical trials: A critical appraisal of guidelines and practice. *Journal of Health Services Research Policy*, 7(1):51–61.

Higgins, J. P., Altman, D. G., Gøtzsche, P. C., Jüni, P., Moher, D., Oxman, A. D., Savović, J., Schulz, K. F., Weeks, L., and Sterne, J. A. (2011). The Cochrane Collaboration's tool for assessing risk of bias in randomised trials. *BMJ*, 343:d5928.

Higgins, J. P., Thomas, J., Chandler, J., Cumpston, M., Li, T., Page, M. J., and Welch, V. A. (2019). *Cochrane Handbook for Systematic Reviews of Interventions*. John Wiley & Sons.

Higgins, J. P. and Thompson, S. G. (2002). Quantifying heterogeneity in a meta-analysis. *Statistics in Medicine*, 21(11):1539–1558.

Higgins, J. P. and Thompson, S. G. (2004). Controlling the risk of spurious findings from meta-regression. *Statistics in Medicine*, 23(11):1663–1682.

Higgins, J. P., Thompson, S. G., and Spiegelhalter, D. J. (2009). A re-evaluation of random-effects meta-analysis. *Journal of the Royal Statistical Society: Series A (Statistics in Society)*, 172(1):137–159.

Hoaglin, D. C. (2016). Misunderstandings about Q and 'Cochran's Q test' in meta-analysis. *Statistics in Medicine*, 35(4):485–495.

Hoenig, J. M. and Heisey, D. M. (2001). The abuse of power: The pervasive fallacy of power calculations for data analysis. *The American Statistician*, 55(1):19–24.

Hoffman, M. D. and Gelman, A. (2014). The No-U-Turn sampler: Adaptively setting path lengths in Hamiltonian Monte Carlo. *Journal of Machine Learning Research*, 15(1):1593–1623.

Hohn, R. E., Slaney, K. L., and Tafreshi, D. (2019). Primary study quality in psychological meta-analyses: An empirical assessment of recent practice. *Frontiers in Psychology*, 9:2667.

Hough, S. L. and Hall, B. W. (1994). Comparison of the Glass and Hunter-Schmidt meta-analytic techniques. *The Journal of Educational Research*, 87(5):292–296.

Hunter, J. E. and Schmidt, F. L. (2004). *Methods of meta-analysis: Correcting error and bias in research findings*. Sage.

Hutton, J. L. (2010). Misleading statistics. *Pharmaceutical Medicine*, 24(3):145–149.

Infanger, D. and Schmidt-Trucksäss, A. (2019). P value functions: An underused method to present research results and to promote quantitative reasoning. *Statistics in Medicine*, 38(21):4189–4197.

Iniesta, R., Stahl, D., and McGuffin, P. (2016). Machine learning, statistical learning and the future of biological research in psychiatry. *Psychological Medicine*, 46(12):2455–2465.

IntHout, J., Ioannidis, J. P., and Borm, G. F. (2014). The Hartung-Knapp-Sidik-Jonkman method for random effects meta-analysis is straightforward and considerably outperforms the standard DerSimonian-Laird method. *BMC Medical Research Methodology*, 14(1):25.

IntHout, J., Ioannidis, J. P., Rovers, M. M., and Goeman, J. J. (2016). Plea for routinely presenting prediction intervals in meta-analysis. *BMJ Open*, 6(7).

Ioannidis, J. P. (2005). Why most published research findings are false. *PLOS Medicine*, 2(8):e124.

Ioannidis, J. P. (2006). Indirect comparisons: The mesh and mess of clinical trials. *The Lancet*, 368(9546):1470–1472.

Ioannidis, J. P. (2012). Why science is not necessarily self-correcting. *Perspectives on Psychological Science*, 7(6):645–654.

Ioannidis, J. P. (2016). The mass production of redundant, misleading, and conflicted systematic reviews and meta-analyses. *The Milbank Quarterly*, 94(3):485–514.

Iyengar, S. and Greenhouse, J. B. (1988). Selection models and the file drawer problem. *Statistical Science*, 3(1):109–117.

Jackson, D. (2013). Confidence intervals for the between-study variance in random effects meta-analysis using generalised Cochran heterogeneity statistics. *Research Synthesis Methods*, 4(3):220–229.

Jackson, D., White, I. R., and Riley, R. D. (2013). A matrix-based method of moments for fitting the multivariate random effects model for meta-analysis and meta-regression. *Biometrical Journal*, 55(2):231–245.

Jordan, A. E., Blackburn, N. A., Des Jarlais, D. C., and Hagan, H. (2017). Past-year prevalence of prescription opioid misuse among those 11 to 30 years of age in the United States: A systematic review and meta-analysis. *Journal of Substance Abuse Treatment*, 77:31–37.

Jöreskog, K. and Sörbom, D. (2006). LISREL 8.80. Chicago: Scientific Software International. *Computer software.*

Jørgensen, L., Paludan-Müller, A. S., Laursen, D. R., Savović, J., Boutron, I., Sterne, J. A., Higgins, J. P., and Hróbjartsson, A. (2016). Evaluation of the Cochrane tool for assessing risk of bias in randomized clinical trials: Overview of published comments and analysis of user practice in Cochrane and non-Cochrane reviews. *Systematic Reviews*, 5(1):80.

Kay, M. (2020). tidybayes: Tidy data and geoms for Bayesian models. R package version 2.1.1. `http://mjskay.github.io/tidybayes`.

Kerr, N. L. (1998). HARKing: Hypothesizing after the results are known. *Personality and Social Psychology Review*, 2(3):196–217.

Kim, E. S. and Menon, V. (2009). Status of women in cardiovascular clinical trials. *Arteriosclerosis, thrombosis, and vascular biology*, 29(3):279–283.

Kirsch, I. (2010). *The emperor's new drugs: Exploding the antidepressant myth.* Basic Books.

Kirsch, I., Moore, T. J., Scoboria, A., and Nicholls, S. S. (2002). The emperor's new drugs: An analysis of antidepressant medication data submitted to the us food and drug administration. *Prevention & Treatment*, 5(1):23a.

Kline, R. B. (2015). *Principles and practice of structural equation modeling.* Guilford.

Knapp, G. and Hartung, J. (2003). Improved tests for a random effects meta-regression with a single covariate. *Statistics in Medicine*, 22(17):2693–2710.

Koffel, E. and Watson, D. (2009). The two-factor structure of sleep complaints and its relation to depression and anxiety. *Journal of Abnormal Psychology*, 118(1):183.

König, J., Krahn, U., and Binder, H. (2013). Visualizing the flow of evidence in network meta-analysis and characterizing mixed treatment comparisons. *Statistics in Medicine*, 32(30):5414–5429.

Kraemer, H. C. and Kupfer, D. J. (2006). Size of treatment effects and their importance to clinical research and practice. *Biological Psychiatry*, 59(11):990–996.

Krahn, U., Binder, H., and König, J. (2013). A graphical tool for locating inconsistency in network meta-analyses. *BMC Medical Research Methodology*, 13(1):35.

Lakens, D., Page-Gould, E., van Assen, M. A., Spellman, B., Schönbrodt, F., Hasselman, F., Corker, K. S., Grange, J. A., Sharples, A., Cavender, C., Hilde, A., Heike, G., Cosima, L., Ian, M., Farid, A., and Anne, S. (2017). Examining the reproducibility of meta-analyses in psychology: A preliminary report. `https://osf.io/xfbjf/`.

Langan, D., Higgins, J. P., Jackson, D., Bowden, J., Veroniki, A. A., Kontopantelis, E., Viechtbauer, W., and Simmonds, M. (2019). A comparison of heterogeneity variance estimators in simulated random-effects meta-analyses. *Research Synthesis Methods*, 10(1):83–98.

Lipsey, M. W. and Wilson, D. B. (2001). *Practical meta-analysis*. SAGE.

Lu, G. and Ades, A. (2009). Modeling between-trial variance structure in mixed treatment comparisons. *Biostatistics*, 10(4):792–805.

Lüdecke, D. (2019). esc: Effect size computation for meta analysis (version 0.5.1). `https://CRAN.R-project.org/package=esc`.

Mahood, Q., Van Eerd, D., and Irvin, E. (2014). Searching for grey literature for systematic reviews: challenges and benefits. *Research Synthesis Methods*, 5(3):221–234.

Makambi, K. H. (2004). The effect of the heterogeneity variance estimator on some tests of treatment efficacy. *Journal of Biopharmaceutical Statistics*, 14(2):439–449.

Mansfield, E. R. and Helms, B. P. (1982). Detecting multicollinearity. *The American Statistician*, 36(3a):158–160.

Mantel, N. and Haenszel, W. (1959). Statistical aspects of the analysis of data from retrospective studies of disease. *Journal of the National Cancer Institute*, 22(4):719–748.

Marsman, M., Schönbrodt, F. D., Morey, R. D., Yao, Y., Gelman, A., and Wagenmakers, E.-J. (2017). A Bayesian bird's eye view of 'Replications of important results in social psychology'. *Royal Society Open Science*, 4(1):160426.

Mattos, C. T. and Ruellas, A. C. d. O. (2015). Systematic review and meta-analysis: What are the implications in the clinical practice? *Dental Press Journal of Orthodontics*, 20(1):17–19.

Mbuagbaw, L., Rochwerg, B., Jaeschke, R., Heels-Andsell, D., Alhazzani, W., Thabane, L., and Guyatt, G. H. (2017). Approaches to interpreting and choosing the best treatments in network meta-analyses. *Systematic Reviews*, 6(1):1–5.

McArdle, J. J. and McDonald, R. P. (1984). Some algebraic properties of the reticular action model for moment structures. *British Journal of Mathematical and Statistical Psychology*, 37(2):234–251.

McAuley, L., Tugwell, P., Moher, D., et al. (2000). Does the inclusion of grey literature influence estimates of intervention effectiveness reported in meta-analyses? *The Lancet*, 356(9237):1228–1231.

McGrayne, S. B. (2011). *The theory that would not die: How Bayes' rule cracked the enigma code, hunted down Russian submarines, and emerged triumphant from two centuries of controversy*. Yale University Press.

McGuinness, L. A. (2019). robvis: An R package and web application for visualising risk-of-bias assessments. `https://github.com/mcguinlu/robvis`.

McGuinness, L. A. and Higgins, J. P. (2020). Risk-Of-Bias VISualization (robvis): An R package and shiny web app for visualizing risk-of-bias assessments. *Research Synthesis Methods*, 12(1).

McNeish, D. (2016). On using Bayesian methods to address small sample problems. *Structural Equation Modeling: A Multidisciplinary Journal*, 23(5):750–773.

McNutt, M. (2014). Reproducibility. *Science*, 343(6168):229.

McShane, B. B., Böckenholt, U., and Hansen, K. T. (2016). Adjusting for publication bias in meta-analysis: An evaluation of selection methods and some cautionary notes. *Perspectives on Psychological Science*, 11(5):730–749.

Meehl, P. E. (1978). Theoretical risks and tabular asterisks: Sir Karl, Sir Ronald, and the slow progress of soft psychology. *Journal of Consulting and Clinical Psychology*, 46(4):806.

Mehta, P. D. and Neale, M. C. (2005). People are variables too: Multilevel structural equations modeling. *Psychological Methods*, 10(3):259.

Mendes, D., Alves, C., and Batel-Marques, F. (2017). Number needed to treat (NNT) in clinical literature: An appraisal. *BMC Medicine*, 15(1):112.

Moher, D., Liberati, A., Tetzlaff, J., Altman, D. G., Group, P., et al. (2009). Preferred reporting items for systematic reviews and meta-analyses: The PRISMA statement. *PLoS Medicine*, 6(7).

Moher, D., Shamseer, L., Clarke, M., Ghersi, D., Liberati, A., Petticrew, M., Shekelle, P., Stewart, L. A., and , P.-P. G. (2015). Preferred reporting items for systematic review and meta-analysis protocols (PRISMA-P) 2015 statement. *Systematic Reviews*, 4(1):1.

Møller, A. P. and Mousseau, T. A. (2015). Strong effects of ionizing radiation from Chernobyl on mutation rates. *Scientific Reports*, 5:8363.

Mosca, L., Hammond, G., Mochari-Greenberger, H., Towfighi, A., and Albert, M. A. (2013). Fifteen-year trends in awareness of heart disease in women: Results of a 2012 American Heart Association national survey. *Circulation*, 127(11):1254–1263.

Muthén, L. K. and Muthén, B. O. (2012). MPlus: Statistical analysis with latent variables–user's guide.

Ngamaba, K. H., Panagioti, M., and Armitage, C. J. (2017). How strongly related are health status and subjective well-being? Systematic review and meta-analysis. *The European Journal of Public Health*, 27(5):879–885.

Nielsen, M. W., Andersen, J. P., Schiebinger, L., and Schneider, J. W. (2017). One and a half million medical papers reveal a link between author gender and attention to gender and sex analysis. *Nature Human Behaviour*, 1(11):791–796.

Nuzzo, R. (2014). Statistical errors: *P* values, the 'gold standard' of statistical validity, are not as reliable as many scientists assume. *Nature*, 506(7487):150–153.

Olkin, I., Dahabreh, I. J., and Trikalinos, T. A. (2012). GOSH–a graphical display of study heterogeneity. *Research Synthesis Methods*, 3(3):214–223.

Olkin, I. and Finn, J. D. (1995). Correlations redux. *Psychological Bulletin*, 118(1):155.

Open Science Collaboration et al. (2015). Estimating the reproducibility of psychological science. *Science*, 349(6251).

O'Rourke, K. (2007). An historical perspective on meta-analysis: Dealing quantitatively with varying study results. *Journal of the Royal Society of Medicine*, 100(12):579–582.

Page, M. J., Sterne, J. A., Higgins, J. P., and Egger, M. (2020). Investigating and dealing with publication bias and other reporting biases in meta-analyses of health research: A review. *Research Synthesis Methods*.

Panageas, K. S., Ben-Porat, L., Dickler, M. N., Chapman, P. B., and Schrag, D. (2007). When you look matters: The effect of assessment schedule on progression-free survival. *Journal of the National Cancer Institute*, 99(6):428–432.

Pastor, D. A. and Lazowski, R. A. (2018). On the multilevel nature of meta-analysis: A tutorial, comparison of software programs, and discussion of analytic choices. *Multivariate Behavioral Research*, 53(1):74–89.

Patsopoulos, N. A., Analatos, A. A., and Ioannidis, J. P. (2005). Relative citation impact of various study designs in the health sciences. *JAMA*, 293(19):2362–2366.

Paule, R. C. and Mandel, J. (1982). Consensus values and weighting factors. *Journal of Research of the National Bureau of Standards*, 87(5):377–385.

Peters, J. L., Sutton, A. J., Jones, D. R., Abrams, K. R., and Rushton, L. (2006). Comparison of two methods to detect publication bias in meta-analysis. *JAMA*, 295(6):676–680.

Peters, J. L., Sutton, A. J., Jones, D. R., Abrams, K. R., and Rushton, L. (2007). Performance of the trim and fill method in the presence of publication bias and between-study heterogeneity. *Statistics in Medicine*, 26(25):4544–4562.

Peters, J. L., Sutton, A. J., Jones, D. R., Abrams, K. R., and Rushton, L. (2008). Contour-enhanced meta-analysis funnel plots help distinguish publication bias from other causes of asymmetry. *Journal of Clinical Epidemiology*, 61(10):991–996.

Peterson, B. G. and Carl, P. (2020). PerformanceAnalytics: Econometric tools for performance and risk analysis. R package version 2.0.4. `https://CRAN.R-project.org/package=PerformanceAnalytics`.

Peto, R. and Parish, S. (1980). Aspirin after myocardial infarction. *The Lancet*, 1(8179):1172–1173.

Piantadosi, S., Byar, D. P., and Green, S. B. (1988). The ecological fallacy. *American Journal of Epidemiology*, 127(5):893–904.

Pigott, T. D. and Polanin, J. R. (2020). Methodological guidance paper: High-quality meta-analysis in a systematic review. *Review of Educational Research*, 90(1):24–46.

Plummer, M. (2019). rjags: Bayesian graphical models using MCMC. R package version 4-10. `https://CRAN.R-project.org/package=rjags`.

Poole, C. and Greenland, S. (1999). Random-effects meta-analyses are not always conservative. *American Journal of Epidemiology*, 150(5):469–475.

Pustejovsky, J. E. and Rodgers, M. A. (2019). Testing for funnel plot asymmetry of standardized mean differences. *Research Synthesis Methods*, 10(1):57–71.

Quintana, D. S. (2015). From pre-registration to publication: A non-technical primer for conducting a meta-analysis to synthesize correlational data. *Frontiers in Psychology*, 6:1549.

Raudenbush, S. (2009). Analyzing effect sizes: Random effects models. In Cooper, H., Hedges, L., and Valentine, J., editors, *The handbook of research synthesis and meta-analysis (2nd Ed.)*. Russell Sage Foundation.

Riley, R. D., Lambert, P. C., and Abo-Zaid, G. (2010). Meta-analysis of individual participant data: Rationale, conduct, and reporting. *BMJ*, 340:c221.

Riley, R. D., Simmonds, M. C., and Look, M. P. (2007). Evidence synthesis combining individual patient data and aggregate data: A systematic review identified current practice and possible methods. *Journal of Clinical Epidemiology*, 60(5):431.e1–431.e12.

Robins, J., Greenland, S., and Breslow, N. E. (1986). A general estimator for the variance of the Mantel-Haenszel odds ratio. *American Journal of Epidemiology*, pages 719–723.

Rosnow, R. L. and Rosenthal, R. (1996). Computing contrasts, effect sizes, and counternulls on other people's published data: General procedures for research consumers. *Psychological Methods*, 1(4):331.

Rosnow, R. L., Rosenthal, R., and Rubin, D. B. (2000). Contrasts and correlations in effect-size estimation. *Psychological Science*, 11(6):446–453.

Rothman, K. J., Greenland, S., and Lash, T. L. (2008). *Modern epidemiology*. Lippincott Williams & Wilkins.

Rothstein, H. R., Sutton, A. J., and Borenstein, M. (2005). *Publication bias in meta-analysis*. John Wiley & Sons.

Röver, C. (2017). Bayesian random-effects meta-analysis using the 'bayesmeta' R package. *ArXiv Preprint 1711.08683*.

Rücker, G. (2012). Network meta-analysis, electrical networks and graph theory. *Research Synthesis Methods*, 3(4):312–324.

Rücker, G. and Schwarzer, G. (2015). Ranking treatments in frequentist network meta-analysis works without resampling methods. *BMC Medical Research Methodology*, 15(58).

Rücker, G. and Schwarzer, G. (2021). Beyond the forest plot: The drapery plot. *Research Synthesis Methods*, 12(1):13–19.

Rücker, G., Schwarzer, G., Carpenter, J. R., Binder, H., and Schumacher, M. (2011). Treatment-effect estimates adjusted for small-study effects via a limit meta-analysis. *Biostatistics*, 12(1):122–142.

Rücker, G., Schwarzer, G., Carpenter, J. R., and Schumacher, M. (2008). Undue reliance on I^2 in assessing heterogeneity may mislead. *BMC Medical Research Methodology*, 8(1):79.

Rücker, G., Krahn, U., König, J., Efthimiou, O., and Schwarzer, G. (2020). *netmeta: Network Meta-Analysis using Frequentist Methods.* R package version 1.2-1.

Salanti, G., Ades, A., and Ioannidis, J. P. (2011). Graphical methods and numerical summaries for presenting results from multiple-treatment meta-analysis: An overview and tutorial. *Journal of Clinical Epidemiology*, 64(2):163–171.

Salanti, G., Del Giovane, C., Chaimani, A., Caldwell, D. M., and Higgins, J. P. (2014). Evaluating the quality of evidence from a network meta-analysis. *PLOS ONE*, 9(7):e99682.

Sanderson, S., Tatt, I. D., and Higgins, J. (2007). Tools for assessing quality and susceptibility to bias in observational studies in epidemiology: A systematic review and annotated bibliography. *International Journal of Epidemiology*, 36(3):666–676.

Saruhanjan, K., Zarski, A.-C., Bauer, T., Baumeister, H., Cuijpers, P., Spiegelhalder, K., Auerbach, R. P., Kessler, R. C., Bruffaerts, R., Karyotaki, E., Berking, M., and Ebert, D. D. (2020). Psychological interventions to improve sleep in college students: A meta-analysis of randomized controlled trials. *Journal of Sleep Research*, e13097.

Schauberger, P. and Walker, A. (2020). openxlsx: Read, write and edit xlsx files. R package version 4.1.5. `https://CRAN.R-project.org/package=openxlsx`.

Scherer, R. W., Meerpohl, J. J., Pfeifer, N., Schmucker, C., Schwarzer, G., and von Elm, E. (2018). Full publication of results initially presented in abstracts. *Cochrane Database of Systematic Reviews*, 1(11).

Schmidt, F. L. and Hunter, J. E. (1977). Development of a general solution to the problem of validity generalization. *Journal of Applied Psychology*, 62(5):529.

Schmucker, C., Schell, L. K., Portalupi, S., Oeller, P., Cabrera, L., Bassler, D., Schwarzer, G., Scherer, R. W., Antes, G., Von Elm, E., and Joerg J, M. (2014). Extent of non-publication in cohorts of studies approved by research ethics committees or included in trial registries. *PLOS ONE*, 9(12):e114023.

Schubert, E., Sander, J., Ester, M., Kriegel, H. P., and Xu, X. (2017). DBSCAN revisited, revisited: Why and how you should (still) use DBSCAN. *ACM Transactions on Database Systems (TODS)*, 42(3):1–21.

Schwarzer, G., Carpenter, J. R., and Rücker, G. (2015). *Meta-analysis with R.* Springer.

Schwarzer, G., Carpenter, J. R., and Rücker, G. (2020). metasens: Advanced statistical methods to model and adjust for bias in meta-analysis. R package version 0.5-0. `https://CRAN.R-project.org/package=metasens`.

Schwarzer, G., Chemaitelly, H., Abu-Raddad, L. J., and Rücker, G. (2019). Seriously misleading results using inverse of Freeman-Tukey double arcsine transformation in meta-analysis of single proportions. *Research Synthesis Methods*, 10(3):476–483.

Schöpfel, J. and Rasuli, B. (2018). Are electronic theses and dissertations (still) grey literature in the digital age? A fair debate. *The Electronic Library*, 36(2):208–219.

Shannon, H. (2016). A statistical note on Karl Pearson's 1904 meta-analysis. *Journal of the Royal Society of Medicine*, 109(8):310–311.

Sharpe, D. (1997). Of apples and oranges, file drawers and garbage: Why validity issues in meta-analysis will not go away. *Clinical Psychology Review*, 17(8):881–901.

Shim, S. R., Kim, S.-J., Lee, J., and Rücker, G. (2019). Network meta-analysis: Application and practice using R software. *Epidemiology and Health*, 41.

Sidik, K. and Jonkman, J. N. (2002). A simple confidence interval for meta-analysis. *Statistics in Medicine*, 21(21):3153–3159.

Sidik, K. and Jonkman, J. N. (2005). Simple heterogeneity variance estimation for meta-analysis. *Journal of the Royal Statistical Society: Series C (Applied Statistics)*, 54(2):367–384.

Sidik, K. and Jonkman, J. N. (2007). A comparison of heterogeneity variance estimators in combining results of studies. *Statistics in Medicine*, 26(9):1964–1981.

Sidik, K. and Jonkman, J. N. (2019). A note on the empirical Bayes heterogeneity variance estimator in meta-analysis. *Statistics in Medicine*, 38(20):3804–3816.

Silberzahn, R., Uhlmann, E. L., Martin, D. P., Anselmi, P., Aust, F., Awtrey, E., Bahník, Š., Bai, F., Bannard, C., Bonnier, E., Carlsson, R., Cheung, F., Christensen, G., Clay, R., Craig, M. A., Dalla Rosa, A., Dam, L., Evans, M. H., Flores Cervantes, I., Fong, N., Gamez-Djokic, M., Glenz, A., Gordon-McKeon, S., Heaton, T. J., Hederos, K., Heene, M., Hofelich Mohr, A. J., Högden, F., Hui, K., Johannesson, M., Kalodimos, J., Kaszubowski, E., Kennedy, D. M., Lei, R., Lindsay, T. A., Liverani, S., Madan, C. R., Molden, D., Molleman, E., Morey, R. D., Mulder, L. B., Nijstad, B. R., Pope, N. G., Pope, B., Prenoveau, J. M., Rink, F., Robusto, E., Roderique, H., Sandberg, A., Schlüter, E., Schönbrodt, F. D., Sherman, M. F., Sommer, S. A., Sotak, K., Spain, S., Spörlein, C., Stafford, T., Stefanutti, L., Tauber, S., Ullrich, J., Vianello, M., Wagenmakers, E. J., Witkowiak, M., Yoon, S., and Nosek, B. A. (2018). Many analysts, one data set: Making transparent how variations in analytic choices affect results. *Advances in Methods and Practices in Psychological Science*, 1(3):337–356.

Simonsohn, U., Nelson, L. D., and Simmons, J. P. (2014a). P-curve: A key to the file-drawer. *Journal of Experimental Psychology: General*, 143(2):534.

Simonsohn, U., Nelson, L. D., and Simmons, J. P. (2014b). P-curve and effect size: Correcting for publication bias using only significant results. *Perspectives on Psychological Science*, 9(6):666–681.

Simonsohn, U., Simmons, J. P., and Nelson, L. D. (2015). Better p-curves: Making p-curve analysis more robust to errors, fraud, and ambitious *p*-hacking, a reply to Ulrich and Miller (2015). *Journal of Experimental Psychology: General*, 144(6):1146–1152.

Simonsohn, U., Simmons, J. P., and Nelson, L. D. (2020). Specification curve analysis. *Nature Human Behaviour*, 4(11):1208–1214.

Smith, M. L. and Glass, G. V. (1977). Meta-analysis of psychotherapy outcome studies. *American Psychologist*, 32(9):752.

Song, F., Loke, Y. K., Walsh, T., Glenny, A.-M., Eastwood, A. J., and Altman, D. G. (2009). Methodological problems in the use of indirect comparisons for evaluating healthcare interventions: Survey of published systematic reviews. *BMJ*, 338:b1147.

Spearman, C. (1904). The proof and measurement of association between two things. *The American Journal of Psychology*, 15(1):72–101.

Stanley, T. D. (2008). Meta-regression methods for detecting and estimating empirical effects in the presence of publication selection. *Oxford Bulletin of Economics and Statistics*, 70(1):103–127.

Stanley, T. D. (2017). Limitations of PET-PEESE and other meta-analysis methods. *Social Psychological and Personality Science*, 8(5):581–591.

Stanley, T. D. and Doucouliagos, H. (2014). Meta-regression approximations to reduce publication selection bias. *Research Synthesis Methods*, 5(1):60–78.

Sterne, J. A., Gavaghan, D., and Egger, M. (2000). Publication and related bias in meta-analysis: Power of statistical tests and prevalence in the literature. *Journal of Clinical Epidemiology*, 53(11):1119–1129.

Sterne, J. A., Hernán, M. A., Reeves, B. C., Savović, J., Berkman, N. D., Viswanathan, M., Henry, D., Altman, D. G., Ansari, M. T., Boutron, I., Carpenter, J. R., Chan, A.-W., Churchill, R., Deeks, J. J., Hróbjartsson, A., Kirkham, J., Jüni, P., Loke, Y. K., Pigott, T. D., Ramsay, C. R., Regidor, D., Rothstein, H. R., Sandhu, L., Santaguida, P. L., Schünemann, H. J., Shea, B., Shrier, I., Tugwell, P., Turner, L., Valentine, J. C., Waddington, H., Waters, E., Wells, G. A., Whiting, P. F., and Higgins, J. P. T. (2016). ROBINS-I: A tool for assessing risk of bias in non-randomised studies of interventions. *BMJ*, 355:i4919.

Sterne, J. A., Savović, J., Page, M. J., Elbers, R. G., Blencowe, N. S., Boutron, I., Cates, C. J., Cheng, H.-Y., Corbett, M. S., Eldridge, S. M., et al. (2019). RoB 2: A revised tool for assessing risk of bias in randomised trials. *BMJ*, 366.

Sterne, J. A. C., Sutton, A. J., Ioannidis, J. P. A., Terrin, N., Jones, D. R., Lau, J., Carpenter, J., Rücker, G., Harbord, R. M., Schmid, C. H., Tetzlaff, J., Deeks, J. J., Peters, J., Macaskill, P., Schwarzer, G., Duval, S., Altman, D. G., Moher, D., and Higgins, J. P. T. (2011). Recommendations for examining and interpreting funnel plot asymmetry in meta-analyses of randomised controlled trials. *BMJ*, 343.

Stijnen, T., Hamza, T. H., and Özdemir, P. (2010). Random effects meta-analysis of event outcome in the framework of the generalized linear mixed model with applications in sparse data. *Statistics in Medicine*, 29(29):3046–3067.

Stinerock, R. (2018). *Statistics with R: A Beginner's Guide*. SAGE, London, UK, 1st edition.

Sweeting, M. J., Sutton, A. J., and Lambert, P. C. (2004). What to add to nothing? use and avoidance of continuity corrections in meta-analysis of sparse data. *Statistics in Medicine*, 23(9):1351–1375.

Tacconelli, E. (2009). Systematic reviews: CRD's guidance for undertaking reviews in healthcare. *The Lancet Infectious Diseases*, 10(4).

Tang, R. W. and Cheung, M. W.-L. (2016). Testing IB theories with meta-analytic structural equation modeling. *Review of International Business and Strategy*, 26(4):472–492.

Terrin, N., Schmid, C. H., Lau, J., and Olkin, I. (2003). Adjusting for publication bias in the presence of heterogeneity. *Statistics in Medicine*, 22(13):2113–2126.

Thalheimer, W. and Cook, S. (2002). *How to calculate effect sizes from published research: A simplified methodology*.

Thompson, B. (2004). *Exploratory and confirmatory factor analysis*. American Psychological Association.

Thompson, S. G. and Higgins, J. P. (2002). How should meta-regression analyses be undertaken and interpreted? *Statistics in Medicine*, 21(11):1559–1573.

Thompson, S. G., Turner, R. M., and Warn, D. E. (2001). Multilevel models for meta-analysis, and their application to absolute risk differences. *Statistical Methods in Medical Research*, 10(6):375–392.

Tipton, E., Pustejovsky, J. E., and Ahmadi, H. (2019). A history of meta-regression: Technical, conceptual, and practical developments between 1974 and 2018. *Research Synthesis Methods*, 10(2):161–179.

Valstad, M., Alvares, G. A., Andreassen, O. A., Westlye, L. T., and Quintana, D. S. (2016). The relationship between central and peripheral oxytocin concentrations: A systematic review and meta-analysis protocol. *Systematic Reviews*, 5(1):1–7.

van Aert, R. C., Wicherts, J. M., and van Assen, M. A. (2016). Conducting meta-analyses based on *p* values: Reservations and recommendations for applying p-uniform and p-curve. *Perspectives on Psychological Science*, 11(5).

van Valkenhoef, G., Lu, G., de Brock, B., Hillege, H., Ades, A., and Welton, N. J. (2012). Automating network meta-analysis. *Research Synthesis Methods*, 3(4):285–299.

Vance, A. (2009). Data analysts captivated by R's power. *New York Times*.

Veroniki, A. A., Jackson, D., Viechtbauer, W., Bender, R., Bowden, J., Knapp, G., Kuss, O., Higgins, J. P., Langan, D., and Salanti, G. (2016). Methods to estimate the between-study variance and its uncertainty in meta-analysis. *Research Synthesis Methods*, 7(1):55–79.

Vevea, J. L. and Woods, C. M. (2005). Publication bias in research synthesis: Sensitivity analysis using a priori weight functions. *Psychological Methods*, 10(4):428.

Viechtbauer, W. (2005). Bias and efficiency of meta-analytic variance estimators in the random-effects model. *Journal of Educational and Behavioral Statistics*, 30(3):261–293.

Viechtbauer, W. (2007). Confidence intervals for the amount of heterogeneity in meta-analysis. *Statistics in Medicine*, 26(1):37–52.

Viechtbauer, W. (2010). Conducting meta-analyses in R with the metafor package. *Journal of Statistical Software*, 36(3):1–48.

Viechtbauer, W. and Cheung, M. W.-L. (2010). Outlier and influence diagnostics for meta-analysis. *Research Synthesis Methods*, 1(2):112–125.

Viechtbauer, W., López-López, J. A., Sánchez-Meca, J., and Marín-Martínez, F. (2015). A comparison of procedures to test for moderators in mixed-effects meta-regression models. *Psychological Methods*, 20(3):360.

Wampold, B. E. (2013). *The great psychotherapy debate: Models, methods, and findings*. Routledge.

Wellek, S. (2017). A critical evaluation of the current "p-value controversy". *Biometrical Journal*, 59(5):854–872.

Whiting, P. F., Rutjes, A. W., Westwood, M. E., Mallett, S., Deeks, J. J., Reitsma, J. B., Leeflang, M. M., Sterne, J. A., and Bossuyt, P. M. (2011). QUADAS-2: A revised tool for the quality assessment of diagnostic accuracy studies. *Annals of Internal Medicine*, 155(8):529–536.

Whittingham, M. J., Stephens, P. A., Bradbury, R. B., and Freckleton, R. P. (2006). Why do we still use stepwise modelling in ecology and behaviour? *Journal of Animal Ecology*, 75(5):1182–1189.

Wicherts, J. M., Veldkamp, C. L., Augusteijn, H. E., Bakker, M., Van Aert, R., and Van Assen, M. A. (2016). Degrees of freedom in planning, running, analyzing, and reporting psychological studies: A checklist to avoid p-hacking. *Frontiers in Psychology*, 7:1832.

Wickham, H., Averick, M., Bryan, J., Chang, W., McGowan, L. D., François, R., Grolemund, G., Hayes, A., Henry, L., Hester, J., Kuhn, M., Pedersen, T. L., Miller, E., Bache, S. M., Müller, K., Ooms, J., Robinson, D., Seidel, D. P., Spinu, V., Takahashi, K., Vaughan, D., Wilke, C., Woo, K., and Yutani, H. (2019). Welcome to the tidyverse. *Journal of Open Source Software*, 4(43):1686.

Wickham, H. and Grolemund, G. (2016). *R for data science: Import, tidy, transform, visualize, and model data*. O'Reilly.

Wiksten, A., Rücker, G., and Schwarzer, G. (2016). Hartung–Knapp method is not always conservative compared with fixed-effect meta-analysis. *Statistics in Medicine*, 35(15):2503–2515.

Williams, D. R., Rast, P., and Bürkner, P.-C. (2018). Bayesian meta-analysis with weakly informative prior distributions. `https://psyarxiv.com/7tbrm/`.

Wolen, A. R., Hartgerink, C. H., Hafen, R., Richards, B. G., Soderberg, C. K., and York, T. P. (2020). osfr: An R interface to the Open Science Framework. *Journal of Open Source Software*, 5(46):2071.

Wołodźko, T. (2020). extradistr: Additional univariate and multivariate distributions. R package version 1.9.1. `https://CRAN.R-project.org/package=extraDistr`.

Xie, Y., Allaire, J. J., and Grolemund, G. (2018). *R Markdown: The definitive guide*. Chapman and Hall/CRC Press.

Yusuf, S., Peto, R., Lewis, J., Collins, R., and Sleight, P. (1985). Beta blockade during and after myocardial infarction: An overview of the randomized trials. *Progress in Cardiovascular Diseases*, 27(5):335–371.

Zhang, J. and Yu, K. F. (1998). What's the relative risk? A method of correcting the odds ratio in cohort studies of common outcomes. *JAMA*, 280(19):1690–1691.

Index

Akaike's Information Criterion, 217, 223, 376
Analysis of Variance, 87, 217, 298, 303, 427
Analysis Plan, 16, 210
Apples and Oranges Problem, 9
Area Under The Curve (AUC), 431
Attenuation, 82

Bayes' Theorem, 356
Bayesian Hierarchical Model, 334, 358, 381
Binomial Test, 257
brms Package, 383, 385
Bubble Plot, 205

Campbell Collaboration, 7, 23
Censoring, 76
Central Limit Theorem, 56
Citation Bias, 229
Cluster Effect, 290
Cochran's Q, 140, 144, 152, 161, 185, 343
Cochrane, 7, 399, 407
Cochrane, Handbook, 7, 14
Cochrane, QUADAS Tool, 408
Cochrane, Risk of Bias Tool, 7, 24, 407
Cochrane, ROBINS-I Tool, 25, 408
Cohen's *d*, 65, 270
Comprehensive R Archive Network (CRAN), 30
Comprehensive Meta-Analysis (CMA), 102
Conditional Probability, 356
Consistency, 329, 332, 333, 337, 343, 350, 368
Continuity Correction, 72, 117
Cook's Distance, 160, 166
Correlation, 54, 61, 63, 127, 424, 426
Correlation, Point-Biserial, 63, 426
Credible Interval, 358, 389
Cumulative Distribution Function (CDF), 402

Data Frame, 38
DBSCAN, 165
DerSimonian-Laird Estimator, 102, 145
dmetar Package, 34, 97, 108, 118, 125, 127, 130, 134, 154, 156, 164, 212, 262, 291, 295, 310, 338, 345, 361, 371, 402, 428, 432
Documentation, R, 45, 107
Double-Counting, 88
Drapery Plot, 180
Dummy Variable, 198, 374

Ecological Bias, 189
Egger's Test, 236
Empirical Cumulative Distribution Function (ECDF), 390
esc Package, 67, 80, 97, 294, 423
Exchangeability Assumption, 100, 333
Exponentiation, 121
extraDistr Package, 385

Factor Analysis, 303, 316
Fidelity, Treatment, 235
File Drawer Problem, 9, 15, 228, 231
Fisher's z, 62, 83, 85, 127, 292
Fisher's Method, 258
Fixed-Effect Model, 93, 95, 112, 198, 307, 401
Fixed-Effects (Plural) Model, 184, 185
Forest Plot, 162, 173, 349, 369, 391
Frequentist Statistics, 334, 335, 356, 381
Function Argument, 31, 107
Function, R, 31

Funnel Plot, 231, 252, 354

Garbage In, Garbage Out Problem, 9
Gaussian Mixture Model, 165
gemtc Package, 360, 373, 381
Generalized Additive Model, 385
Gibbs Sampler, 358, 360, 383
Glass' Delta, 112
Graph Theory, 330, 337

HARKing, 230
Hedges' *g*, 68, 80, 83, 85, 97, 108, 423
Heterogeneity, 99, 100, 104, 139, 149, 183, 184, 282, 343
History of Meta-Analysis, 6, 61, 62, 258

I^2, Higgins & Thompson's, 145, 149, 152, 155, 204, 282, 295
Incidence Rate Ratio, 76, 125
Influential Case, 153, 156, 183
Interaction (Regression), 207, 217
Inverse-Variance Weighting, 97, 102, 115, 125, 141
IPD Meta-Analysis, 5

JAGS, 360

K-Means, 165
Knapp-Hartung Adjustment, 104, 113
Kolmogorov-Smirnov Test, 269

Language Bias, 229
Leave-One-Out Method, 156, 159, 162
Likelihood Ratio Test, 217
Limit, 246, 250
Log-Incidence Rate Ratio, 78
Log-Odds Ratio, 74
Log-Risk Ratio, 71
Logarithm, Natural, 60, 123, 174
Logit-Transformation, 60, 132

Machine Learning, 164
Mantel-Haenszel Method, 72, 97, 115, 119, 126
Markdown, R, 413, 415
Markov Chain Monte Carlo, 358, 365, 366, 383, 387
MARS Statement, 14
Maximum Likelihood, 102, 133, 209, 217, 307, 313
Mean Path Length, 346
Mean, Arithmetic, 53, 59, 130, 423
meta Package, 32, 93, 105, 145, 173, 190, 212, 291, 383, 423
Meta-Regression, 17, 187, 197, 373
metafor Package, 32, 103, 163, 212, 214, 276, 291, 383
metaSEM Package, 308, 309
metasens Package, 251
Minimal Parallelism, 346
Mixed-Effects Model, 132, 187, 197, 198
Mixed-Treatment Comparison Meta-Analysis, 329
Moderator Analysis, 183, 298
Multi-Collinearity, 210, 213
Multi-Model Inference, 212, 220
Multilevel Meta-Analysis, 89, 187, 287
Multiple Publication Bias, 229
Multivariate Meta-Analysis, 303, 309

netmeta Package, 335
Network Graph, 344, 364
No-U-Turn Sampler (NUTS), 383
Node Splitting, 333, 352, 368
Non-Centrality Parameter, 264
Number Needed to Treat, 430

Occam's Razor, 296
Odds, 60, 73
Odds Ratio, 73, 76, 115, 430
Open Science Framework (OSF), 17, 230, 413, 417
openxlsx Package, 38
Ordinary Least Squares (OLS), 201, 209
Outcome Reporting Bias, 229
Outlier, 153, 183, 262, 282
Overfitting, 209, 210

P-Curve, 254, 281
P-Hacking, 229, 254, 260
P-Score, 348
P-Value, 180, 254, 428
Package, R, 31

Parsimony, 210
Path Diagram, 304, 326
Paule-Mandel Estimator, 103, 119, 126
PerformanceAnalytics Package, 214
Permutation, 210, 219
Person-Time, 76, 125
PET-PEESE, 246, 281
Peter's Test, 240
Peto Method, 97, 116
Peto Odds Ratio, 116
PICO, 13
Pipe, R, 45
Pooled Standard Deviation, 64
Position Matching, 107
Posterior Distribution, 357, 383, 390
Potential Scale Reduction Factor, 367, 388, 399
Power, 187, 256, 259, 265, 399
Power Analysis, 17, 399
Power Approach Paradox, 400
Prediction Interval, 150, 315
Preprint, 17
Preregistration, 16, 420
Prior Distribution, 357, 359, 382
PRISMA Statement, 14, 17, 23
Project, R, 413
Proportion, 60, 132
Protocol, 17
Psychometric Meta-Analysis, 86
Publication Bias, 228

Questionable Research Practice (QRP), 229, 254

R Studio, 29
Random-Effects Model, 6, 17, 93, 99, 139, 184, 185, 198, 288, 307, 404
Range Restriction Correction, 84
Reporting Bias, 229
Research Question, 12
Researcher Agenda Problem, 9
Restricted Maximum Likelihood Estimator, 102, 103, 128, 217
Review Manager (RevMan), 102, 178, 407
Review, Narrative, 4
Review, Systematic, 4
Risk of Bias, 24, 176, 206, 373, 407
Risk Ratio, 70, 75, 115
rjags Package, 360
robvis Package, 407
Root Mean Square Error of Approximation (RMSEA), 325

Sampling Error, 55, 95, 99, 141, 149, 159, 197, 288
Selection Model, 272
semPlot Package, 326
Sidik-Jonkman Estimator, 102, 103
Small-Study Effect, 230, 254
Sparse Data, 115
STAN, 385
Standardized Mean Difference, 6, 54, 63, 65, 68, 80, 97, 108, 112, 239, 246, 248, 384, 423, 428
Step Function, 273
Step-Wise Regression, 211
Stouffer's Method, 258
Structural Equation Model, 303
Study Search, 11, 18, 230
Subgroup Analysis, 17, 183, 197, 298, 303, 406
SUCRA Score, 371
Surface Under the Cumulative Ranking (SUCRA) Score, 348
Survival Analysis, 76

t-Distribution, 104, 259, 264
Three-Parameter Selection Model, 275, 281
tidybayes Package, 391
tidyverse Package, 32, 237
Time-Lag Bias, 229
Transitivity Assumption, 332, 333
Trim and Fill Method, 242, 254

Uninformative Prior, 359, 383
Unit-of-Analysis Problem, 88, 289, 434
Unrealiability Correction, 81

Version, R, 34

Wald-Type Test, 104, 202, 314
Weakly Informative Prior, 383
Weight, 96, 102, 111, 155, 167, 200, 273
Weighted Least Squares (WLS), 201, 241,
309
WEIRD Populations, 15
WinBUGS, 377

Zero Cell Problem, 72, 75, 117

Printed in the United States
by Baker & Taylor Publisher Services

Printed in the United States
by Baker & Taylor Publisher Services